Cell Engineering

Cell Engineering

Volume 3

The titles published in this series are listed at the end of this volume.

CELL ENGINEERING

Vol. 3: Glycosylation

Edited by

Mohamed Al-Rubeai

School of Chemical Engineering,
The University of Birmingham,
Edgbaston, Birmingham, U.K.

KLUWER ACADEMIC PUBLISHERS
DORDRECHT / BOSTON / LONDON

A C.I.P. Catalogue record for this book is available from the Library of Congress.

ISBN 1-4020-0733-7

Published by Kluwer Academic Publishers,
P.O. Box 17, 3300 AA Dordrecht, The Netherlands.

Sold and distributed in North, Central and South America
by Kluwer Academic Publishers,
101 Philip Drive, Norwell, MA 02061, U.S.A.

In all other countries, sold and distributed
by Kluwer Academic Publishers,
P.O. Box 322, 3300 AH Dordrecht, The Netherlands.

Printed on acid-free paper

Printed in the Netherlands.

Glycosylation

Cell Engineering
Volume 3

With contributions by:

Professor Mike Betenbaugh, John Hopkins University, Baltimore, MD,
U.S.A.
Professor Michael Butler, The Glycobiology Institute, Oxford, U.K.
Dr. Harald Conradt, Protein Glycosylation Department, GBF,
Braunschweig, Germany
Dr. Eckart Grabenhorst, Protein Glycosylation Department, GBF,
Braunschweig, Germany
Dr. David Harvey, The Glycobiology Institute, Oxford, U.K.
Dr. Tsu-An Hsu, National Health Research Institutes, Taipei, Taiwan
Professor Roy Jefferis, University of Birmingham, Birmingham, U.K.
Dr. Nigel Jenkins, Eli Lilly & Company, Indianapolis, IN, U.S.A.
Dr. Lynne Krummen, Genentech, Inc., San Francisco, CA, U.S.A.
Dr. S.M. Lawrence, John Hopkins University, Baltimore, MD, U.S.A.
Dr. Cheng-Kang Lee, National Taiwan University of Science and
Technology, Taipei, Taiwan.
Dr. John Lund, University of Birmingham, Birmingham, U.K.
Dr. Tony Merry, The Glycobiology Institute, Oxford, U.K.
Dr. Lucia Monaco, San Raffaele Scientific Institute, Milan, Italy
Dr. Manfred Nimtz, Protein Glycosylation Department, GBF,
Braunschweig, Germany
Dr. V. Restelli, University of Manitoba, Winnipeg, Manitoba, Canada
Dr. Myriam Taverna, Université de Paris Sud, Chatenay-Malabry,
France
Dr. Thuy Tran, Université de Paris Sud, Chatenay-Malabry, France
Dr. Stefanie Weikert, Genentech, Inc., San Francisco, CA, U.S.A.
Dr. Jyh-Ming Wu, National Taiwan University of Science and
Technology, Taipei, Taiwan.
Dr. Kevin Yarema, Johns Hopkins University, Baltimore, U.S.A.

Contents

1   ANALYSIS OF GLYCANS OF RECOMBINANT   1
    GLYCOPROTEINS
    T. Merry, M. Taverna, T. Tran, D. Harvey

2   THE EFFECT OF CELL CULTURE PARAMETERS ON PROTEIN   61
    GLYCOSYLATION
    V.Restelli and M.Butler

3   GLYCOSYLATION OF RECOMBINANT IgG ANTIBODIES AND   93
    ITS RELEVANCE FOR THERAPEUTIC APPLICATIONS
    R. Jefferis

4   CELLULAR MODIFICATION FOR THE IMPROVEMENT OF THE   109
    GLYCOSYLATION PATHWAY
    L. Krummen and S. Weikert

5   CONTROLLING CARBOHYDRATES ON RECOMBINANT   131
    GLYCOPROTEINS
    N. Jenkins, J. Lund, and L. Monaco

6   TARGETING OF GENETICALLY ENGINEERED   149
    GLYCOSYLTRANSFERASES TO IN VIVO FUNCTIONAL
    GOLGI SUBCOMPARTMENTS OF MAMMALIAN CELLS
    E. Grabenhorst, M. Nimtz, and H.S. Conradt

7   A METABOLIC SUBSTRATE-BASED APPROACH TO   171
    ENGINEERING NEW CHEMICAL REACTIVITY INTO
    CELLULAR SIALOGLYCOCONJUGATES
    K.J. Yarema

8   ADDRESSING INSECT CELL GLYCOSYLATION   197
    DEFICIENCIES THROUGH METABOLIC ENGINEERING
    S.M. Lawrence and M.J. Betenbaugh

9   ASPARAGINE-LINKED GLYCOSYLATIONAL   215
    MODIFICATIONS IN YEAST
    J.-M. Wu, C.-K. Lee, T.-A. Hsu

Index   233

1. ANALYSIS OF GLYCANS OF RECOMBINANT GLYCOPROTEINS

T. MERRY[1]*, M. TAVERNA[2], T. TRAN[2], D. HARVEY[3]

[1] *Glycan Consultancy,Charlbury , Oxford , OX7 3HB England*
[2]*Laboratoire de chimie analytique, Faculté de Pharmacie, Université de Paris Sud, Rue JB Clement, 92290 Chatenay-Malabry, France*
[3]*Oxford Glycobiology Institute, University of Oxford, South Parks Road, Oxford OX2 6HE England*
* *Corresponding Author: E-mail: tony@merryone.freeserve.co.uk*

1.   Introduction

An important aspect of recombinant protein analysis is the consideration of post translational modifications. This review will focus on the analysis of glycosyla-tion, which is one such major modification. A number of studies have shown that glycosylation depends on both the cell line used for the culture and other factors in the culture medium (Doyle, de la Canal et al., 1986); (Page, Killian et al., 1990) (Patel, Parekh et al., 1992; Borys, Linzer et al., 1993) (Gawlitzek, Valley et al., 1995) (Andersen, Bridges et al., 2000; Yang and Butler, 2000). It is important to consider these factors and also to be able to monitor glycosylation as this may affect several properties of the glycoprotein under production, most significantly the half life in the circulation, a factor that is an important aspect in considering the pharmokinetics of recombinant glycoproteins (Baynes and Wold, 1976); (Elbein, 1991; Goochee, 1992; Kobata, 1992; Dwek, 1995; Rice, Chiu et al., 1995) such as the immunoglobulins ((Elbein, 1991; Kunkel, Jan et al., 1998; Tang, Nesta et al., 1999).

The study of the glycan chains (oligosaccharides) attached to glycoproteins presents a number of analytical problems that generally makes their analysis more difficult than that of the peptides to which they are attached. Predominant among these problems are the branched nature of many carbohydrates and the ability of constituent monosac-charide units to form bonds to several positions on the adjacent unit. The following analytical techniques have all been applied to the study of glycans.

1.   Nuclear magnetic resonance (NMR)
2.   Analysis by high pH anion exchange chromatography (HPAEC)
3.   Analysis by sodium dodecyl sulphate polyacrylamide gel electrophoresis (SDS-PAGE)
4.   Analysis by capillary electrophoresis (CE)
5.   Analysis by gel permeation chromatography (GPC)
6.   Analysis by high-performance liquid chromatography (HPLC) following fluorescent labeling
7.   Analysis by mass spectrometry (MS)

M. Al-Rubeai.(ed.). Cell Engineering. 1-60.
© 2002 *Kluwer Academic Publishers. Printed in the Netherlands.*

Each technique has its own merits and all have been successfully employed for the characterisation of recombinant glycoproteins. The choice of technique is, therefore, dictated by a number of factors including level of information required, available expertise and budget. The speed of analysis and interpretation of data may also be considerations.

Definitive characterisation has traditionally been performed by NMR (Anderson, Atkinson et al., 1985); (Anderson, Atkinson et al., 1985); (De Beer, Van Zuylen et al., 1996). However, this technique is not generally applicable to analysis in a routine manner as it requires relatively large amounts of material (in the milligram range). In addition, access to the sophisticated equipment and the expertise required for the interpretation of data mean that the technique is only available to specialised dedicated laboratories. Another complex technique which has also been applied extensively to studies of glycan structure is that of mass spectrometry in various forms, several of which are discussed below. Electrophoretic techniques have several advantages in terms of speed and generally do not require sophisticated instrumentation. The most widely used are those employing polyacrylamide gels or capillary electrophoresis. Chromatographic techniques offer a number of advantages and various forms have been widely used. In particular, the use of fluorescent labels such as 2-aminobenza-mide (2-AB) and HPLC column systems calibrated with standards allows the assignment of structures and comparative profiling. In combination with exogly-cosidase digestion and mass spectrometry, complete analysis of complex glycans may be achieved. Monosaccharide analysis is a traditional means of analysis and is still widely used in quality control although it is now becoming replaced by methods that give more information on glycan structure.

2.   Analysis Of Intact Glycoproteins

2.1. ANALYSIS BY CAPILLARY ELECTROPHORESIS

The emergence of capillary electrophoresis (CE) with the various modes in which separation can be performed has brought new possibilities in the field of microheterogeneity of glycoproteins evaluations. CE method, combining the advantages of high efficiency and resolution, accurate quantification, short analysis time and automation, has gained interest not only in lot-to-lot routine control of glycoproteins produced by DNA technology, but also in process monitoring, purity assessment and product quality evaluation.

In this chapter, applications of recombinant glycoprotein analyses with the aim to separate either the glycoforms of the intact glycoprotein or oligosaccharides derived from glycoproteins will be described. A great number of publications have reviewed the potential of this technique for the analysis of recombinant glycoproteins (Chiesa, Oefner et al., 1995; El Rassi and Mechref, 1996; Taverna, Tran et al., 1998).

2.1.1. Glycoform Analysis

This part will consider the main strategies for analysing the glycoforms of the intact glycoproteins by means of the most employed modes afforded by CE. This approach represents the most straightforward method to establish a pattern of glycoprotein heterogeneity; and information on the identity, the heterogeneity, the purity of the glycoprotein are readily obtained.

Due to their high resolving power, capillary zone electrophoresis (CZE) and capillary isoelectric focusing (CIEF) are the most employed for sensitive and rapid screening techniques for recombinant glycoproteins. Owing to their ease of automation and their selectivity which facilitate the development of validatable routine analysis, CZE and CIEF appear well suited for the quality control environment.

Generally, when the analyses of an unknown glycoprotein is required, the first and the simplest mode that should be tried to separate the different glycoforms is CZE in uncoated capillaries, which can be carried out with a reasonable number of parameters to optimize. CIEF can then be employed either to further improve the resolution or to achieve separations that could not be possible using CZE mode. This mode is particularly suitable for separation of glycoproteins bearing complex type oligosaccharides, or which glycoforms exhibit different degree of sialylation, sulfatation or phosphorylation. The practical application of different microscale techniques of CE for analyzing the intact recombinant glycoproteins (Taverna *et al.*, 2001).

2.1.2. Capillary Zone Electrophoresis (CZE)

CZE separation is based on charge-to-mass ratio. In the case of glycoforms, CZE is well suited to the separation of variants with different glycosylation degrees and different number of sialic acid residues. It has been shown that neutral glycoforms can also be separated by careful selection of the separation conditions (using the complexation of sugars with borate).

It is therefore recommended to firstly use a wide pH range to ensure that at least one peak can be visualized, improvement of the glycoform separation is then accomplished by testing pHs closer to the isoelectric point of the protein where differences in their charge to mass ratio is more pronounced. When glycoforms differ by only slight variations in their glycosylation, reduction of the electroosmotic flow (by adding alkylamines to the separation buffer) may be necessary to achieve a complete resolution of the glycoforms. Watson and Yao (Watson and Yao, 1993) have first reported the well-known separation of the glycoforms of recombinant human erythropoietin (rHuEPO) into six well resolved peaks using an uncoated silica-fused capillary in a tricine buffer (pH 6.2) with the addition of 2.5 mM 1,4-diaminobutane (DAB) and 7M urea. Sialic acid residues are negatively charged at neutral pH values and thus, glycoforms eluted in a predictable manner in order of increasing numbers of sialic acid residues.

Although CZE has been the most studied mode for glycoforms separation, the major problem encountered is the adsorption of glycoproteins to the capillary wall. In this case, permanent coated or dynamically coated capillaries should be preferred (the latter ones offering the advantages to be cheaper, more stable and more compatible with the diode array detection). Several papers have also reported the use of neutral polymers like cellulose derivatives (*e.g.* methyl cellulose, hydroxyethyl cellulose (HEC), hydroxypropylmethyl cellulose (HPMC)) to pre-vent protein adsorption by dynamically coating the silica wall. The modified cellulose are generally used at very low concentrations. Phosphate ions are believed to strongly bind to the capillary surface converting acidic silanols to protonated silica-phosphate complex. In this case, capillaries are also called phosphate deactivated fused-silica.

Thus, CZE may be used for on-line analysis to monitor bioproduction of pharmaceutical glycoproteins at different stages of the purification or production process such as cultivation step, downstream process. CZE may also contribute to assess the recombinant glycoprotein identity, its purity, heterogeneity, quantity and stability.

2.1.3. Process Control

Reif et al. (Reif and Freitag, 1995) have reported the control of the cultivation process of recombinant antithrombin III (r-AT III). The downstream process was monitored by CZE using 50 mM phosphate buffer with 0.1% hydroxypropylmethylcellulose (HPMC) (pH 2.0) at 20°C, and an uncoated capillary under voltage ramping. The detection limit was found to be 50µg/ml using detection at 200nm increasing the sensitivity. One advantage of using CZE is that analysis time of 10 min is regarded as acceptable, especially as the down-streamm process takes several hours.

Somerville et al. (Somerville, Douglas et al., 1999) reported the CE separation of glycosylated and non-glycosylated recombinant human granulocyte macrophage colony stimulating factor (rHuGM-CSF) in 40 mM phosphate buffer, pH 2.5 and containing 0.03% HPMC. Glycosylated G-CSF eluted as a double peaks compared to the non-glycosylated recombinant human granulocyte colony stimulating factor (G-CSF) that eluted as a single and sharp peak. This technique enabled a detection limit of 10 µg/mL for the rHuGM-CSF. The doublet peaks could be due to the neuraminic acid residue present at the threonine site on the glycosylated molecule. This method has been used to reveal distinct differences between the bacculovirus expressed products and the comercially available glycosylated rHuGM-CSF, since bacculovirus expressed G-CSF gave a peak shape similar to that obtained for the non-glycosylated form although it is glycosylated. The difference in the glycosylation pathway between insect and mammalian cell explained the absence of the doublet peaks. These results have confirmed other reports indicating that insect cells are unable to process complex carbohydrate structures.

Using the phosphate deactivated fused-silica capillary method, an interesting application of CZE method to detect the effect of changes in the fermentation conditions on recombinant human growth hormone (rhGH) production from E. Coli has been reported. The rhGH and its variants could be resolved in less than 20 min from very crude mixtures of E. Coli using a 250 mM phosphate (pH 6.8)-1% (v/v) propylene glycol buffer whereas three different HPLC separations were required to detect the natural rhGH and its variants (McNerney, Watson et al., 1996).

2.1.4. Purity Testing

Pedersen et al. (Pedersen, Andersen et al., 1993) have studied microheterogeneity of the proteinase A for the evaluation of product purity and also of the suitability of the host organism. The CZE analysis of the proteinase A and the variant glycoform was performed, using an untreated silica capillary with 100 mM acetate-phosphate buffer (pH 3.2). Both molecules resolved into three peaks that probably correspond to charge heterogeneities atributable to differences in the phosphorylation level of the carbohydrate group at Asn68. This study revealed that approximately 70% of the product was native proteinase A presenting two glycosylation sites Asn68 and Asn269, whereas the remaining 30% was a proteinase A variant glycoform lacking the carbohydrate moiety at Asn269.

2.1.5. Batch-to- Batch Consistency

Hoffstetter-Kuhn *et al.* (Hoffstetter-Kuhn, Alt et al., 1996) have described the use of CZE based on complex formation of borate with carbohydrates in a borate buffer at pH 9.4 to monitor batch-to-batch consistency of monoclonal antibody (mAb). It was found that lower temperatures, higher borate concentrations and higher pH values improved the separation by stabilizing the complex, giving three peaks detected by UV absorbance at 200 nm. Firstly, linearity of the peak areas was measured up to a protein concentration of 0.1% w/v, with correlation coefficients better than 0.999. On the other hand, the separation method was validated for reproducibility (n=6) as the individual peaks were quantified with fairly good precision with relative standard deviations of the peak area ranging from 5.5% to 7.3% (5.1% for peak 3, 5.5% for peak 2 and 7.3% for peak 1). Electropherograms of different batches of the mAb showed two profiles quite similar while the third is markedly different. Although all three peaks were detected in the latter profile, the relative peak areas deviated significantly. This technique was therefore useful for stability testing of galenical formulations.

2.1.6. Stability and Quantification

The precedent method described by Hoffstetter-Kuhn *et al.* (Hoffstetter-Kuhn, Alt et al., 1996) showed that separation profiles changed distinctly after antibody storage in glass vials for three months at different temperature. Pronounced degradation (only one of the initial three peaks could be detected clearly) was observed at 37°C and even more after storage at 37°C for six months. Other stability testing of the final product formulations such as recombinant human tumor necrosis factor beta (rHuTNF-β) (Yao *et al.*, 1995) (Yao, Loh et al., 1995) and immunoglobulin G (IgG) (Klyushnichenko and Kula, 1997) (Klyushnichenko and Kula, 1997) have been successfully reported using borate as CZE buffer.

Bietlot *et al.* (1997) (Bietlot and Girard, 1997) developed a CZE method for quantitative recombinant human erythropoietin (rHuEPO) determination in final drug formulations. Large amounts of human serum albumin (HSA) are generally added to rHuEPO as a protein excipient. A complete separation of the two proteins without affecting the resolution pattern of rHuEPO into several glycoform populations could be achieved only through the addition of 1 mM nickel chloride to a 200 mM phosphate buffer, pH 4 with a fused silica capillary. Indeed, metal ions are known to interact with (glyco)proteins, to affect their properties and conformation, and then to alter their electrophoretic mobility. Then, it appears that the presence of nickel ions in the buffer decreases the electrophoretic mobility of HSA. This method was linear over the concentration range of 0.03-1.92 mg/ml, with limits of detection and of quantitation of

0.01 and 0.03 mg/ml, respectively. The precision of the method was evaluated from intra- and inter-day triplicate injections of rhEPO standard solutions and formulations over four batches. This method was found to be useful for quantitative measurement rhEPO in formulations (components of within- and between-batch variances were less than 5%). In addition, this method was found to be useful to reflect variations in manufacturing processes. Clear differences emerged first in the number and relative amounts of the resolved glycoforms, and second in peak shape and relative proportions of the four largest peaks (see Figure 1).

2.1.7. Natural Versus Recombinant Glycoprotein Comparisons

CZE has been shown to be useful to compare interleukin-2 (IL-2) in its native state and Escherichia coli (E. Coli) derived recombinant IL-2 (rIL-2) (Knuver-Hopf and Mohr, 1995) Using a coated capillary and 100 mM pH 2.5 phosphate buffer at 10kV, CZE separation of natural interleukin-2 (nIL-2) shows three different forms (non-glycosylated, glycosylated-monosialylated and glycosylated-disialylated). CZE analysis of rIL-2 exhibited two peaks with approximately the same electrophoretic mobility as the first peak in the electropherogram of nIL-2 under the same conditions. The authors suggested that one of those two peaks represents a rIL-2 form with a conformation slightly modified. This work demonstrated the high selectivity of CZE for separation of proteins with a single charge difference.

The potential of CZE for the routine analytical characterisation of glycoform heterogeneity of rtPA has been demonstrated by Thorne, 1996. An excellent precision of the method was determined thanks to the high migration time reproducibilities (relative standard deviation is less than 0.2%) and full protein recovery (resulting of the

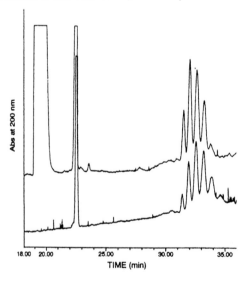

Figure 1. Comparison of the profile of rhEPO formulations from two manufacturers. Conditions: fused capillary 50μm×47cm, 200mM sodium phosphate buffer, pH 4.0, 1mM nickel chloride hexahydrate, 8kV, 20°C, 200nm.

addition of 0.01% (v/v) Tween 80 to the running buffer). These authors showed that CZE was a rugged technique using ω-amino acids buffer with PAA and PVA coated capillaries for the separation of rtPA variants.

2.1.8. Capillary Isoelectric Focusing (CIEF)

CIEF permits to analyze the charge heterogeneity of glycoforms on the basis of their degree of sialylation, phosphorylation and/or sulfatation.

The separation of charged analyte molecules takes place in a pH gradient created in a capillary by ampholyte mixtures under the influence of an electric field. Since the migration of charged molecules stop when their net charge is zero, they focalise in narrow zones in the region of pH corresponding to their isoelectric point (pI). A mobilization step is thus required to drive the focused zones past the detector window. Two CIEF techniques have been essentially developed:

In the two-step method, focusing and mobilization take place sequentially; the latter is achieved independently after the focusing step. Pressure mobilization is the simplest method. However if complete resolution of glycoforms is not attained, chemical mobilization may be preferred. Capillary with nearly no EOF have to be employed, using neutral coated capillaries such as polyacrylamide or polyvinyl alcohol coated capillaries.

In the one-step method, the components are mobilized simultaneously with the focusing, by means of the EOF which is maintained constant by the addition of polymers such as cellulose derivatives in buffers. Employing either uncoated or coated capillaries, faster separation are obtained using this approach. The stability of the coating is of prime concern as it will influence the velocity of the mobilization and thereby the reproducibility of the migration times.

For glycoproteins the pI depends not only on the number and type of charged amino acids of the protein backbone but also on the number of sialic acid residues of the glycan moiety. CIEF today offers a high resolving power (0.01-0.02 units of pH) and the advantages of automatization and rapidity. In addition, the concen-trating effect arising from the focusing step enables the detection of components present in small quantities. This mode of separation is particularly suitable for separation of glycoproteins bearing complex type oligosaccharides and exhibiting a good solubility in water. However, CIEF suffers from two main drawbacks including problems associated with reliability of pI markers, and with protein solubility.

The determination of pI values by CIEF is generally based on the comparison of migration times of the unknown peaks to those of pI standards added to the sample mixture (internal standards). Although protein with known pI may be used as pI markers, the main problem is their instability. Another approach consisted in the use of synthesized small molecular weight substituted aromatic aminophenols (Rodriguez-Diaz, 1997) synthetic (oligo)peptides (Shimura, Wang et al., 2000) with UV absorption which assure a pI reproducibility of 0.06%.

One of the main problem encountered in CIEF is precipitation of proteins at a pH close to their pI that may occur during the focalisation step, resulting in clogging the small-diameter capillaries used, in irreproducibility of migration times and peak area. A number has discussed about protein solubilization (Rabilloud, 1996); (Liu, 1996). In a recent review, we have made up a list of different solubilizers employed for CIEF of glycoproteins (Taverna *et al.*, 2001). In addition reviewed in details the desirable properties of a carrier ampholyte and its influence on the protein solubility during the focusing step.

Another problem encountered with CIEF of glycoproteins is the presence of salts in the samples. These salts can affect the pH gradient and can have adverse effect on reproducibility (Grossman, Wilson et al., 1988). Thus, a small change of pI values of glycoforms often can not be properly identified because of the effect of salt on the separation pattern. To counteract these problems, it is usually recommended to desalt the sample prior to separation (Wu, 1995). On-line (Clarke *et al.*, 1997) (Clarke, Tomlinson et al., 1997) and off-line (Cifuentes, Moreno-Arribas et al., 1999) desalting have been tested for CIEF of erythropoietin glycoforms. It has been then concluded that for sample with high salt content, off-line desalting before introducing the sample into the capillary was preferred to an on-line desalting step.

2.1.9. One-Step CIEF

One-step CIEF was introduced as early as 1992 (Krull and Mazzeo, 1992) permitting to reduce run times from 30 min for a two-step CIEF to 5 min through reversing the polarity (cathode at the end of the capillary) and shortening the separation distance.

Since then, this approach has been exploited by Moorhouse *et al.* (Moorhouse, Eusebio et al., 1995) to further improve the separation of the rt-PA glycoforms in less than 10 min. A neutral, coated capillary of 50 μm I.D. was used with HPMC added to reduce EOF to a constant and reproducible value. A 50:50 mixture of two different ampholytes (pH=5-8 and pH=3-10) at a 3% level, 4M urea, 7.5% N,N,N'N'-tetramethylethylenediamine (TEMED) and 0.1% HPMC ensure the best resolution and the solubility of the glycoprotein throughout the focusing, resolving the sample into ten glycoforms. With the optimal conditions (ratio of the two ampholytes, sample concentration, the concentration of urea, HPMC and TEMED) inter-assay precision for migration time (<5%) and for normalized areas (<10%) was adequate. These authors have validated the above method with minor modifications (Moorhouse, Rickel et al., 1996) in a series of experiments examining accuracy, precision, specificity and ruggedness of the method.

Lee (Lee, 1997) developed a rapid (<5 min), simple and reproducible one-step CIEF to resolve isoforms and glycoforms of monoclonal antibodies and to determine their pI. The method permitted the monitoring of the lot-to-lot consistency and of the purification of MAbs. Fused-silica capillaries of 50 μm I.D. were used, with reverse polarity. Samples and standards were dissolved in ampholyte-HPMC mixtures, with or without addition of urea which was needed to solubilize MAbs. Analyses both with and without urea, using two different sets of reagents and capillaries and performed on three consecutive days, showed that the migration times of isoforms were highly reproducible. Linearity of pI calibration was also investigated: the use of urea resulted in correlation coefficients slightly lower than in absence of urea, probably due to the pI modification of protein standards under (partially) denaturing conditions. The microheterogeneity fingerprints of MAbs can thus be determined and monitored in a simple, rapid and economical way.

More recently, our group developed a one-step CIEF method to resolve the glycoforms of the heterogeneous recombinant human immunodeficiency virus envelope glycoprotein (rgp 160) in 5 min (Tran, 2000). A typical glycoform pattern was obtained using an optimized mixture of ampholytes (narrow and wide pH range) and a combination of sucrose and 3-(cyclohexylamino)-1-propane-sulfonic acid which was shown to be the most efficient additive to avoid protein precipitation (see Figure 2). However, although the reproducibility of rgp160 separation depicted in Figure 2B, C and D was then confirmed, spikes appeared sometimes, indicating that the protein insolubility phenomenon was not completely resolved. The profiles obtained under optimal conditions for five standard markers with pI values ranging from 2.75 to 9.45 is also

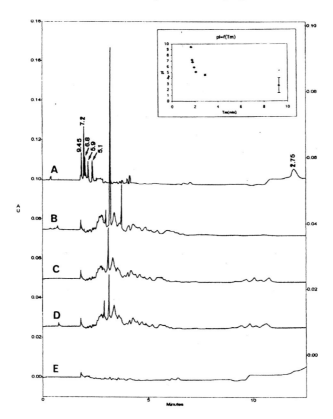

Figure 2. (A) One-step CIEF electrophoregram of protein standards. (B-D) Reproducibility of the separation of rgp 160 by the one-step CIEF. (E) Backgroung electrophoregram. **Inset:** Calibration curve (pI versus migration times) made from the mean of six analyses of the protein standards, bars indicate the corresponding SD. **Conditions**: PVA coated capillary 50μm×47cm; Carrier ampholyte solution: 5% of pH 3.5-5 ampholines, pH 5-8 ampholines and pH 3.5-10 ampholytes in the ratio (71/12/17: v/v/v), 1% TEMED, 0.085M CAPS, 6% saccharose in 0.1% HPMC;Concentration: 0.3-0.8 μg/μl of rgp 160 and 0.01-0.4 μg/μl of pI standards markers; Anolyte: 100 mM H₃PO₄; Catyholyte: 20 mM NaOH; Focusing and mobilization at –20 kV.

shown in Figure 2. Although the calibration curve (isoelectric point *versus* migration time displayed in the inset of Figure 2) showed a non-linear relationship, an adequate linearity could be yielded over a narrow range of pH as previously discussed by Schwer (Schwer, 1995). However, this relation permitted to exhibit the acidic character of the different glycoforms of rgp 160 (pI from 4.00 to 4.95). Equivalent resolution was observed using PVA or PAA coated capillaries but PAA capillary showed longer migration time and a profile more extended, indicating that PVA capillary exhibited a higher residual EOF. It has been shown that not only the intra-day reproducibility but also the long term stability were lower in the case of the PAA capillary. The application of this method to a comparison of glycoform patterns of the rgp 160 of two sub-populations (or clades) of the virus HIV-1 suggested that one clade exhibited not only a lower microheterogeneity but had also more acidic glycoforms.

2.1.10.   Two-Step CIEF

A two-step method with pressure mobilization was utilized for the routine analysis of rtPA (Thorne, Goetzinger et al., 1996) Using PAA coated capillary, 50 μm I.D., ampholine as carrier ampholyte and denaturing conditions produced by the addition of urea, rtPA was resolved into at least eight species, with pI values ranging from 6.4 to 9.2. The developed method is shown to be validatable: acceptable total protein recovery from the capillary (93%, n=3) and method reproducibility (RSD 2-3%, n=17 for the four major peaks) were demonstrated. Migration time precision was shown to be considerably improved if the glycoform peaks are bracketed with peptide pI markers, and their migration time calculated relative to the marker peaks.

Two-step CIEF has been used for monitoring recombinant human interferon-γ (IFN-γ) N-glycosylation during perfused-fluidized-bed and stirred-tank batch culture of CHO cells.(Kopp, Schluter et al., 1996). At least 11 differently sialylated glycoforms over a pI range of 3.4 to 6.4 have been resolved. Desialylation of the protein of a non specific neuraminidase resulted in a major shift in the molecular pI to a narrow peak grouping with pI values between 8.3 and 9.6. It has been shown that the degree of sialylation had a marked effect on the apparent pI of rIFN-γ. For quantitative analysis, the relative proportion of acidic rIFN-γ glycoforms increased after 210 h of culture, indicating an increase in N-glycan sialylation during establishment of the perfusion culture.

Two-step CIEF has been used for characterizing the purified recombinant antithrombin III (rATIII) produced in BHK cells by its isoelectric point (Reif and Freitag, 1995). Using a dextran-coated capillary 50μm I.D. and a mixture of four differents ampholytes (2% Ampholine pH 4-6, 0.5% Pharmalyte pH 3-10, 0.5% Pharmalyte pH 2.5-5, 0.5% Pharmalyte pH 4-6.5) in the presence of 0.01% HPMC, 0.1% TEMED and 0.001% triton X-100, hydrodynamic mobilisation allowed a pattern of six fractions focused in the pH range 4.7-5.3. It is important to note that desalting of the sample was crucial for the correctness of the determined pI.

In-process control in the production of ATIII and human clotting factor IX (FIX), two human plasma glycoproteins from the blood clotting cascade, has been investigated by CIEF (Buchacher, Schulz et al., 1998; Hunt, Hotaling et al., 1998) CIEF of the two glycoproteins turned out to require rather different measuring conditions: rATIII gave best results with a linear PAA-coated capillary, a mixture of ampholytes of ranges 3-10 and 3-5, as well as TEMED and HPMC as additives. FIX was separated in a CElect-H150 coated capillary using a mixture of ampholytes 2.5-5, 4.2-4.9 and 3-10 with only HPMC as additive. In both cases a cathodic mobilizing reagent was used to achieve mobilization, but the anolyte-catholyte couple was different as a result of extensive optimization experiments. These methods showed six isoforms of rATIII and four of rFIX, permitting the determination of their isoelectric points ranging from pH 4.3-5.2 and 4.1-4.5, respectively. Since both glycoproteins are also produced by recombinant mammalian cells, the newly developed capillary methods can be useful in the routine control process of the biotechnological production.

Hunt *et al.* (Hunt, Hotaling et al., 1998) have developed a CIEF method for determining the identity and the charge distribution of recombinant monoclonal antibody C2B8, by using a bioCAP LPA capillary 50μm I.D., a mixture of 2% of Pharmalyte pH 8-10.5, Bio-Lyte pH 7-9 and Bio-Lyte pH 3-10 containing 0.5% TEMED and 0.2% HPMC under chemical mobilization. The validation was performed in accordance with the guidelines of International Conference on Harmonization of Technical Requirements for Registration of Pharmaceuticals for Human Use, studied parameters such as linearity, accuracy, limits of detection and quantitation, repeatability, intermediate precision, specificity, robustness (evaluated with respect to the lot-to-lot variability of the reagents, the focusing and mobilizing voltage, capillary-to-capillary variability), use of alternate capillaries and alternate instruments, sample stability, ability of the method to monitor stability and the system suitability. The method was proven to be satisfactory with respect to each parameter studied.

Recently, Cifuentes *et al.* (Cifuentes, Moreno-Arribas et al., 1999) explored the effect of several experimental conditions which can affect the performance of CIEF separation including the presence of salts in the samples, the addition of urea, the pH range of the carrier ampholytes and the time length and applied voltage of the focusing step. The optimized CIEF method allowed the separation and quantitation in 12 min of at least seven peaks representing glycoforms of rhEPO with apparent pI in the range 3.78-4.69. In comparison with the well-known CZE method using putrescine and urea in tricine and sodium acetate buffer (Watson and Yao, 1993)these authors concluded that CZE should be the method of choice when strict control are required, and CIEF should be most convenient when rapid quality control have to be performed.

Tang *et al.* (Tang, Nesta et al., 1999) have evaluated the potential of one-step CIEF and two-step CIEF for routine analysis of recombinant Immunoglobulin G (rIgGs). It has been found that two-step CIEF yielded much higher resolution and reproducibility than the one-step method, particularly for glycoforms having high pIs (see Figure 3). Quantitative and qualitative rIgGs focusing profiles were obtained. Peak area response *versus* sample concentration and pI *versus* migration time exhibited linear relationships. Good reproducibility was obtained for peak areas (RSDs 1% intra-day, RSDs 8% inter-day) and migration times (RSDs 1% intra-day, RSDs 3% inter-day). No decrease in the peformance of the capillary was observed over 150 runs. These data demonstrated the superiority of CIEF for routine analysis of rIgGs.

2.2. EXAMINATION OF INTACT GLYCOPROTEINS BY MASS
SPECTROMETRY

If the glycoprotein contains a single glycosylation site, or very little diversity in its constituent glycan population, glycoforms can frequently be resolved by MALDI or electrospray (ESI) mass spectrometry (Figure 4).

The development of MALDI in the late 1980's (Karas and Hillenkamp, 1988) provided a technique for obtaining accurate molecular weights of glycoproteins up to a mass of about 200 kDa. In order to obtain a spectrum, the sample, typically in the low pmole range, is mixed, in solution, with a large excess (about 5000 fold) of a suitable solid matrix dissolved in 1 - 2 ml of solvent, allowed to dry and introduced into the mass spectrometer. Ionization is by irradiation with a laser, usually at 337 nm or 266 nm. The

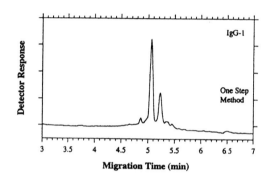

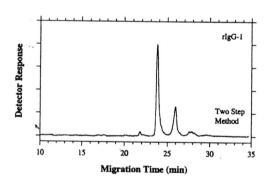

Figure 3. Comparison of the one-step and two-step CIEF methods for the separation of rIgG-1. **Conditions:** two-step CIEF: DB-1 dimethyl siloxane capillary 50µm×47cm (40 cm effective length); Catholyte 20mM NaOH in 0.4% MC; Anolyte 120 mM H₃PO₄ in 0.4% MC; 2% of 30/73 mixture of pharmalyte pH 3-10 and 8-10.5 in in 0.4% MC; 30°C; Focusing: 30 kV for 8-20 min; Mobilisation: 0.5 psi and 30 kV, 280 nm; one-step CIEF: neutral PAA-coated capillary 50µm×37cm(5.5 cm effective length); Catholyte 20mM NaOH; Anolyte 10 mM H₃PO₄; 4% of pharmalyte pH 3-10; 5% TEMED in 0.4% HPMC; 23°C; -15 kV; 280 nm

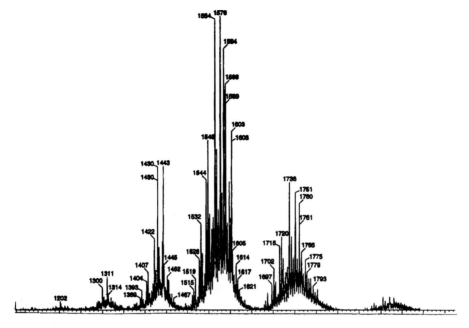

Figure 4. Glycoforms resolved by MALDI-mass spectrometry.

function of the matrix is to dilute the sample, minimise cluster formation, absorb the laser energy and ionize the sample. However, even today, the detailed mechanism of MALDI ionization is poorly understood. Proteins and glycoproteins are most success-fully ionised with sinapinic acid (3,5-dimethoxy-4-hydroxycinnamic acid) (Beavis and Chait, 1989) whereas lighter compounds give better responses with a-cyano-4-hydroxy-cinnamic acid (HCCA) (Beavis, Chaudhary et al., 1992) or 2,5-dihydroxybenzoic acid (2,5-DHB) (Strupat, Karas et al., 1991). Poor resolution generally restricts the use of MALDI on linear TOF instruments to studies of glycoproteins with masses below about 20 kDa if glycoform resolution is required. Above this mass, even structures differing by the mass of a monosaccharide residue may not be resolved. Early results from ribonuclease B (around 15 kDa), a glycoprotein containing five glycoforms each differing by one mannose residue, illustrate the problem (Harvey, 1992). Above about 20 kDa, only broad peaks are generally produced with little or no resolution of glycoforms. However, with the advent of high-resolution (20,000) reflectron-TOF instruments, it is now possible to resolve glycoforms at very much higher masses. As with the electrospray example illustrated above, MALDI can be used at higher masses to determine the presence or absence of N-linked glycosylation as illustrated by a study of carbohydrate-deficient glycoprotein syndrome (Wada, Gu et al., 1994). Resolution in this study was, however, inferior to that obtained in a parallel electrospray study by Yamashita et al. (Yamashita, Ohkura et al., 1993).

Electrospray mass spectrometry has made a significant contribution to the examin-ation of intact glycoproteins. In this technique, a solution of the sample molecules is ejected from a capillary needle carrying an electrical potential of 3-5 kV. The solvent is evaporated from the resulting spray of fine, charged droplets to leave charged sample ions of the type [M + X]nn+, where X = H or a metal, or [M - H]-. Larger molecules attract several charges depending on factors such as the proton affinity and number of chargeable sites, to give an envelope of ions differing by one charge state. Deconvolution of the spectrum leads to a very accurate measurement of the molecular weight. Furthermore, because the ions are multiply charged, they appear in the mass spectrum at relatively low m/z values allowing quite modest instruments such as quadrupoles to be used for their measurement even though masses of up to 100 kDa may be involved.

Subtraction of the mass of the protein (calculated from its amino-acid sequence) and any other post-translational modification, gives the mass of the attached glycans. Because N- and O-linked sugars are composed of relatively few isobaric structures (e.g. hexose, HexNAc etc), it is then a relatively simple matter to calculate the composition of the sugar by use of the residue masses listed in and even to propose a probable structure. For example, a neutral mammalian N-linked sugar of nominal mass 1641 reduces to a composition of (Hex)5(HexNAc)4 for which a biantennary structure is the most likely. It must be emphasised, however, that this proposed structure must be confirmed by additional information such as a composition and linkage analysis by GC/MS or by techniques such as MS/MS, NMR or exoglycosidase digestion.

Proteins and glycoproteins give exceptionally good electrospray spectra although the complexity of some glycoprotein structures often reduces the mass range at which glycoforms can be resolved. Hen ovalbumin, for example with a single N-glycosylation site containing about 20 glycans and a mass of only 44 kDa is only just resolved by electrospray (Duffin, Welply et al., 1992). With glycoproteins having simpler glycosylation patterns, it is sometimes possible to resolve glycoforms and detect site occupation when several sites are potentially glycosylated as illustrated by studies on human serum transferrin present in carbohydrate-deficient glycoprotein syndrome where compounds containing zero, one and two biantennary glycans have been resolved at 75 kDa (Yamashita, Ohkura et al., 1993).

2.3. MONOSACCHARIDE COMPOSITIONAL ANALYSIS FOR DETECTION OF PROTEIN GLYCOSYLATION

Monosaccharide compositional analysis (Lee, Loganathan et al., 1990; Townsend and Hardy, 1991; Anumula, 1994; Anumula and Du, 1999) represents one of the basic techniques for showing protein glycosylation and giving some information on the types of glycan present. It may be considered analogous to amino acid compositional analysis of proteins and is the technique which has been employed for the longest period of time in analysis of oligosaccharides. It is therefore a generally accepted technique and is still widely used for characterisation of recombinant glycoproteins. However, there are several technical problems associated with accurate analysis. Even when performed at high accuracy it has been shown that it may not give a true reflection of the type of glycosylation present on the protein and two forms of a glycoprotein which have different glycans present may have very similar monosaccharide composition.

To analyse the monosaccharide composition of a glycoprotein it is first hydrolysed under acid conditions. Sialic acids may be specifically cleaved by hydrolysis under relatively mild conditions followed by subsequent use of stronger conditions to release other monosaccharides. (Manzi, Diaz et al., 1990; Mawhinney and Chance, 1994; Rohrer, 2000) Use of appropriate standards in known amounts allows quantitation of each monosaccharide present. Allowance must be made for losses during preparation, the hydrolysis procedures and for accurate work an internal standard is usually added.

Monosaccharide compositional analysis was originally performed by tandem gas chromatography and mass spectrometry with ionisation detection GC/MS (1-3). Subsequently techniques were developed using column chromatography and HPAEC-PAD has been widely used   Higher sensitivity may be obtained by introduction of a fluorescent label such as 2-amino benzoic acid and such derivatised glycans may be analysed by reverse phase chromatography. A typical separation of standard monosac-charides is shown in Figure 5.

The monosaccharides derived from the glycoprotein may then be analysed and quantified by reference to the standard monosaccharides as shown for this example of recombinant IgG.

Although compositional analysis is a valuable technique for detection and monitoring of glycosylation, it should be noted that different glycans may give very similar values for composition so additional analysis of the glycan type is desirable (Patel, Parekh et al., 1992)

Compositional analysis (Lee, Loganathan et al., 1990; Townsend and Hardy, 1991); (Anumula, 1994; Anumula and Du, 1999) represents one of the basic techniques for demonstrating protein glycosylation and giving some information on the types of glycan present. It may be considered analogous to amino acid compositional analysis of proteins and is the technique which has been employed for the longest period of time in analysis of oligosaccharides. It is, therefore, a generally accepted technique and is still widely used for characterisation of recombinant glycoproteins. However, there are several technical problems associated with accurate analysis. Even when performed at

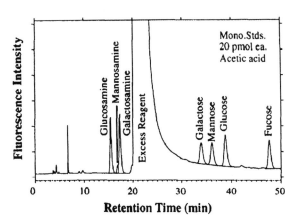

Figure 5. Analysis of standard monosaccahrides labelled with2-amino benzoic acid (taken from Anumula KR. Anal Biochem. 1994 220(2):275-83.

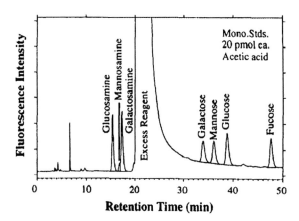

Figure 6. Analysis of monosaccahrides from IgG (taken from Anumula KR. Anal Biochem. 1994 220(2):275-83.

high accuracy it may not give a true reflection of the type of glycosylation present as different glycoforms may have very similar monosaccharide composition.

To analyse the monosaccharide composition of a glycoprotein it is first hydrolysed under acid conditions. Sialic acids may be specifically cleaved by hydrolysis under relatively mild conditions followed by subsequent use of stronger conditions to release other monosaccharides (Manzi, Diaz et al., 1990); (Mawhinney and Chance, 1994; Rohrer, 2000).Use of appropriate standards in known amounts allows quantitation of each monosaccharide present. Allowance must be made for losses during preparation, the hydrolysis procedures and, for accurate work, an internal standard is usually added.

Monosaccharide compositional analysis was originally performed by GC/MS with electron-impact (EI) ionisation (see below). Subsequently, column chromatography and HPAEC with pulsed amperometric detection (PAD) have been widely used. Higher sensitivity may be obtained by introduction of a fluorescent label such as 2-AB and such derivatised glycans may be analysed by reversed phase chromatography. Although compositional analysis is a valuable technique for detection and monitoring of glycosylation, it should be noted that different glycans may give very similar values for composition so additional analysis of the glycan type is desirable.

2.4. DETERMINATION OF GLYCOSYLATION SITE OCCUPANCY

Identification of site occupancy generally depends on enzymatic or chemical cleavage of the protein chain, separation of the resulting peptides and glycopeptides by HPLC and detection by mass spectrometry. A variety of mass spectrometric techniques can be used, either off-line as in the case of MALDI, or on-line as with electrospray (Huddleston, Bean et al., 1993). Earlier methods have been reviewed by Carr et al. (Carr, Barr et al., 1990). Conditions can usually be found that induce cleavage between the glycosylation sites to leave a mixture of peptides and glycopeptides with each glycosylation site, or group of sites in the case of O-linked glycans, localised to a single glycopeptide. Assuming that the sequence of the glycopeptide is known, the masses of the carbohydrates can be obtained by difference and their probable compositions deduced .

Enzymatic removal of the sugars with endoglycosidases such as protein-N-glycosidase-F (PNGase-F) or endoglycosidase-H (endo-H) and revaluation of the residual peptide mixture by combined liquid chromatography/MS (LC/MS) reveals the glycopeptides by their shift in the chromatogram (Liu, Volk et al., 1993). Both enzymes leave a modified protein and, thus, partial site occupancy can be determined. Endo-H, cleaves the sugar between the two GlcNAc residues of the core and, thus, leaves the reducing-terminal GlcNAc residue, together with any substituents, attached to the peptide. PNGase-F cleaves the sugar at the amide bond to leave an aspartic acid residue rather than asparagine and this can be detected by a mass increment of 1 Da (Ferranti, Pucci et al., 1995). The extent of acid formation gives an indication of site occupancy. If the incubations are performed in water containing H218O2, the labelled oxygen is incorporated into the asparagine resulting in doublet peaks separated by two mass units in the resulting mass spectra (Gonzalez, Takao et al., 1992; Küster and Mann, 1999), thus aiding detection.

In the approach taken by Huddleston and Carr (Carr, Huddleston et al., 1993; Huddleston, Bean et al., 1993), and which has now become a standard technique, electrospray spectra were acquired with a high ion-source cone voltage at the low mass end of a spectrum and a low voltage at the end. This procedure gives diagnostic glycan fragments such as those at m/z 163 (hexose), 204 (HexNAc) and 366 (Hex-HexNAc) whose single-ion plots can be used to map the glycopeptides on the total ion trace. Hunter and Games (Hunter and Games, 1995), however, have pointed out that the ion at m/z 204 derived from HexNAc may also arise from peptides and recommend the use of m/z 366 from HexHexNAc as being more specific.

Site analysis of O-linked sugars has not received as much attention as the N-linked carbohydrates, due largely to the clustered nature of many O-linked sites and to the absence of a general O-glycosidase. These glycans are usually released by b-elimination with sodium hydroxide, a procedure that leaves an unsaturated propyl substituent from threonine and dehydroalanine from serine-linked glycans (Greis, Hayes et al., 1996). However, some peptides are degraded by the base treatment necessary for b-elimination, thus reducing the general applicability of the method. In order to avoid degradation, Rademaker et al. (Rademaker, Haverkamp et al., 1993) have released sugars from serine-containing peptides with the milder reagent, ammonium hydroxide. Under these conditions, the reduced serine reacted with the ammonia to produce an aminated amino acid that exhibited a mass increment of 17 Da. Site analysis has also been performed by partial deglycosylation with trifluoromethanesulfonic acid to the level of a single GalNAc residue at each site followed by cleavage of the protein with the Arg-C-specific endopeptidase, clostripain. The resulting glycopeptide was analysed be MALDI mass spectrometry employing post-source decay (PSD) fragmentation (Müller, Goletz et al., 1997). However, there appears to be a tendency for GalNAc residues to be lost under PSD conditions from some glycopeptides (Alving, Körner et al., 1998), thus limiting the usefulness of the technique.

3.   Detailed Examination of Glycan Structure

Glycoproteins are generally too large and heterogeneous for structural studies of their attached glycans to be carried out directly and, consequently, they are typically broken down into smaller units before this type of analysis can be attempted. Two methods are in common use, production of small glycopeptides by chemical or enzymatic proteolysis or cleavage of the glycans, again either by chemical means or by the use of endoglycosidases.

3.1. ANALYSIS OF GLYCOPEPTIDES

Glycopeptides obtained by protease, usually tryptic, cleavage can be separated by HPLC and then examined by mass spectrometry to reveal the glycoform population as well as to define the site occupancy. The amino groups of the peptide can be protonated, allowing matrices such as HCCA to be used to improve detection limits by MALDI. One disadvantage of this approach is that both [M + H]+ and [M + Na]+ ions are produced, with the shorter peptides giving rise to higher proportions of [M + Na]+. Salt formation can also occur at the acidic site of the amino acid, thus complication the spectra further. This approach, combined with the first example of MALDI analysis in this area, was used by Sutton et al. (Sutton, O'Neill et al., 1994) to examine N-linked glycosylation from the two sites in the glycoprotein, recombinant human tissue inhibitor of metalloproteinase (TIMP) using a-cyano-4-hydroxycinnamic acid as the MALDI matrix. Further structural elucidation was performed by exoglycosidase digestion combined with MALDI analysis.

Pronase treatment has also been used in an attempt to reduce the protein to a single amino acid at the reducing terminus of the oligosaccharide. The basic amino group of this amino acid can then be protonated by electrospray, fast atom bombardment (FAB) or MALDI ionization (Wang, Chen et al., 1987; Bock, Schuster-Kolbe et al., 1994). However, complete digestion with pronase is often difficult, with some peptides being particularly resistant and the method has not been widely adopted.

3.2. STRUCTURAL ANALYSIS BY EXOGLYCOSIDASE DIGESTION

The large number of isomeric and branched structures encountered with carbohydrates makes detailed structural analysis a daunting task if viewed from a purely chemical perspective. However, because glycans are assembled and degraded enzymatically, the number of structures that actually exist naturally is reasonably small and, furthermore, the existence of the degradative enzymes can be used analytically to determine their structures. The technique, which is equally applicable to glycopeptides or released glycans, involves sequential removal of monosaccharide or small oligosaccharides from the reducing terminus with exo- or endo-glycosidases, respectively, with monitoring of the products either by chromatography or by mass spectrometry. This measurement will give the number of residues released and the specificity of the enzyme will reveal the type of monosaccharide and, in many cases, its linkage. Suitable enzymes have been described in a number of recent reviews (Montreuil, Bouquelet et al., 1986; Dwek, Edge et al., 1993; Küster, Krogh et al., 2001) and several examples of the technique are given below. Alternatively, digestions can be performed in parallel with mixtures of exoglycosidases such that digestion of individual samples proceeds to the point where the enzyme required for the next stage is missing (Edge, Rademacher et al., 1992). This procedure considerably reduces the number of steps that have to be performed in comparison with the isolation of individual compounds. It also requires less glycoprotein.

3.3. EXAMINATION OF GLYCAN STRUCTURE FROM RELEASED GLYCANS

Glycans may be released from glycoproteins either by chemical or enzymatic methods. Each technique has its strengths and limitations.

3.3.1. Chemical Release

N- (Takasaki, Misuochi et al., 1982; Patel, Bruce et al., 1993) and O-linked (Patel and Bhatt, 1993) glycans can both be removed from glycoproteins with hydrazine. This reagent cleaves peptide bonds, including that between the N-linked glycan and asparagine. O-Linked glycans are specifically released at 60oC, whereas 95oC is needed to release the N-linked sugars. Although all types of glycan can be released in this way, unlike the situation with some enzymatic release methods, the technique has several major disadvantages. Because all peptide bonds are destroyed, all information relating to the protein, such as the site of glycan attachment, is lost. Secondly, the acyl groups are cleaved from the N-acetylamino sugars and sialic acids. Normally, these acyl groups are replaced chemically in a reacetylation step on the assumption that they were originally acetyl. Although this assumption is true for the N-acetylaminohexoses, it is not always true for the sialic acids that frequently contain N-glycoyl groups and, thus, this information is also lost. The reacetylation step also often adds a small amount of acetyl substitution to hydroxyl groups. Thirdly, the reducing terminus of some of the glycans contains residual hydrazide or amino groups. Acetylation of this amino group produces a compound with a mass increment that is equivalent to that between a hexose and an N-acetylaminohexose and, thus, can lead to the apparent appearance of additional GlcNAc-containing glycans in a profile. Furthermore, these by-products contain a blocked reducing terminus that cannot subsequently be labelled with a fluorescent tag. Bendiak and Cumming (Bendiac and Cumming, 1985) have examined the hydrazinolysis/ reacetylation reaction in detail, and have calculated that as much as 25% of the total glycans are converted into products that bear nitrogen-containing groups at the reducing terminus. They further conclude that these compounds can never be converted into the parent sugar. Finally, if the hydrazinolysis conditions are too vigorous, then the N-acetylamino group can be removed from the reducing terminus. There have also been reports of the reducing-terminal GlcNAc residue being removed from, for example, plant glycans that contain an a1-3-linked fucose residue (Costa, Ashford et al., 1997). Unfortunately the automated system for hydrazinolysis release (Oxford GlycoSciences GlycoPrep) is no longer commercially available but it is possible to perform the same procedures manually as a suitable supply of anhydrous hydrazine is available from Glyco Inc.

3.3.2.   Enzymatic Release

Several enzymes are available for releasing N-glycans. The most popular is PNGase-F (Tarentino, Gómez et al., 1985), which cleaves the intact glycan as the glycosylamine and leaves aspartic acid in place of the asparagine at the N-linked site of the protein. The released glycosylamine readily hydrolyzes to the glycan except if the reaction is performed in ammonium-containing buffers (Küster, Wheeler et al., 1997) when a considerable amount of residual glycosylamine has been detected by MALDI mass spectrometry. Under these conditions, the glycan can be regenerated by incubation with dilute acetic acid.

PNGase-F releases most glycans except those that contain fucose α1-3 linked to the reducing-terminal GlcNAc (Tretter, Altmann et al., 1991) as commonly found in plants. In these situations, PNGase-A is usually effective. This enzyme is also capable of releasing glycans as small as GlcNAc, whereas PNGase-F appears to require at least two GlcNAc residues for effective release. In other respects, the glycan specificity of these enzymes appears to be similar although the rates of release of related glycans vary greatly as the result of differing protein structure (Altmann, Schweizer et al., 1995). Costa et al. (Costa, Ashford et al., 1997) have found an almost complete resistance towards PNGase-F of glycans from the 31 and 15 kDa subunits of the aspartic proteinase, cardosin A from Cynara cardunculus L. due to the presence of an α1-3-linked fucose residue. However, PNGase-A also failed to release the glycans from the intact glycoproteins, which had to be partially digested with pronase before incubation with PNGase-A was successful. Glycan profiles measured from the 2-AB derivatives were similar to those obtained by automated hydrazinolysis, although the latter technique was observed to cause substantial decomposition of the glycans at 95oC. N-Linked glycans from the Fe(III)-Zn(II) purple acid phosphatase of the red kidney bean (Phaseolus vulgaris) have been identified, following cyanogen bromide and trypsin digestion to isolate the glycosylation site (Stahl, Thurl et al., 1994) prior to incubation with PNGase-A. Glycosylation of soybean peroxidases have also been studied and found to contain mannose-type glycans with xylose attached to the core (Gray and Montgomery, 1997).

Endo-H is another popular enzyme for releasing N-linked glycans. It hydrolyses the bond between the two GlcNAc residues of the chitobiose core, leaving the core GlcNAc with any attached fucose attached to the protein. Information on the presence of core fucosylation is, thus, not available from the spectra of the resulting glycans. Another potential disadvantage of this enzyme is that it only releases high-mannose and hybrid, glycans although this property does yield some structural information.

For release of O-glycans the so-called O-glycanase enzyme is too specific for general use and a modified hydrazinolysis procedure should be employed if free glycans with a reducing-terminal are required. Otherwise, β-elimination is more appropriate.

3.3.3.   Enzymatic Release of N-linked Glycans from Protein Bands on SDS PAGE Gels

An 'in-gel' release method has been developed to release N-glycans directly from proteins on SDS PAGE gels. The sugars are released using peptide-N-glycosidase F

(PNGase F) (Kuster, Wheeler et al., 1997; Wheeler and Harvey, 2001) which hydrolyses the β-aspartylglycosylamine bond between asparagine and N-acetylglucosamine.

Once released the N-glycan pool can be analysed directly by MALDI TOF MS and after fluorescent labelling, by both MS or HPLC. The two methods are complementary, MS gives compositional analysis while HPLC gives structural information with respect to the type of monosaccharides in the oligosaccharide chains, their linkage and arm specificity. The preliminary assignments made to components of the glycan pool resolved by HPLC are confirmed by analysing the products of digestions of the intact pool with enzyme arrays (Guile, Rudd et al., 1996) either by MS or HPLC.

3.3.4. Labelling of Released Glycans

Carbohydrates are difficult to detect at low concentration because of the absence of a suitable chromophore or other tag. Consequently, most detection methods require prior labelling of the glycans. Fortunately, the glycans released by hydrazinolysis contain a unique reducing-terminal site that can be used for such labelling reactions. It is desirable that the derivatisation is non-selective in order for quantitative analysis and it should not cause structural changes such as desialylation. The incorporation of tritium into the C1 position of the reducing terminal monosaccharide by reduction with sodium borotritide fulfils these requirements most effectively and many studies have been performed by this technique (Takasaki, Mizuochi et al., 1982; Endo, Amano et al., 1986; Mizuochi, 1993). Tritiation does, however, require the use of relatively large amounts or radioactivity and the use of scintillation counting for high sensitivity work.

Fluorescent labels may also be introduced into the C1 position by reductive amination reactions, usually with aromatic amines, and a number of these labels including 2-AB (Bigge, Patel et al., 1995), 2-aminopyridine (2-AP), 2-aminoacridone (2-AMAC) (Okafo, Burrow et al., 1996; Okafo, Langridge et al., 1997), 3-(acetyl-amino)-6-aminoacridine (AA-Ac) (Charlwood, Birrell et al., 2000), and 2-aminoben-zoic acid (2-AA) have been described. The derivatisation employs conjugation via Schiff's base formation with the free reducing terminus with the product being stabilised by reduction with sodium cyanoborohydride. Such derivatives are ideal for detection following HPLC separations. For mass spectrometric detection, derivatization can be used to increase sensitivity by introducing either chemical groups with a high proton affinity or groups with a constitutive charge. Examples include the use of 4-aminobenzoic acid 2-(diethylamino)ethyl ester (ABDEAE) to introduce a tertiary amino group with greatly increased sensitivity over that of the free carbohydrate (Yoshino, Takao et al., 1995; Mo, Sakamoto et al., 1999) and the use of trimethyl-(p-aminophenyl)ammonium chloride (TMAA) (Dell, Carman et al., 1987; Okamoto, Takahashi et al., 1995) to introduce a charged trimethylammonium group. Other derivatives of the type used for introducing fluorophores such as 2-AB , 2-AMAC or AA-Ac do not, in general, produce a significant increase in MALDI mass spectral sensitivity although they aid proton attachment for electrospray.

3.3.5.   Analysis of Released Carbohydrates by Polyacrylamide Gel Electrophoresis

Another type of electrophoresis technique which is well suited to batch to batch analysis of recombinant glycoprotein glycans is separation by polyacrylamide gel electrophoresis (Jackson, 1991; Jackson, 1994; Jackson, Pluskal et al., 1994; Starr, Masada et al., 1996; Klock and Starr, 1998) (PAGE). The analysis of carbohydrates by PAGE is an attractive technique for its speed and simplicity (Jackson, 1996).

Two major types of analysis are commonly performed by SDS-PAGE are mono-saccharide composition (Starr, Masada et al., 1996) or profiling of glycans (Jackson, 1994; Jackson, 1994). The use for monosaccharide compositional analysis provides a technique which is simpler and quicker to perform than the more traditional approaches of GC-MS or HPLC based techniques. The major monosaccharides commonly found in recombinant glycoproteins, N-acetyl glucosamine, mannose, galactose, fucose, N-acetyl

run as borate complexes. Quantitation is also possible but attention has to be paid to the stoichiometry of the fluorescent labelling techniques employed. For precise measurements a known amount of standard should always be labelled along with the samples under study. Optimisation studies should be performed to ensure that a reproducible and quantifiable amount of the label is incorporated and that incorporation into the different monosaccharides is equivalent. A study was performed with the label 2-AA in which the effect of all variables in the reductive amination were examined an optimum conditions for labelling of oligosaccharides were determined (Bigge, Patel et al., 1995). However these will all have the same reducing terminal monosaccharide. If a standard is used then it should have the same terminal monosaccharide e.g. chitotrose for a standard for N-linked glycans. When considering efficiency of labelling of different monosac-charides it is important to recognise that this may vary and that use of standard mono-saccharides labelled under identical conditions to the sample is essential to ensure this.

The use of ANTS for quantifying monosaccharides has been described by Starr et al and protocols are available (Starr, Masada et al., 1996). This technique is increasingly being used for the routine determination of monosaccharides in recombinant glycoproteins where it is more a questions of monitoring the relative amounts of known mono-saccharides rather than trying to determine the composition in an unknown sample.

3.3.6.   Oligosaccharides

The analysis of oligosaccharides by PAGE is more complex and the wide variety of possible structures that might be present requires consideration of the properties influencing type of separation that is required. In many cases the two major properties influencing separation namely the charge and size of the molecules can produce opposite effects which may mean that bands overlap each other and can considerably complicate the analysis of the electrophoretic separation. Nevertheless it has been possible to obtain good separations for glycans derived from a number of recombinant glycoproteins although the technique does not give resolution which is possible with most HPLC separations it is often sufficient to give a profile which is characteristic and can be used for batch to batch comparison. In this respect it may even be an advantage to have a separation technique which only resolves major glycans for this particular application as the patterns are easier to interpret and comparison of different profiles is simplified. Of prime concern is the reproducibility and the ability to detect 'significant' changes in glycosylation patterns between glycoproteins.

Unfortunately there are number of properties of carbohydrates which complicate the analysis by such a method. Firstly many carbohydrates, with the notable exception of those containing sialic acid residues or those with phosphate or sulphate substitutions, do not carry any net charge and therefore would not migrate under electrophoresis in most buffer systems. Secondly carbohydrates possess no chromogenic groups or selective staining properties which may make them difficult to detect. It has however proved possible to overcome such limitations, generally by a derivatisation procedure, which addresses both of these problems.

In 1991 Jackson (Jackson, 1991) reported the development of a technique utilising the flourophore ANTS (8-aminonapthalene-1,3,6-trisulphonate) which both provides the necessary charge and also enables detection by fluorescence. The derivatisation employs conjugation via a Schiff's base on carbohydrates which have a free aldehyde reducing terminus which is then stabilised by reduction in a reductive amination to give a product suitable for electrophoresis on polyacrylamide gels. He also described the preparation of gels with a high percentage of monomer which is necessary to prepare gels with smaller pore size than generally used for protein work and which are essential to give good resolution of such relatively small molecules. Since then the use of other flourophores has been described in particular AMAC (2-aminoacridone) and 2-AA (2-amino anthanilic acid). These are attached through reductive amination reactions. Each of the labels have specific properties and these are shown in Figure 7.

In a review by Jackson in 1997 (Jackson, 1996) he describes the principles of the separation of carbohydrates by polyacrylamide gel electrophoresis and of the use of different buffer systems. He also shows how separation of small sugars, such as mono, di or tri-saccharides or oligosaccharides can be carried out by the use of suitable buffers. For ANTS labelled glycans stacking buffers containing chloride and running buffers containing glycine may be used. In the case of AMAC the lack of charge means that a borate buffer must be used to complex neutral glycans. He then gives numerous examples of separation of protein glycans labelled with ANTS and detailed protocols are provides for the analysis of differing amounts of glycoprotein labelled with ANTS as well as details of imaging by Polaroid photography and by Charge Coupled Device (CCD) camera systems.

The technology has also been reviewed by Starr et al (Starr, Masada et al., 1996) They present data on the kinetics of labelling with ANTS which showed that reproducible labelling could be obtained if the amount of glycan was less than 20nmol the collection of quantitative data using a CCD camera showed between 5 and 500pmol per band could be detected with a %CV of less than 10%. Data is shown for glycans released from glycoproteins, glycolipids and free glycosaminoglycans and protocols are given for each type of glycconjugate. Monosaccharide analysis can also be performed by using a borate buffer - the formation of the borate complex with different monosaccharides allows them to be separated in the gel. They then showed how further information can be obtained by use of specific exoglycosidases to digest the glycans before analysis on the gel and enables some sequence information to be determined.

Type of Label	Formula	Advantages	Disadvantages
ANTS 8-aminonapthalene-1,3,6-trisulphonate	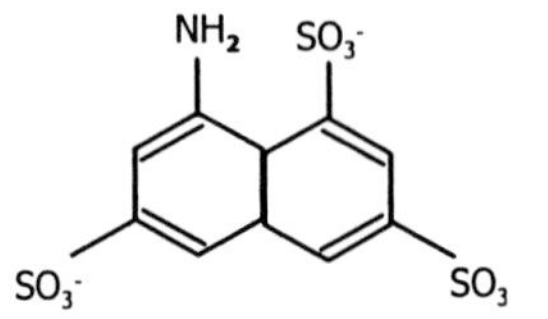	Charge gives glycans high mobility High fluoresence intensity	High mobility may mask glycan properties
AMAC 2-aminoacridone	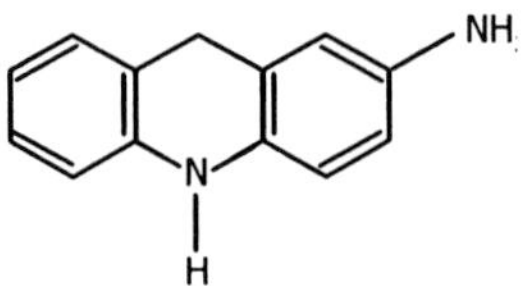	No charge	Separations based on glycan charge (substituents or borate complexes)
2-AA 2-amino-anthanilic (benzoic) acid	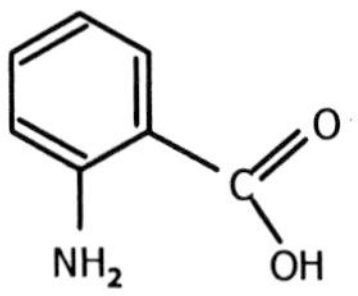	Small size Single negative charge Glycan properties predominate in separation	Lower fluorescence intensity

Figure 7. Fluorescent labels used for glycan electrophoresis.

Information can also be obtained by the use of enzymes in "fingerprinting" of glycan pools from glycoproteins to show different types of glycan present and to detect modifications such as sulphation and phosphorylation. They conclude that the technique is ideal for applications such as process development and quality control where monitoring of glycosylation is required.

The technique of fluorophore-assisted carbohydrate electrophoresis (sometimes referred to as FACE™) is finding growing utility in the field of analysis of glycoprotein glycosylation particularly where routine and comparative analysis is require such as in the monitoring of batch to batch consistency. Indeed the use of this type of technology has been recently approved by the FDA for pre-release analysis of batches of recombinant proteins, and there is no doubt that they will be more widely used in future. Masada et al (Starr, Masada et al., 1996) investigated the use of the technique in a quality control environment and determined the effect of a number of variables in sample preparation, electrophoresis ands gel formulation on the analysis of N-linked oligosaccharides in a quantitative manner. Enzymatic release with the enzyme PNGaseF was investigated on a number of stand glycoproteins and repeatability of released measured.

This was followed by studies on the consistency of analysis of the glycans under different electrophoretic conditions and with different batches of gels. They concluded that using relative luminance of bands gave good consistence with <5% CV for most standard glycoproteins but that temperature was an important parameter for consistency.

Detailed practical protocols for analysis of N-linked oligosaccharides, O-linked oligo-saccharides and imaging are also provided in the article by Klock and Starr (Klock and Starr, 1998). This article covers all aspects from the release of glycans from the protein and labelling. Guidance on problems that may occur in running the gels and in imaging them are also provided.

The analysis of IgG glycans labelled with 2-AA has been reported by Frears (Frears, Merry et al., 1999) smaller dye molecule in combination with Tris Glycine buffer gives excellent resolution of closely related neutral structures such as the bi-antennary glycans in IgG. Although not yet applied to recombinant proteins this would be a very good means of assessing the glycosylation of monoclonal IgG hybridomas on a routine basis and given the growing number of such antibodies being produced for therapeutic purposes this technique shows great promise for monitoring their glycosylation patterns (see Figure 8).

It seems that there are an increasing number of applications for this technology for routine use where the simplicity of the technique and the ability to provide comparative data in an easily understood form make this an attractive option in a monitoring environment. If required the techniques can also be supplemented with more sophisti-cated electrophoretic, HPLC or mass spectrometry techniques.

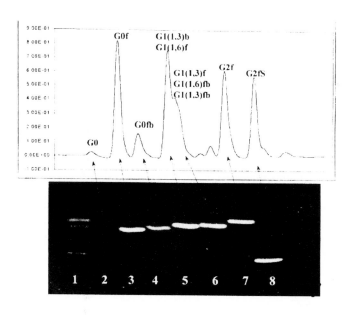

Figure 8. Correlation of peaks from normal phase HPLC with bands on SDS-PAGE. Bands of 2-AA labelled glycans were collected from HPLC and run on SDS PAGE. Taken rom Frears, E. R., A. H. Merry, et al. (1999).

3.3.7.   Analysis of Released Glycans by Capilliary Electrophoresis

To monitor routinely the glycosylation of recombinant glycoproteins different profiling methods have been developed. Apart the HPLC ones, capillary electrophoresis has proven its efficiency and has been successfully applied to perform the oligosaccharide mapping of glycoproteins. Three modes of CE are frequently encountered for oligosaccharide mapping: capillary zone electrophoresis (CZE) ; Micellar electrokinetic capillary chromatography (MEKC) and capillary gel electrophoresis (CGE)

Two major problems are encountered when dealing with CE of carbohydrates derived from glycoproteins. First, with the exception of complex oligosaccharides which may be charged through sialylation and/or sulfatation, other oligosaccharides such as high mannose chains are neutral precluding their direct separation in electrophoretic systems. Second, carbohydrates weakly absorb UV and they may be detected only at non specific low wavelengths (typically 200 nm).

However, neutral carbohydrates can be converted to charged species through chemical derivatization with a suitable tag which confers to the oligosaccharide a constant charge whatever the structure is and which increase the sensitivity of UV or fluorescence detection Alternatively, it is possible to exploit the ability of carbohydrate to form charged complexes with borate ions or with metal ions. Pre-column derivatization with a charged chromophore is now the most employed strategy.

Basically, three different types of derivatization reactions may be employed to label the oligosaccharide. A wide variety of derivatization reagents by reductive amination has been suggested for glycan labelling. The mechanism is based on the reaction of the amino group in the primary amine function of the derivatization reagent with the aldehyde group at the reducing termini of the carbohydrates. Other derivatization methods such as condensation of the carboxyl group in acidic saccharides with the amino group of sulfanilic acid or 7-amino-naphtalen-1,3-disulfonic acid can be employed for the introduction of these tags through the amide function (Mechref and el Rassi, 1994). A different reaction scheme involving condensation of the carbonyle group of the reducing termini of the carbohydrates with 1-phenyl-3-methyl-5-pyrazolone (PMP), yielding a bis-PMP derivative, is an attractive method for tagging reducing carbohydrates, especially sialylated oligosaccharides since it proceeded quantitatively under mild conditions and avoid loss of the sialic acid residue (Honda, Togashi et al., 1997).

Reaction efficiency is an important factor when choosing a labelling reagent, since high efficiency is especially required for the labelling of sialic acid-bearing carbohydrate chains that are labile under acidic conditions. A high reaction efficiency should make it possible to use milder derivatization conditions without loss of sialic acids or other labile residues during derivatization.

3.3.7.1.   CZE

The mobility of oligosaccharides in CZE can be related to their charge to mass ratio. Thus complex oligosaccharides can been separated according to their degree of sialylation, sulfatation, or number of antenna. However neutral oligosaccharides such as oligomannose type ones cannot be separated each from other and comigrate as a single peak.

CZE as borate complexes in alkaline borate buffer have been the major separation modes for underivatized carbohydrates. Sugar-borate complexes adsorb at 190-200 nm more readily than free sugars. In addition, separations especially those of neutral oligosaccharides are generally enhanced by borate complexation which in turns is dependant on both the pH and the borate concentration of the electrolyte. Greve et al. (Greve, Hughes et al., 1996) optimized the separation of glycans from the fusion protein (CTLA4Ig), a homodimer with a MM of 92 000 Da which is produced by recombinant techniques using a 50 mM boratebuffer with the addition of 30mM phytic acid. Oligosaccharides released from CTLA4Ig were evaluated under normal and thermally stressed conditions. The thermally stressed sample displayed a different pattern of peaks than the unstressed sample indicating selectivity with respect to oligosaccharide composition.

Introduction of chromophore or fluorescent tag by precapillary derivatization offers the advantage of enhancing the selectivity in CZE and at the same time sensitivities of detection. The choice of the appropriate tag is important not only for the spectral properties of the oligosaccharides but also for the electric charge of the derivatives. If the oligosaccharides are labelled with an anionic tag such as APTS (aminopyrene trisulfonate), ANTS (aminonaphtalene trisulfonate), or aminobenzoic acid, it is drawn to the anode and *vice versa* for a positively charged derivatizing agent (*e.g.* 2-AP, 3-acetylamino-6-aminoacridone).

To attach a charge to the carbohydrates to induce their electrophoretic migration, very acidic groups such as sulfonic acids are suitable since they remained ionized over a wide pH range, thus allowing modulation of the separation.

Guttman *et al* (Guttman and Starr, 1995) have compared the performance of slab gel electrophoresis and capillary electrophoresis at low pH for the profiling of N-linked oligosaccharides released from HIV envelope recombinant glycoprotein (gp 120). Using CE with a neutral coated capillary, oligosaccharides were separated as their ANTS-derivatives and monitored with laser induced fluorescence (He-Cd laser with anexcitation wavelength of 325 nm). The use of the acetate buffer pH 4.75 as running buffer resulted in very high reproducibility of the migration times (RDS<1%).With this method, an excellent resolution was attained without the use of a sieving matrix. CE was able to separate at least a dozen peaks while only nine to ten bands were present on the PAGE-line (See Figure 9).

Although not yet applied to recombinant glycoproteins, derivatization of oligosaccharide with 2-aminopyridine (2-AP)-is widely employed. Honda et al (Honda, Makino et al., 1990) have first reported the separation of labeled ovalbumin glycans by two modes including size separation using acidic buffer and separation at alkaline pH based on complexation of sugars with borate. Using these methods, five and nine 2-AP oligosaccharides derived from ovalbumin could be separated respectively. To be able to potentially map all the tremendous number of existing glycoprotein glycans, Zieske et al (Zieske, Fu et al., 1996) extended the dual mode analysis to a three dimensional map byusing a different combination of buffer systems. One dimension used sodium acetate buffer (200mM, pH 4.0) for electroendosmotic flow-assisted zone CE. A second dimension involved separation based on configurational difference of the hydroxyl groups using borate complexation electrophoresis in a polyethylene glycol-containing buffer

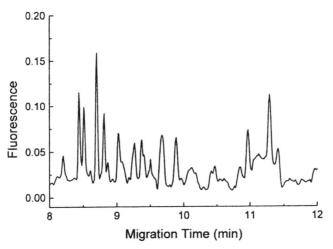

Figure 9. CZE/LIF separation of oligosaccharides released from HIV recombinant glycoprotein gp120 and derivatized with ANTS (314)

(500 mM sodium borate buffer, pH 8.5 containing 1% polyethyleneglycol). A third dimension developed specifically for neutral oligosacharides, using a sodium phosphate buffer (pH 2.5), has been shown to resolve neutral oligosaccharides which could not be separated by the other two dimensions. Thus, a three-dimensional map has been generated for the structures of 20 identified N-oligosaccharides, from which co-migrating unkown oligosaccharides could be characterized. Finally, the multi-dimensional mapping of the relative the migration times of the oligosaccharides can be employed to obtain structural informations on the separated oligosaccharides.

3.3.7.2. MECC mode

In micellar electrokinetic capillary chromatography (MECC), surfactants which may be anionic or cationic are added to the running buffer at a concentration exceeding the critical micelle concentration. Hydrophobic compounds may interact with the core of the charged micelles and this confers an electrophoretic mobility to uncharged oligosaccharides. To perform oligosaccharide mapping using MEKC, two main strategies may be employed. The oligosaccharides can be derivatized using a tag which converts the carbohydrate to hydrophobic derivatives that may include into ionic micelles, permitting their separation by MECC. Altenatively underivatized oligosaccharides can be analysed by exploiting their complexation with divalent cations.

An MEKC method for the profiling of underivatized oligosaccharides was successfully applied to establish the glycosylation pattern of two variants of the recombinant tissue plasminogen activator (rt<u>PA)</u> which differ by the presence or the absence of glycans at one of the glycosylation sites (Taverna, Baillet et al., 1995) rtPA produced from CHO cells consists of a mixture of two variants namely Type I and Type II. A 50mM phosphate buffer (pH 7.0) containing 50mM of sodium dodecyl sulfate (SDS) and 10mM of magnesium chloride was employed for this separation. The addition of Mg^{++}

cations to SDS solutions provided an effective means to enhance the selectivity of separation via complexation of Magnesium with both sugars and micelles. CE oligosaccharide mapping of rtPA displayed more than 20 structures separated, and two trisialylated structures differing by the type of linkage between the sialic acid and the galactose residue could be resolved. The method was also applied to compare the distribution of the oligosaccharides at each glycosylation site of rtPA. Oligosaccharides were released from the three glycopeptides, carrying individual glyco-sylation sites, using Glycanase F digestion and were subjected to CE analysis. As shown in Figure 10, 5 main regions of peaks corresponding to neutral oligosaccharide (mainly of oligomannose type); mono to tetrasialyaled structures could be distinguished. The porofile obtained for the Asn-184 and Asn448 glycosylation sites indicate that these two site carry nearly the same population of oligosaccharides (mainly complex type) (344).

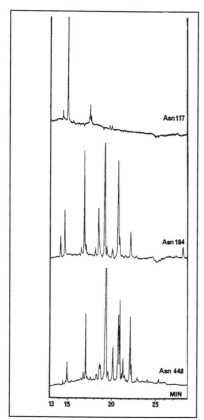

Figure 10. Oligosaccharide mapping of the three glycosylation sites Asn117; Asn184; Asn448 of the recombinant tissue plasminogen activator by CE.

The Camilleri group has introduced a neutral reagent 2-aminoacridone (2-AMAC) as a sensitive fluorophore for the analyses of glycans by MEKC using boric acid/taurodeoxycholate buffer system (Camilleri, Tolson et al., 1998). The migration behavior of linear oligosaccarides has been found to be related to size: solutes with the largest hydrodynamic volume migrated first. This methodology was used to obtain preliminary information on the mixture of oligosaccharides in human IgG monoclonal antibodies (Harland, Okafo et al., 1996) and in a recombinant soluble form of the complement receptor 1 (327). The mAbs, an IgG1 and IgG3 were purified from two Epstein Barr virus-transformed human lymphooblastoïd lines, oligosaccharides were then released by hydrazinolysis and derivatized with 2-AMAC. The detection of the oligosaccharides was performed using an helium/cadnium laser with an excitation wavelength of 442nm. The running buffer was a 500mM borate buffer with 80mM taurodeoxycholate pH8.5. Separation were performed on a fused silica capillary. 0.1 to 0.5 of carbohydrate from 5 to 25 ug of protein were sufficient for these fingerprinting.

3.3.7.3. CGE

Capillary gel electrophoresis is performed through the addition of an hydrophilic polymer to the running buffer. Oligosaccharides are generally derivatized with a ionic fluorophore which imparts the same charge to all the oligo-saccharides. Coated capillary with no electrosomotic flow are employed. Due to the presence of a sieving medium (e;g; agarose or linear polyacrylamide), the electrophporetic mobility of the oligo-saccharides is proportional to the log of their molecular mass.

Guttman et al. (Guttman and Starr, 1995) has reported the use of APTS to derivatize sialylated asparagin-linked oligosaccharides released from bovine fetuin which have been then separated by CGE. Derivatization at 37°C with malic, citric and malonic acids produced high labeling yields with retention of more than 90% of sialic acid residues on the oligosaccharides (Guttman, Chen et al., 1996). With 25 mM acetate buffer containing 0.4% polyethylene oxide, pH 4.75, a complete separation of two primary corresponding to trisialylated triantennary and tetraantennary trisialylated structure. Further, Guttman (Guttman, 1997) has described the oligosaccharide sequencing by specific enzymatic digestion of N-linked glycan pool from fetuin using carefully designed exoglycosidase matrix containing different mixtures of exoglycosidases, in conjonction with separation of the combined digests by CGE. Comparison of the migration times of the exoglyco-sidase digest fragments to the maltoligosaccharides of known size, enabled calculation of migration shifts, due to cleavage based on the actual exoglycosidases used. Hence, the particular sequence of each oligosaccharide in a released glycan pool could be proposed with high confidence based on the migration shifts of the various oligosaccharide structures. Additionally, high-sensitivity laser-induced fluorescence detection enabled complete sequence information from picomolar amounts of glycoproteins.

Chen et al. (Chen and Evangelista, 1998)have described a method for the analysis of N-oligosaccharides derived from glycoproteins including sialic acid and fucose residues containing carbohydrate chains. N-linked glycans released from several glycoproteins have been derivatized with APTS under mild citric acid-catalyzed reductive amination to preserve the structural integrity of N-oligosaccharides e.g. to reduce desialylation and to minimize the loss of fucose residues. The glycan fingerprinting has been performed by CE-LIF with 25 mM acetate buffer containing 0.4% polyethyleneoxide, pH 4.75. The profiles of heavily sialylated N-linked oligosaccharides derived rhEPO have been reported and yielded high resolution. The rhEPO profile was complex, containing multiple sialic acids in the oligosaccharides. Mild acid hydrolysis (0.1N TFA, 80°C, 1h) of the APTS-derivatized complex carbohydrates yielded species with substantially slower electrophoretic mobility a much less complicated pattern.

Similarly, N-linked oligosaccharides have been enzymatically released from an IgG antibody, derivatized with APTS to monitor the variation of glycosylation during manu-facture. The antibody, rituximab, is a genetically engineered mouse/human chimeric antibody to human CD20 antigen approved for the treatment of low grade non-Hodgkins lymphoma. The glycans present on rituximab are neutral complex biantennary oligosac-charides with zero, one and two terminal galactose residues (G0, G1 and G2, respective-ly). The oligosaccharides derivatized with 8-APTS were separated by CGE with laser induce fluorescence using a argon-ion laser with an excitation wavelength of 488 nm. The background electrolyte: carbohydrate separation gel buffer is provided by Beckman.

Capillaries with a variety of hydrophilic polymeric coatings (e.g. polyvinyl alcohol or polyacrylamide) could can be employed for this separation.. Figure 11 shows that the glycan are baseline resolved from each other. In addition two positional isomers (G1) were also resolved. The method was validated and the performance of this assay in terms of accuracy and precision were very satisfactory. The overall precision determined as the interlaboratory reproducibility was better than 0.2, 0.1 and 0.9% for G0, G1 and G2 respectively (319).

Using the same method, Raju et al. (Raju, Briggs et al., 2000) have undertaken a detailed structural study of N-linked oligosaccharides present in the IgGs of 13 different animal species in order to determine the extent of the variation of glycosylation. The N-linked glycans of IgGs have been released by PNGase F, derivatized with 9-aminopyrene 1,4,6-trisulfonic acid and analyzed by CE-LIF. The results indicated that glycosylation of IgGs is species-specific and revealed the necessity for appropriate cell line selection to express recombinant IgG for human therapy.

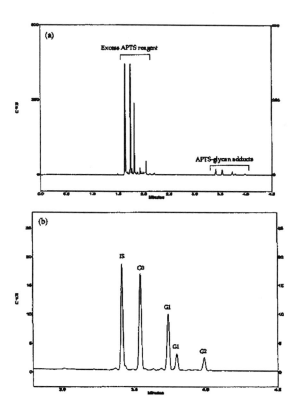

Figure 11. Typical electropherogram obtained by CGE showing (a) the glycan distribution on a sample of rituximab; (b) an expanded-scale view showing the glycans of interest.

4.   Analysis of Released Glycans by Liquid Chromatographc Techniques

The development of HPLC analysis for glycans presented several unique problem. In particular, glycans generally do not possess a strong chromophore or fluorophore for detection. In addition, they may be neutral and characteristically have have very similar monosaccharide composition and physicochemical properties. Therefore, their study has required the development of techniques both for their derivatisation, and separation which differ from those used for peptides.

Upon the release of the glycan by hydrazinolysis, endo-glycosidase or endo-glycopeptidase, a reducing terminus is generated. This enables the derivatisation of the glycans, by the relatively simple reaction of reductive amidation (Bigge, 1995).

It is desirable that the derivatisation is non-selective in order for quantitative analysis and it should not cause structural changes such as desialylation. The incorporation of tritium into the C1 position of the reducing terminal monosaccharide fulfils these requirements most effectively, and many studies have been performed by this technique (Takasaki, Mizuochi et al., 1982; Endo, Amano et al., 1986; Mizuochi, 1993) This technique does however require the use of relatively large amounts or radioactivity and the use of scintillation counting for high sensitivity work. Fluorescent labels may also be introduced into the C1 position by similar reductive amidation reactions and a number of these have now been described including 2-amino pyridine, 2-aminoacridone, 3-(acetylamino)-6-aminoacridine (AA-Ac), 2-aminoanthanilic acid and 2-aminobenzoic acid.

A number of chromatographic systems have been developed for separation including gas-liquid chromatography (Lowe and Nilsson, 1984) size exclusion chromatography on polyacylamide based beads, notably the BioGel P4 series (Yamashita, Hitoi et al., 1983; Kobata, Yamashita et al., 1987), ion exchange chromatography on standard matrices (Honda, Takahashi et al., 1981; Freeze and Wolgast, 1986; Jilge, Unger et al., 1989) and the development of specialised matrices for anion exchange (HPAEC) (Townsend and Hardy, 1991; Hardy and Townsend, 1994). They have been used in a large number of studies on protein glycosylation by all these techniques have certain drawbacks. Low pressure size exclusion chromatography require great care in column packing and run times are long. HPAEC requires the use of high pH and high salt buffers which have to be removed for further analysis of separated glycans, for example exoglyco-sidase sequencing. Use of labelled glycans with a fluorescent tag such as 2-AB (Bigge, 1995) with suitable HPLC matrices (Guile, 1996; Guile, 1998) provides a convenient and sensitive means of profiling and sequencing of glycoprotein glycans (Rudd, 1999; Rudd, 1997; Guile, 1996) which can now be performed by any well equipped laboratory.

The analysis of glycans by this technique falls into a number of stages including release of glycans form the glycoprotein. The techniques may be adapted according to the source and purity of the (glyco)proteins under study and of the techniques available for their analysis as will be discussed below.

4.1. ANALYSIS BY GEL PERMEATION CHROMATOGRAPHY

This technique originally developed by Yamasita and Kobata (Kobata, Yamashita et al., 1987) allows the separation of glycans on the basis of size. It has been traditionally employed in the analyses of many glycoproteins and the original determination of the glycosylation of such important recombinant glycoproteins as erythropoetin (Takeuchi, Inoue et al., 1989) and tissue plasminogen activator (Parekh, Dwek et al., 1989; Rudd, Joao et al., 1994) were performed by this technique. Most studies employed labelling by means of tritium introduced into the C1 position of the glycan by reductive amidation (Yamashita, Mizuochi et al., 1982; Kobata, Yamashita et al., 1987). The most widely used medium was BioGel P4 and the technique can be coupled with exoglycosidase digestion for complete glycan characterisation.

However the relatively long time required for analysis and the lack of resolution has led to this technique being largely superseded by HPLC techniques.

4.2. ANALYSIS BY HIGH PH ANION EXCHANGE CHROMATOGRAPHY (HPAEC) WITH PULSED AMPEROMETRIC DETECTION (PAD)

The problems associated with the separation and detection of glycans led to the development of novel matrices and detection techniques. One such combination that is widely employed is that of separation on a unique anion exchange bead matrix HPAEC (Paskach, Lieker et al., 1991) combined with electrochemical detection (PAD (Neuburger and Johnson, 1987; Hardy and Townsend, 1988; Hardy, Townsend et al., 1988; Townsend, Hardy et al., 1988; Hardy and Townsend, 1994)). This technique has been commercialised by Dionex. The matrix may be used for both monosaccharide and glycan analysis, (Freeze and Wolgast, 1986; Townsend, Hardy et al., 1988; Spellman, 1990; Goodarzi and Turner, 1995) and gives high resolution. The PAD detection technique allows non-derivatised glycans to be visualised with reasonable sensitivity.

Many recombinant glycoproteins have been analysed by this technique including prorenin (Aeed, Guido et al., 1992), follitropin, lutropin and choriogonadotropin (Amoresano, Siciliano et al., 1996), urinary plasminogen activator (Buko, Kentzer et al., 1991), human E-selectin (Burrows, Franklin et al., 1995) and recombinant acetylcholin-esterase (Chitlaru, Kronman et al., 1998). The technique does allow the

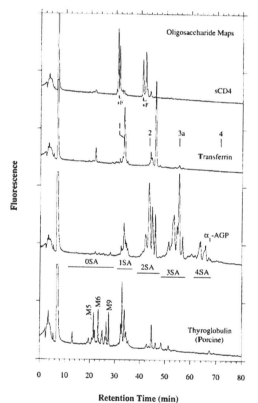

Figure 12. Separation of standard glycoprotein glycans and recombinant soluble CD4 glycans by HPAEC-PAD (From KR Anumula and ST Dhume Glycobiology 1998 8: 685-94)

rapid analysis of glycans and is widely used for routine glycan profiling. A typical chromatogram showing glycan separation id shown in Figure 12.

For details of methodologies reviews by Rohrer (Rohrer, 1995) and Townsend and Hardy (Townsend and Hardy, 1991; Hardy and Townsend, 1994) are recommended. Despite the convenience of the system there are some considerations that should be taken into account when considering this approach.

Drawbacks with the system are that the complex nature of interaction with the matrix makes prediction of glycan elution position difficult and a number of factors such as the carbon dioxide concentration of buffers are critical for reproducible chromatography. The detection sensitivity may vary to some extent depending on the type of glycan so precise quantitation can be difficult. Non-glycan components may also give a strong response and complicate analysis of the chromatograms. The high salt concentrations required for the separation and the high pH for detection make the use of the technique in a preparative manner difficult as removal of salt must be performed before concentration. This may also complicate subsequent analysis by, for example, mass spectrometry although with modifications this can be performed. (Richardson, Cohen et al., 2001) (Thayer, Rohrer et al., 1998). Nevertheless this technique is widely used in the routine analysis of glycosylation of recombinant glycoproteins.

4.3. SEPARATION OF RELEASED GLYCAN POOLS BY HPLC

Three types of HPLC matrices , weak anion exchange, normal phase amide and reversed phase C18 are commonly used for separating released oligosaccharides into pools according to their different characteristics (Guile, Wong et al., 1994; Guile, Rudd et al., 1996; Rudd, Guile et al., 1997). All of these HPLC systems can be used preparatively, thus individual peaks can be collected, the volatile buffers removed, and the glycans analysed by mass spectrometry or capillary electrophoresis to validate assignments (Rudd, Guile et al., 1997; Rudd, Mattu et al., 1999).

4.3.1. Weak Anion Exchange (WAX) Chromatography

WAX chromatography separates N-and O-linked sugars on the basis of the number of charged groups they contain. For example, Fig.?? shows the separation of a library of N-linked sugars from fetuin by PNGaseF using the chromatographic method described in (Guile, 1996; Rudd, 1997; Guile, 1994).

In the predominant separation mode is by charge but the size of a glycans will also have some small effect, thus within one charge band larger structures elute before smaller ones. The double peaks within the mono- and di-sialylated fetuin bands contain tri- and bi- antennary structures in which tri-antennary glycans elute first (see Figure 13).

4.3.2. Analysis by HPLC Following Fluorescent Labelling

The use of labelled glycans with a fluorescent tag such as 2-AB (Bigge, Patel et al., 1995) with suitable HPLC matrices   provides a convenient and sensitive means of profiling and sequencing of glycoprotein glycans which can now be performed by any well equipped laboratory. (Guile, Wong et al., 1994; Guile, Rudd et al., 1996)

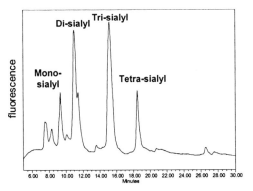

Figure 13. Separation of bovine serum fetuin glycans labelled with 2-amino benzamide on weak anion exchange chromatography.

(Anumula, 2000; Rudd, Guile et al., 1997) 2AB has a high molar labelling efficiency, and also labels all the oligosaccharide components of a glycan pool non-selectively so, when a fluorescence detector is used, they are detected in their correct molar proportions. 2-AB is also compatible with a range of separation techniques such as normal phase (NP), reverse phase (RP) and anion exchange HPLC, MALDI TOF and electrospray mass spectrometry (ES-MS).

The analysis of glycans by this technique falls into a number of stages including release of glycans form the glycoprotein, the 2-AB labelling of the glycans, the initial HPLC profiling, enzymatic sequencing of glycans and conformation of structures by mass spectrometry which are summarised in the diagram shown in Figure 1. The techniques may be adapted according to the source and purity of the (glyco)proteins under study and of the techniques available for their analysis as will be discussed below HPLC techniques for glycan analysis are now very sensitive and can be readily combined with further analysis by a number of techniques. In order to produce reliable gradients and good resolution however, good quality pumping systems with mixing at high pressure are required. A number of different column systems which rely on different properties of the glycan for separation are currently used. Where the glycans have been labelled with 2-AB they can be separated on an amide-silica column in normal phase chromatography as described by Guile et al (Guile, Wong et al., 1994; Guile, Rudd et al., 1996) Glycans labelled with 2-AA (Anumula and Du, 1999) or aminoacridone (Camilleri, Tolson et al., 1998; Charlwood, Birrell et al., 2000) may also be analysed on this column system. This gives separation based on hydrophilic interactions so the number of hydroxyl groups and therefore the size predominates but other features can contribute and closely related structures can be resolved.   Standardisation with a ladder of glucose oligomers allows glucose unit values to be calculated for each peak which can be predictive of the type of glycan as the residue, linkage and position in the glycan can all contribute to the glucose unit value (Guile, Rudd et al., 1996; Rudd, Guile et al., 1997). It is possible to combine data either on a different system such as reversed phase HPLC or by obtaining accurate values by mass spectrometry which can be helpful in characterising peaks. This may be off-line performed by MALDI-TOF or on-line by using a microbore column coupled to an electrospray mass spectrometer e.g. QTOF. At the present time however peak identification is usually carried out firstly by the use of specific exoglycosidases to progressively digest structures with the chromatogram. Wheen the digested glycan pool is compared with the orginal profile a change in position of a peak indicated that it has been digested by the particular exoglycosidase used. From knowledge of likely structures based on their elution positions and from their behaviour with carefully chosen glycosidase digests a full sequence can in many cases be determined. This is shown in the example of CD8 portrayed in Fig 17. Digestions with mixtures of exoglycosidases can be carried out on a pool of glycans and the data interpreted to show most of the structures present. This considerably reduces the number of steps which have to be performed in comparison with isolation of individual peaks and also requires less glycoprotein.

An alternative system is to use pyridylamino (PA) derivatives have been characterised on a number of different types of HPLC column including anion exchange reverse phase on octadecylsilica and normal phase on amide-silica columns to give a multidimensional separation which can allow identification of a particular glycan as described by Takahashi (Takahashi, 1996).

The strategy of 2-AB labelling provides a rapid and robust method of N- and O-glycan analysis which can be applied routinely to sugars released from picomoles of glycoproteins from a variety of sources. If the protein is in a purified form the glycans may be released directly from the protein solution. For more complex mixtures the proteins may be separated by SDS-PAGE and glycans released from the bands corres-ponding to the protein(s) under study. The technology is being further refined to achieve the high sensitivity required to analyse sub-picomole levels of glycoproteins in 2D gels.

An advantage of the 'in-gel' release method is that the protein remains in the gel after the sugars have been removed, allowing 'in-gel' proteolysis of the protein with trypsin and analysis of the peptide fragments by nanospray MS to enable identification of the protein. This is achieved by using a protein database to compare the molecular weights of the tryptic fragments with those of the predicted sequences of fragments from tryptic digests of known proteins (Kuster, Wheeler et al., 1997; Wheeler and Harvey, 2001)

The steps involved in the analysis of N-linked glycans include (a) releasing the sugars from the protein in the gel, followed by (b) labelling (c) separating and (d) characterising the components of the released glycan pool. The technology to release O-glycans from proteins in gels is still not available, since there is as yet no generic O-glycanase.

4.3.3. Normal Phase (NP) High Performance Liquid Chromatography (HPLC)

A sensitive and reproducible NP HPLC technology has been developed (Guile, 1996; Rudd, 1997; Guile, 1994) using the GlycoSep N column (Glyko Ltd.). This system can resolve sub-picomole quantities of fluorescently labelled neutral and acidic N- (and O-) linked glycans simultaneously and in their correct molar proportions. Elution positions are expressed as glucose units (gu) by comparison with the elution positions of glucose oligomers (dextran ladder). The contribution of individual monosaccharides to the overall gu value of a given glycan can be calculated from the elution positions of standard sugars and these incremental values may be used to predict some possible structures for unknown glycans on the basis of their gu values. The GlycoSep N column is able to resolve many arm specific isomers to give a further level of specificity (Rudd, 1999; Rudd, 1997). The resolving power and reproducibility of predictive NP-HPLC, and the ability to analyse sialylated and neutral sugars in one run, make it particularly useful for profiling both N- (and O-) glycan pools and for making comparisons between different samples. Figure 14 shows the NP HPLC for glycans released from IgG.

4.3.4. Reverse Phase (RP) HPLC

Reverse Phase HPLC, using C-18 columns, separates sugars on the basis of hydrophobicity. Samples applied in aqueous buffer to the RP column are eluted with increasing concentrations of organic solvent. The standard dextran ladder is not well resolved over the length of the reverse phase gradient, therefore the elution positions of sugars separated by RP are measured in comparison with an arabinose ladder and are assigned arabinose units (au). Closely related glycans that co-elute on NP HPLC can often be separated by RP (Figure 15) and can be used, for example, to distinguish between bisected and non-bisected structures which elute very close to each other on NP HPLC. HPLC profiling of 2AB labelled glycans requires only 5% of the material needed for one MALDI MS run of the same material.

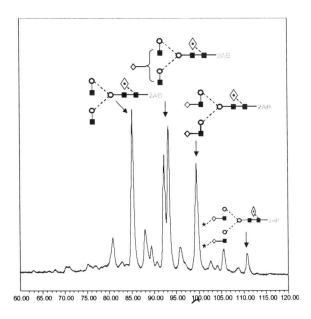

Figure 14. Analysis of IgG glycans labelled with 2-AB by NP-HPLC. The major types of biantennary complex glycans are shown.

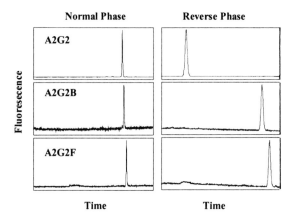

Figure 15. Analysis of closely related bi-antenarry complex glycan structure by normal and reverse phase.

4.3.5. Simultaneous Sequencing of Oligosaccharides Using Enzyme Arrays

The preliminary assignment of structures from the initial NP-HPLC run can be confirmed rapidly by sequencing all of the oligosaccharides in a glycan pool simultaneously using enzyme arrays. Enzymatic analysis of oligosaccharides using highly specific exoglycosidases is a powerful means of determining the sequence and structure of glycan chains. However, until recently, it was necessary to isolate single sugars from the glycan pool for digestion with exoglycosidases. The high resolving power of the NP HPLC system allowed a new approach to be developed (Rudd, Guile et al., 1997; Rudd, Mattu et al., 1999) This involves the simultaneous analysis of the total glycan pool by digesting aliquots with a set of enzyme arrays (see Figure 16).

After overnight incubation the products of each digestion are analysed by NP HPLC or by MALDI TOF MS. On the HPLC system, structures are assigned to each peak from the known specificity of the enzymes and the pre-determined incremental values of individual monosaccharide residues. To illustrate this technique, the rapid profiling and simultaneous analysis of the major N-glycans attached to the leucocyte antigen CD8 is shown.

5.   Analysis of Released Carbohydrates by Mass Spectrometry

5.1. FAST ATOM BOMBARDMENT (FAB)

Larger carbohydrates could not be analysed as intact molecules before about 1981 when FAB was introduced (Barber, Bordoli et al., 1981) as a technique for ionizing molecules of intermediate mass (up to about 10 kDa). In this method, the sample, dissolved in a liquid matrix of low volatility, such as a mixture of glycerol and thioglycerol and bombarded with either a beam of neutral argon or xenon atoms or with a beam of caesium ions. Unlike electron-impact ionization, which produces odd-electron

CD8 N glycans

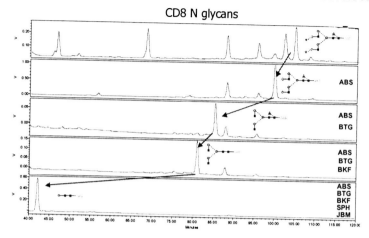

Figure 16. Exoglycosidase Sequencing of glycans from recombinant CD8. ABS Arthrobacter Ureafaciens sialidase; BTG Bovine Testes galactosidase : BKF Bovine Kidney Fucosidase ; SPH Streptococcus Pneumoneae Hexosaminidase ; JBM Jack Bean mannosidase.

ions, FAB usually results in the production of more stable "pseudo molecular ions" that contain an additional hydrogen ([M + H]+) or sodium atom ([M + Na]+) and have an even electron configuration. With all but the smallest molecules, derivatization, usually by permethylation (Hakomori, 1964; Ciucanu and Kerek, 1984) or peracetylation (Dell, Carman et al., 1987; Dell, 1990; Dell, Khoo et al., 1993) is necessary in order to reduce hydrophobicity and promote diffusion of the molecules to the matrix surface where they are ionized. However, permethylation needs somewhat vigorous conditions, is rarely quantitative, and requires sodium-removal from the product. An alternative strategy is to use derivatisation of the reducing terminus by reductive amination with compounds containing hydrophobic groups (Poulter and Burlingame, 1990).

Glycan profiling by FAB has made a substantial contribution to our knowledge of glycan structure with the identification and confirmation of several novel structural types such as tyvelose- (Reason, Ellis et al., 1994) and GalNAc-containing (Chan, Morris et al., 1991) N-linked glycans. However, it is now used by only a few laboratories on account of its relatively poor sensitivity and signal-to-noise ratio. Anionic sugars such as those containing sialic acids generally present no particular problems but, although sulphated oligosaccharides give good negative ion FAB spectra, the sulphate group is frequently not seen in spectra recorded in the positive ion mode (Chan, Morris et al., 1991).

FAB spectra usually contain large numbers of fragment ions that can provide much structural information (Dell, Khoo et al., 1993). However, their presence often prevents the determination of a true glycan profile. Fragment ions are usually produced by glycosidic cleavage, with Y and B ions (see below) being particularly abundant, especially adjacent to HexNAc residues where ions of the type [Gal-GlcNAc + H]+ (m/z 464) are produced. These ions tend to eliminate methanol if there is a free methoxy group in the 3-position of the GlcNAc residue but, otherwise, there is little in the fragmentation spectrum that yields much more information of linkage.

5.1.1. Linkage Analysis by FAB Mass Spectrometry

Linkage information can, however, be obtained directly from FAB spectra following additional chemical modification such as periodate oxidation (Angel and Nilsson, 1990). This reaction causes oxidative cleavage of the carbon-carbon bond between carbon atoms bearing cis-hydroxy groups with the formation of a di-aldehyde. Reduction of this di-aldehyde with sodium borodeuteride followed by methylation gives a product whose mass differs from that of the original permethylated oligosaccharide. Thus, for hexoses, the presence of 1 and 4 linkages causes cleavage between the 2- and 3-carbon atoms resulting in a shift of the residue weight of the hexose moiety from 204 mass units to 208 units. Although the presence of linkages at the 1 and 2 positions also cause a similar mass shift due to cleavage of the 3 - 4 bond, the linkage can be recognised by the presence of a prominent fragment ion produced by loss of methanol in the spectrum of the unoxidised oligosaccharide. Hexoses containing 1 and 3 linkages produce no periodate cleavage product because of the absence of a cis-diol group whereas linkage at the 1 and 6 positions cause cleavages between both C-3 and C-4 and between C-4 and C-5 resulting in loss of the carbon atom at C-3. A terminal hexose gives rise to a prominent ion at m/z 179. Other mass shifts are listed in the paper by Angel and Nilsson (Angel and Nilsson, 1990).

5.1.2. Determination of Anomericity by FAB Mass Spectrometry

Determination of anomericity is a topic not easily addressed directly by mass spectrometry. However, Khoo and Dell (Khoo and Dell, 1990) have made use of the property of chromium trioxide in acetic acid to oxidise specifically b-pyranoses to keto esters. a-Pyranoses remain essentially intact under these conditions. The carbohydrates were first deuteroacetylated and then heated with the reagent for 2 hours at 50oC before being examined by FAB mass spectrometry with thioglycerol as the matrix. Oxidation produced a mass increment of 14 mass units and the difference in mass recorded before and after oxidation defined the number of b-anomers. Deuteroacetylation rather than acetylation provided differentiation between the reaction product and native acetate groups. The method was fairly specific but it was noted that some 1-6 linked sugars underwent oxidation at the 6-position.

5.2. MATRIX-ASSISTED LASER DESORPTION/IONIZATION

The ability of MALDI mass spectrometry to ionize underivatised N-linked carbohydrates was first demonstrated by Mock et al. in 1991 (Mock, Davy et al., 1991) and the method is now one of the most powerful available for glycan analysis. The technique, as applied to carbohydrate analysis, has been reviewed (Harvey, 1999). ([M + Na]+) ions are produced from most carbohydrates but [M + H]+ ions are sometimes present in the spectra of those derivatised at the reducing terminus. Detection is usually with a TOF analyser, although magnetic sector instruments fitted with array detectors have been used (Harvey, Rudd et al., 1994).

MALDI provides much improved resolution over HPLC, in terms of the number of glycans resolved, particularly for larger structures, and the resulting signal strength

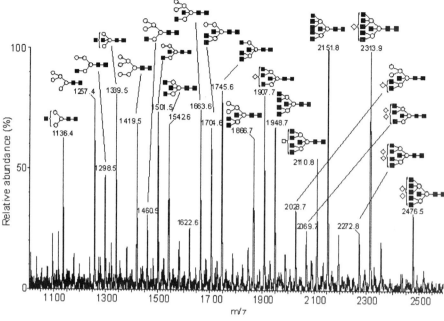

Figure 17. MALDI profile

appears to reflect sample concentration reasonably accurately (Harvey, 1993; Naven and Harvey, 1996) providing that the glycan mass is not too high (Chan and Chan, 2000). Furthermore, unlike FAB spectra, MALDI spectra are generally not complicated by the presence of fragment ions except when sialic acids are present. Sialylated glycans have a tendency to eliminate sialic acid, both in the ion source and in the flight tube of TOF instruments (see Figure 18).

Ions formed in the flight tube of reflectron-TOF mass spectrometers give rise to unfocused PSD or metastable ions in the MALDI spectra. Their mass, which is instrument-dependent, can be calculated from the masses of their respective parent and focused fragment ions (Harvey, Küster et al., 1999). The problem of sialic acid loss has been addressed in several ways. Tsarbopoulos et al., (Tsarbopoulos, Bahr et al., 1997) for example, advocate acquisition of spectra in linear mode to avoid separation of molecular and fragment ion peaks whereas Powell and Harvey (Powell and Harvey, 1996) have stabilised the sialic acids by forming methyl esters. Because sialylated glycans form both negative ([M − H]−) and positive ([M + Na]+) ions, thus splitting the signal, methyl ester formation has the advantage of converting all ionization into the positive mode with restoration of the quantitative relationship between neutral and acidic compound types. Without such derivatization, however, sialylated glycans can be examined as negatively charged ions. Further complications with mass spectra of all anionic glycans arise from their tendency to form salts with alkali metals, thus giving rise to multiple peaks in the spectra. However, addition of a small amount of salt to the MALDI matrix promotes quantitative salt formation and largely prevents sialic acid loss.

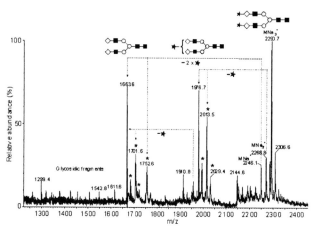

Figure 18. Loss of sialic acid on MALDI mass spectrometry.

The first MALDI matrix to be used for neutral oligosaccharides, 3-amino-4-hydroxybenzoic acid, was reported by Mock et al. in 1991 (Mock, Davy et al., 1991). However, this matrix has now been superseded by others, of which 2,5-DHB is the most widely used (Stahl, Steup et al., 1991). When this matrix:sample solution evaporates, DHB tends to crystallise from the periphery of the target spot in the form of long needles that point towards the centre of the target leaving much of the target blank. Methods used to improve crystallization include the addition of 10% 2-hydroxy-5-methoxybenzoic acid to produce a mixture known as "super DHB" (Karas, Ehring et al., 1993), addition of 1% 1-hydroxyisoquinoline (Mohr, Börnsen et al., 1995), addition of spermine (Mechref, Baker et al., 1998) or simple recrystallization of the DHB target spot from ethanol (Harvey, 1993). Other matrices suitable for the ionization of carbohydrates include 2-(4'-hydroxyphenylazo)benzoic acid (HABA) (Mohr, Börnsen et al., 1995), 3-aminoquinoline (3-AQ) (Metzger, Woisch et al., 1994), 5-chloro-2-mercaptobenzithiazole (Xu, Huang et al., 1997), b-carbolines (Nonami, Fukui et al., 1997; Nonami, Tanaka et al., 1998), osazones (Chen, Baker et al., 1997), ferulic acid (Kim, Shin et al., 1998), 6-aza-2-thiothymine (Geyer, Schmitt et al., 1999) and various polyhydroxyacetophenones (Krause, Stoeckli et al., 1996). The latter compounds appeared particularly effective for ionising sialylated glycans. Osazones and, in particular, arabinosazone, provide good ionization of sulphated glycans with a very low matrix background (Wheeler and Harvey, 2001). The more basic matrices such as b-carbolines and 3-AQ are capable of producing [M + H]+ as well as [M + Na]+ ions. Although the latter matrix has been reported to be superior to 2,5-DHB for ionization of sialylated sugars (Stahl, Thurl et al., 1994), Papac, Wong, and Jones (Papac, Wong et al., 1996) have found that it sublimes too rapidly to be of great practical use and we have noted Schiff base formation from reducing sugars in an acid environment (Harvey and Hunter, 1998). Although less fragmentation has been reported from this matrix than from 2,5-DHB, it appears to be more sensitive to the presence of contaminants (Stahl, Linos et al., 1997). The use of 6-aza-2-thiothymine has enabled exoglycosidase digestions to be performed on the MALDI probe in the presence of the matrix (Geyer, Schmitt et al., 1999), thus allowing sequential digests to be undertaken. Although other methods for on-target enzymolysis have been reported (Küster, Naven et al., 1996; Colangelo and Orlando, 1999), the matrix, DHB, was in each case, added after exoglycosidase digestion.

Although MALDI is reasonable tolerant to the presence of small amounts of buffer salts and other contaminants, larger amounts cause problems and can inhibit crystal formation. Several recent micro-purification techniques have, thus, been developed for clean-up of MALDI samples. Salts, for example, may be removed by drop-dialysis (Marusyk and Sergeant, 1980; Görisch, 1988) on a 500 Da cut-off membrane and both salts and residual peptides can be removed with a Nafion membrane in its hydrogen form (Börnsen, Mohr et al., 1995). Rouse and Vath (Rouse and Vath, 1996) have removed contaminants by adding small amounts of resins to the sample:matrix solution on the MALDI target to adsorb salts and other interfering material. Huang et al. (Huang and Riggin, 2000) report that detergents, residual glycoproteins and other hydrophobic contaminants can be removed with Supelco SP20SS resin. Bioaffinity clean-up was employed by Wang et al. (Wang, Sporns et al., 1999) in a technique that involved derivatization of the glycans with EZ-Link biotin hydrazide from Pierce using the reductive amination procedure. These compounds were then extracted by binding to avidin that had been immobilised on a small piece of Cannon NP transparency film type E attached directly onto the MALDI probe.

5.3. ELECTROSPRAY

Electrospray has had much less impact on the examination of neutral oligosaccharides than the desorption techniques such as FAB or MALDI, mainly because the neutral oligosaccharides lack a site with any appreciable proton affinity. Thus, ion formation usually requires the presence of a metal such as sodium or potassium which is often added to the electrospray solvent in the form of sodium acetate (Duffin, Welply et al., 1992). Addition of acetic or formic acid can produce protonation but not with the exclusion of the metal-adducted species. Doubly-charged ions are usually more abundant than their singly-charged analogues with spectra being dominated by species such as MH_2^{2+}, $MHNa_2^+$ and MNa_2^{2+} particularly when the glycans contain a reducing-terminal derivative that can be protonated (Charlwood, Tolson et al., 1999). Two major advantages of electrospray over MALDI mass spectrometry is its ability to detect sialylated glycans without desialylation (Huang and Riggin, 2000) and its ability to be coupled directly with HPLC.

The ready coupling of electrospray with liquid chromatography and HPLC, together with the high solubility of oligosaccharides in the normal electrospray solvents, provides great impetus for development in this area (Camilleri, Tolson et al., 1998; Charlwood, Langridge et al., 1999; Charlwood, Tolson et al., 1999). Dionex high pH ion exchange chromatography is frequently used for oligosaccharide analysis and effluents from these columns have also been examined (Conboy and Henion, 1992). Anionic micromembrane suppression was used by these authors to remove the high base concentrations and the resulting solution was examined by ionspray (pneumatically assisted electrospray).

5.4. PLASMA DESORPTION (PD)

Like MALDI, this technique is also performed with a TOF instrument, but generally has lower resolution and is now regarded as having only historical interest. Nevertheless, it has given results with intractable compounds such as heparin fragments (McNeal, Macfarlane et al., 1986). The sample, on a thin metal target is placed into the spectrumeter and bombarded with fission fragments from 252californium. Fission of this isotope ejects two fragments in diametrically opposed directions; one fragment ionises the sample and the other starts the timing clock for the TOF process. A major disadvantage of the method is the considerable time, often several hours, necessary to acquire a spectrum.

5.4.1. Information From Electron Impact Ionization

Although electron impact ionization is not capable of ionising large, intact glycans, it has, nevertheless made a substantial contribution to the structural analysis of small carbohydrates and their hydrolysis products. In order to achieve ionization, molecules in the vapour state are bombarded with electrons to form unstable M^- species. In nearly all cases, chemical derivatization such as permethylation or trimethylsilylation (Sweeley, Bentley et al., 1963) is required to produce the necessary volatility. The technique is ideally coupled with on-line GC/MS and, when combined with specific derivatization techniques, as described below, provides the main methods for the determination of constituent monosaccharide composition and linkage of larger carbohydrates. Compounds as involatile as monosaccharide phosphates can be examined if suitably derivatised (Harvey and Horning, 1973). Although monosaccharides are the most frequent targets, a few investigators have reported GC/MS analysis of polysaccharides containing as many as six or seven monosaccharide residues (Karlsson, Carlstedt et al., 1989). However, column temperatures in the region of 400oC are needed for the larger compounds and, above this temperature, most compounds decompose.

5.5. COMPOSITIONAL ANALYSIS BY METHANOLYSIS AND GC/MS

The nature of constituent monosaccharides is usually not apparent from direct mass or fragmentation spectra and must be obtained by other means. Most commonly, this involves cleavage of the polysaccharide into its constituent monosaccharides and examination by GC/MS. Hydrolysis with aqueous acids, although efficient, is problematical on account of the acid lability of some monosaccharides (Biermann, 1989; Merkle and Poppe, 1994). Unwanted side reactions such as deacylation also occur. Consequently, methanolysis, a milder procedure, is often used in its place. The technique involves heating the oligosaccharide with methanol containing hydrogen chloride to produce methyl glycosides which are then converted into their trimethylsilyl ethers for GC/MS evaluation (Fergusson, 1993). The method provides both a retention time and a mass spectrum to aid identification. All commonly-occurring monosaccharides are separable on gas-liquid chromatographic (GLC) columns of low polarity and, although some sugars give multiple GLC peaks, identification is usually unambiguous.

5.6. LINKAGE ANALYSIS BY GC/MS

Only limited information on linkage is normally obtained from the mass spectra of intact carbohydrates. However, mass spectrometry still provides the best method for determination of this property using a method, first developed in 1968 (Hellerqvist, Lindberg et al., 1968) and commonly referred to as "methylation analysis" (Hellerqvist, 1990). Many variants exist, for example the "reductive cleavage" modification introduced by Rolf and Grey (Rolf and Gray, 1982) but all depend initially on complete derivatisation of the intact carbohydrate, usually accomplished by permethylation (Hakomori, 1964; Ciucanu and Kerek, 1984). Next, the carbohydrate is hydrolysed to the constituent monosaccharides and the newly introduced hydroxyl groups are derivatised with another reagent such as acetic anhydride. The products, which are specifically labelled with acetyl groups at the points of linkage, are then examined by GC/MS, in this case as partially methylated alditol acetates, or PMAAs, following reduction with sodium borohydride to their alditols in order to avoid multiple GLC peak formation as the result of equilibration between the anomeric forms. Among modifications of the basic technique, Anumula and Taylor (Anumula and Taylor, 1992) have described a procedure that allows all stages of the method to be performed in two days in a "single pot" with minimal sample loss. Most PMAAs give unique GC/MS properties but the method is unable to distinguish between a few sugars such as 4-linked aldohexopyranose from 5-linked aldohexofuranose because each product contains the same substitution as the result of acetyl group addition both at the linkage and ring-closure positions. The reducing terminal sugar can be specifically labelled with deuterium if the oligosaccharide is reduced with sodium borodeuteride before methylation. The review by Hellerqvist (Hellerqvist, 1990) lists ions diagnostic of specific PMAAs.

5.6.1. Mass Spectrometric Fragmentation of Carbohydrates

Modern mass spectrometers are now able to produce fragmentation spectra from small amounts of complex carbohydrates with great reproducibility and such spectra often provide sufficient information to characterise a carbohydrate without recourse to laborious wet chemical procedures. Carbohydrates show two general types of fragmentation, cleavages between the monosaccharide rings and cleavages across the rings (Figure 19).

The former type of fragmentation, known as glycosidic cleavage, produces the most abundant fragment ions and provides information on sequence and branching (Reinhold, Reinhold et al., 1995). A hydrogen migration, usually involving a hydroxylic hydrogen atom (Hofmeister, Zhou et al., 1991), always accompanies the glycosidic bond cleavage. Cross-ring cleavages occur mainly from [M + metal]+ rather than from [M + H]+ ions (Orlando, Bush et al., 1990) and yield more information on linkage. They are generally weak in the fragmentation spectra recorded at low energies but become more abundant as the energy rises (Lemoine, Fournet et al., 1993; Harvey, 1997). Unfortunately, because of the branched nature of many carbohydrates, several cleavage pathways can often lead to fragment ions of the same mass. Such "internal cleavage fragments" considerably hinder the interpretation of many spectra. Another recently-reported phenomenon is that of "internal residue loss" (Kovácik, Hirsch et al., 1995) in which

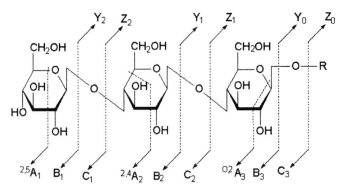

Figure 19. Types of glycan fragmentation seen in mass spectrometry.

migration of constituent monosaccharides occurs during fragmentation of $[M + H]+$, but not $[M + Na]+$ ions (Brüll, Kovácik et al., 1998), with loss of residues that previously did not occupy terminal positions. Fucose appears to be particularly prone to such migrations (Mattu, Royle et al., 2000), especially when present in glycans that have been derivatised at the reducing terminus.

The accepted nomenclature for describing fragment ions derived from carbohydrates is that devised by Domon and Costello (Domon and Costello, 1988). Ions retaining charge on the reducing terminus are labelled X (cross-ring), Y ($C1\tilde{O}O$ glycosidic) and Z ($O\tilde{O}Cx$ glycosidic) or A, B and C if the charge is located at the non-reducing end (Figure 4). Subscript numerals denote the bond or sugar ring broken, starting from the reducing end for the X, Y and Z fragments and the other terminus for the others. The heaviest chains from branched compounds are additionally given a, b etc. suffixes with a assigned to the heaviest chain. Bonds involved in cross-ring cleavage fragments are denoted by a superscript prefix consisting of the lowest numbered carbon (or oxygen) atom of the two bonds cleaved. In the spectra of most carbohydrates the B and Y ions are usually the most abundant.

Fragmentation obtained from reflectron-TOF instruments after MALDI ionization is generally that from spontaneous cleavages occurring within the flight tube (post-source decay, PSD) (Spengler, Kirsch et al., 1995) although this can be enhanced by incorporating a collision cell (Mechref, Baker et al., 1998). This fragmentation gives mainly B-, C- and Y-ions, tends to be weak and provides only moderate mass measurement accuracy. With instruments equipped with time-lag focussing (also known as delayed extraction) (Vestal, Juhasz et al., 1995), increasing the delay reduces the PSD ion intensity as a greater proportion of the fragmentation occurs within the ion source (Naven, Harvey et al., 1997). Such in-source fragmentation has also been reported from magnetic sector instruments following MALDI ionization (Harvey, 1995). Much improved fragmentation is obtained by collision-induced decomposition (Gillece-Castro and Burlingame, 1990; Lemoine, Strecker et al., 1991), particularly with tandem quadrupole-TOF (Q-Tof) instruments (Charlwood, Birrell et al., 1999; Charlwood, Birrell et al., 1999; Harvey, 2000). The electrospray ion source of the Q-Tof instrument produces

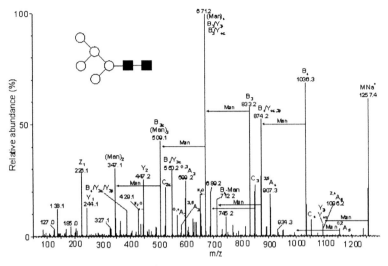

Figure 20. Glycan Fragmentation in Q-Tof.

mainly [M + 2H]2+ ions from glycans derivatised at their reducing termini with fragment ions consisting almost exclusively of glycosidic fragments. With high cone voltages, however, prominent [M + Na]+ ions can be produced, the fragmentation of which gives rise to both glycosidic and cross-ring fragments (Harvey, 2000) (Fig. 20).

Unfortunately, considerable in-source fragmentation occurs under these conditions and, furthermore, there is a decrease in ion production with increasing molecular weight. Both problems can be overcome by using recently-developed MALDI ion sources with these Q-TOF instruments (Harvey, 2000).

More detailed structural analysis of glycans can be performed by utilising successive stages of fragmentation in an ion trap instrument (Weiskopf, Vouros et al., 1997; Weiskopf, Vouros et al., 1998). Either residues, such as those containing HexNAc that direct fragmentation can be removed to allow detailed structure of the remainder of the glycan to be examined or ions corresponding to selected regions of the glycan can be removed and fragmented further. A library of carbohydrate epitopes has been constructed to aid the latter approach (Tseng, Hedrick et al., 1999).

5.7. PROTOCOLS FOR HIGH SENSITIVITY GLYCAN ANALYSIS BY HPLC AND MASS SPECTROMETRY

Küster et al. (Küster, Wheeler et al., 1997; Küster, Hunter et al., 1998) have developed a method whereby glycoproteins were separated or isolated by one dimensional SDS-PAGE gels prior to glycan release from the reduced and alkylated glycoprotein by cleavage with PNGase-F from within the gel. The glycans were extracted and subsequently desalted by passage through a mixed bed resin column of AG-3 (removal of anions), AG-50 (removal of cations), and C18 (removal of organic material) packed into a gel-loader pipette tip. Samples were examined directly by MALDI or HPLC combined with exoglycosidase digestion. The AG-3 resin was found to be essential for the production of MALDI signals, but unfortunately it removed most acidic glycans. For retention of sialic acid-containing glycans, the methyl ester derivatization method developed by Powell and Harvey (Powell and Harvey, 1996) was adopted and for examination of sulphated glycans, the sample could be desalted with porous graphatized carbon (Wheeler and Harvey, 2001). Sensitivity was such that only 100 pmoles of glycoprotein applied to the gel was sufficient for a complete glycan analysis. More recently, it has been found that more thorough washing of gel and clean-up with a Nafion-117 membrane enables sialylated glycans to be detected without the need for permethylation.

Charlwood et al. (Charlwood, Birrell et al., 2000) have extended this method to the analysis of N-linked glycans separated by 2D SDS-PAGE gels. In the first of three protocols, PNGase-F-released glycans were extracted from the gels and desalted with GlycoClean-H resin from Oxford GlycoSciences (OGS) before being labelled with AA-Ac and further cleaned with a Waters OASIS cartridge. In the second procedure, the glycoproteins underwent in-gel tryptic digestion and glycans were released from the resulting glycopeptides after extraction from the gel. In the third, the gel-separated glycoproteins were blotted onto polyvinylidine difluoride (PVDF) membranes and the excised spots were subjected to PNGase-F digestion to release the glycans. In all cases, the released glycans were examined by both MALDI and HPLC analysis. The best results were obtained from glycoproteins containing neutral glycans when these were released from within the gel. These glycans were detected from as little as 0.5 mg of glycoprotein (ovomucoid). Glycans containing sialic acid were not recovered as effectively and better results were obtained either by release from the tryptic peptides or by desialylation prior to in-gel release. Recoveries from the PVDF membranes were poor.

Another sensitive method for N-glycan analysis is that reported by Papac et al. (Papac, Briggs et al., 1998) in which MALDI spectra of sialylated N-linked glycans were obtained from as little as 0.1 mg of recombinant tissue plasminogen activator. The method relied on the absorption of up to 50 mg of the glycoprotein directly onto PVDF (Immobilon P) membranes of a 96-well MultiScreen IP plate (pore size 0.45 mm), where the glycoprotein was reduced and alkylated before the glycans were released with PNGase-F. Tris-acetate buffer (10 mM, pH 8.5) rather than the more usual sodium phosphate was used for the incubation in order to avoid any lactone formation from sialic acid residues (Papac, Wong et al., 1996). Samples were desalted with AG50W-X8 resin, a process that was found to cause about 3% loss of sialic acid. MALDI analysis was performed with THAP/ammonium citrate (negative ion) or with super-DHB.

6.   Conclusions

Glycosylation analysis is an important aspect of the production of recombinant glycoproteins both in the initial characterisation and product registration and also in batch analysis of product prior to release. As shown in this article a number of techniques are now available and a choice of the most suitable may be made based on the requirements and available resources. Although some of the techniques may be unfamiliar to many who are more used to protein analysis, they are now fully developed and validated and can be incorporated into the routine analysis with a minimum of

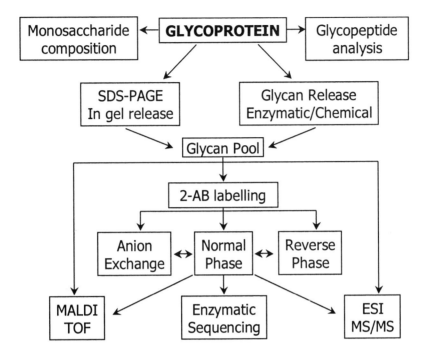

Figure 21. Scheme for Glycosylation Analysis.

training. Fig 21 shows a general scheme for recombinant glycoprotein analysis using techniques described in this chapter.

The importance of glycosylation analysis is growing as it is increasingly being required by both the FDA and European regulatory authorities. It enables monitoring to ensure all aspects of product consistency in particular with respect to pharmacokinetics. Its value may be gauged from a recent patent case which hinged on the glycosylation of recombinant EPO. The engineering of glycosylation can even lead to an improved versions of the therapeutic as in the case of EPO from Amgen and investment in glycosylation analysis can readily be justified.

6.1. ACKNOWLEDGEMENTS

The authors would like to thank Max Crispin, David Neville for their advice and comments on the manuscript.

7. Abbreviations

AA-Ac, 3-acetamido-6-aminoacridine; 2-AB, 2-aminobenzamide; ABDEAE, 4-aminobenzoic acid 2-(diethylamino)ethyl ester; aFGF, acidic fibroblast growth factor;

2-AMAC, 2-aminoacridone; ANTS, 8-aminonapthalene-1,3,6-trisulphonate; 2-AP, 2-aminopyridine; 3-AQ, 3-aminoquinoline; AT, antithrombin; au, arabinose units; BSA, bovine serum albumin; CCD, charge coupled device; CE, capillary electrophoresis; CHO, Chinese hamster ovary; CMC, critical micelle concentration; CV, coefficient of variation; CZE, capilliary zone electrophoresis; Da, Dalton; DAB, diaminobutane; DAPS, N-dodecyl-N,N-dimethyl-3-amino-1-propanesulfonate; DHB, dihydroxybenzoic acid; DNA, deoxyribonucleic acid; DSPA, Desmodus salivary plasminogen activator; EI, electron impact; Endo-H, endoglycosidase-H; EOF, electro-osmotic flow; EPO, erythropoietin; ESI, electrospray; FAB, fast-atom bombardment; FACE, fluorophore-assisted carbohydrate electrophoresis; FDA, Food and Drug Administration; FSCE, free-solution capillary electrophoresis; Gal, galactose; GalNAc, N-acetylgalactosamine; GC/MS, gas chromatography/mass spectrometry; GH, growth hormone; GLC, gas-liquid chromatography; GlcNAc. N-acetyl-glucosamine; GM-CSF, granulocyte-macrophage colony stimulating factor; GPC, gel permeation chromatography; gu, glucose unit; HABA, 2-(4'-hydroxyphenylazo)benzoic acid; HCCA, a-cyano-4-hydroxycinnamic acid; HCV, hepatitis A virus; Hex, hexose; HexNAc, N-acetylhexosamine; HPAEC, high pH anion exchange chromatography; HPCE, high-performance capillary electrophoresis; HPLC, high performance liquid chromatography; HPMC, hydroxypropylmethylcellulose; HAS, human serum albumin; HU, human; IFN, interferon; IGF, insulin-like growth factor; IgG, immunoglobulin G; IL, interleukin; LC/MS, liquid chromatography/mass spectrometry; mAb, monoclonal antibody; MALDI, matrix-assisted laser desorption/ionization; MEKC, micellar electrokinetic capillary chromatography; MS, mass spectrometry; m/z, mass to charge ratio; NMR, nuclear magnetic resonance; NP, normal phase; OGS, Oxford GlycoSciences; PAA, polyacrylamide; PAD, pulsed amperometric detection; PAGE, polyacrylamide gel electrophoresis; PEG, polyethylene glycol; PD, plasma desorption; PMAA, partially methylated alditol acetates; PNGase, peptide-N-glycosidase; PSD, post-source decay; PVA, polyvinyl alcohol; PVDF, poly(vinylidine difluoride); Q, quadrupole; Q-Tof, quadrupole-time-of-flight; r, recombinant; rh, recombinant human; RP, reversed phase; SDS, sodium dodecyl sulphate; THAP, trihydroxyacetophenone; TIMP, tissue inhibitor of metalloproteinase; TMAA, trimethyl-(p-aminophenyl)ammonium chloride; TNF, tumor necrosis factor; TOF, time-of-flight; TPA, tissue plasminogen activator; UV, ultraviolet; WAX, weak anionic exchange.

8.   References

Aeed, P.A., D.M. Guido, et al. (1992). "Characterization of the oligosaccharide structures on recom-binant human prorenin expressed in Chinese hamster ovary cells." Biochemistry 31: 6951-61.

Altmann, F., S. Schweizer, et al. (1995). "Kinetic comparison of peptide: N-glycosidases F and A reveals several differences in substrate specificity." Glycoconjugate Journal 12: 84-93.

Alving, K., R. Körner, et al. (1998). "Nanospray-ESI low-energy CID and MALDI post-source decay for determination of O-glycosylation sites in MUC4 peptides." Journal of Mass Spectrometry 33: 1124-33.

Amoresano, A., R. Siciliano, et al. (1996). "Structural characterisation of human recombinant glyco-hormones follitropin, lutropin and choriogonadotropin expressed in Chinese hamster ovary cells." Eur J Biochem 242: 608-18.

Andersen, D. C., T. Bridges, et al. (2000). "Multiple cell culture factors can affect the glycosylation of Asn-184 in CHO-produced tissue-type plasminogen activator." Biotechnol Bioeng 70: 25-31.

Anderson, D. R., P. H. Atkinson, et al. (1985). "Major carbohydrate structures at five glycosylation sites on murine IgM determined by high resolution 1H-NMR spectroscopy." Arch Biochem Biophys 243: 605-18.

Angel, A. S. and B. Nilsson (1990). "Analysis of glycoprotein oligosaccharides by fast atom bombard-ment mass spectrometry." Biomedical and Environmental Mass Spectrometry 19: 721-30.

Angel, A.-S. and B. Nilsson (1990). "Linkage positions in glycoconjugates by periodate oxidation and fast atom bombardment mass spectrometry." Methods in Enzymology 193: 587-607.

Anumula, K.R. (1994). "Quantitative determination of monosaccharides in glycoproteins by high-performance liquid chromatography with highly sensitive fluorescence detection." Anal Biochem 220: 275-83.

Anumula, K. R. (2000). "High-sensitivity and high-resolution methods for glycoprotein analysis." Anal Biochem 283(1): 17-26.

Anumula, K. R. and P. Du (1999). "Characterization of carbohydrates using highly fluorescent 2-aminobenzoic acid tag following gel electrophoresis of glycoproteins." Anal Biochem 275: 236-42.

Anumula, K. R. and P. B. Taylor (1992). "A comprehensive procedure for preparation of partially methylated alditol acetates from glycoprotein carbohydrates." Anal Biochem 203: 101-8.

Barber, M., R. S. Bordoli, et al. (1981). "Fast atom bombardment of solids (FAB): A new ion source for mass spectrometry." Journal of the Chemical Society, Chemical Communications: 325-7.

Baynes, J. W. and F. Wold (1976). "Effect of glycosylation on the in vivo circulating half-life of ribonuclease." J Biol Chem 251: 6016-24.

Beavis, R. C. and B. T. Chait (1989). "Cinnamic acid derivatives as matrices for ultraviolet laser de-sorption mass spectrometry of proteins." Rapid Communications in Mass Spectrometry 3: 432-5.

Beavis, R. C., T. Chaudhary, et al. (1992). "α-cyano-4-hydroxycinnamic acid as a matrix for matrix-assisted laser desorption mass spectrometry." Organic Mass Spectrometry 27: 156-8.

Bendiac, B. and D. A. Cumming (1985). "Hydrazinolysis-N-reacetylation of glycopeptides and glycoproteins. Model studies using 2-acetamido-1-N-(L-aspart-4-oyl)-2-deoxy-a-D-glucopyranosylamine." Carbohydrate Research 144: 1-12.

Biermann, C. J. (1989). Hydrolysis and other cleavage of glycosidic linkage. Analysis of Carbohydrates by GLC and MS. G. D. McGinnis. Boca Raton, FL, CRC Press: 27-41.

Bietlot, H. P. and M. Girard (1997). "Analysis of recombinant human erythropoietin in drug formulations by high-performance capillary electrophoresis." J Chromatogr A 759: 177-84.

Bigge, J. C., T. P. Patel, et al. (1995). "Nonselective and efficient fluorescent labeling of glycans using 2-amino benzamide and anthranilic acid." Anal Biochem 230: 229-38.

Bock, K., J. Schuster-Kolbe, et al. (1994). "Primary structure of the O-glycosidically linked glycan chain of the crystalline surface layer glycoprotein of Thermoanaerobacter thermohydrosulfuricus L111-69 Galactosyl tyrosine as a novel linkage unit." Journal of Biological Chemistry 269: 7137-44.

Börnsen, K. O., M. D. Mohr, et al. (1995). "Ion exchange and purification of carbohydrates on a Nafion(r) membrane as a new sample pretreatment for matrix-assisted laser desorption-ionization mass spectrometry." Rapid Communications in Mass Spectrometry 9: 1031-4.

Borys, M.C., D.I. Linzer, et al. (1993). "Culture pH affects expression rates and glycosylation of recombinant mouse placental lactogen proteins by Chinese hamster ovary (CHO) cells." Biotechnology (NY) 11: 720-4.

Brüll, L. P., V. Kovácik, et al. (1998). "Sodium-cationized oligosaccharides do not appear to undergo 'internal residue loss' rearrangement processes on tandem mass spectrometry." Rapid Communications in Mass Spectrometry 12: 1520-32.

Buchacher, A., P. Schulz, et al. (1998). "High-performance capillary electrophoresis for in-process control in the production of antithrombin III and human clotting factor IX." J Chromatogr A 802: 355-66.

Buko, A. M., E. J. Kentzer, et al. (1991). "Characterization of a posttranslational fucosylation in the growth factor domain of urinary plasminogen activator." Proc Natl Acad Sci U S A 88: 3992-6.

Burrows, S. D., S. G. Franklin, et al. (1995). "Biological and biophysical characterization of recombinant soluble human E-selectin purified at large scale by reversed-phase high-performance liquid chromatography." J Chromatogr B Biomed Appl 668: 219-31.

Camilleri, P., D. Tolson, et al. (1998). "Direct structural analysis of 2-aminoacridone derivatized oligosaccharides by high-performance liquid chromatography/mass spectrometric detection." Rapid Commun Mass Spectrom 12: 144-8.

Carr, S. A., J. R. Barr, et al. (1990). "Identification of attachment sites and structural classes of asparagine-linked carbohydrates in glycoproteins." Methods Enzymol 193: 501-18.

Carr, S. A., M. J. Huddleston, et al. (1993). "Selective identification and differentiation of N- and O-linked oligosaccharides in glycoproteins by liquid chromatography-mass spectrometry." Protein Science 2: 183-96.

Chan, A. L., H. R. Morris, et al. (1991). "A novel sialylated N-acetylgalactosamine-containing oligosaccharide is the major complex-type structure present in Bowes melanoma tissue plasminogen activator." Glycobiology 1: 173-85.

Chan, P. K. and T.-W. D. Chan (2000). "Effect of sample preparation methods on the analysis of dispersed polysaccharides by matrix-assisted laser desorption/ionization time-of-flight mass spectrometry." Rapid Communications in Mass Spectrometry 14: 1841-7.

Charlwood, J., H. Birrell, et al. (2000). "Analysis of oligosaccharides by microbore high-performance liquid chromatography." Anal Chem 72: 1469-74.

Charlwood, J., H. Birrell, et al. (1999). "Carbohydrate release from picomole quantities of glycoprotein and characterisation of glycans by high-performance liquid chromatography and mass spectrometry." J Chromatogr B Biomed Sci Appl 734: 169-74.

Charlwood, J., H. Birrell, et al. (2000). "A probe for the versatile analysis and characterization of N-linked oligosaccharides." Anal Chem 72: 1453-61.

Charlwood, J., H. Birrell, et al. (1999). "A chromatographic and mass spectrometric strategy for the analysis of oligosaccharides: determination of the glycan structures in porcine thyroglobulin." Rapid Commun Mass Spectrom 13: 716-23.

Charlwood, J., J. Langridge, et al. (1999). "Profiling of 2-aminoacridone derivatised glycans by electrospray ionization mass spectrometry." Rapid Commun Mass Spectrom 13: 107-12.

Charlwood, J., D. Tolson, et al. (1999). "A detailed analysis of neutral and acidic carbohydrates in human milk." Anal Biochem 273: 261-77.

Chen, F. T. and R. A. Evangelista (1998). "Profiling glycoprotein n-linked oligosaccharide by capillary electrophoresis." Electrophoresis 19: 2639-44.

Chen, P., A. G. Baker, et al. (1997). "The use of osazones as matrices for the matrix-assisted laser desorption/ionization mass spectrometry of carbohydrates." Analytical Biochemistry 244: 144-51.

Chiesa, C., P. J. Oefner, et al. (1995). "Micellar electrokinetic chromatography of monosaccharides derivatized with 1-phenyl-3-methyl-2-pyrazolin-5-one." J Capillary Electrophor 2: 175-83.

Chitlaru, T., C. Kronman, et al. (1998). "Modulation of circulatory residence of recombinant acetylcholinesterase through biochemical or genetic manipulation of sialylation levels." Biochem J 336 (Pt 3): 647-58.

Cifuentes, A., M. V. Moreno-Arribas, et al. (1999). "Capillary isoelectric focusing of erythropoietin glycoforms and its comparison with flat-bed isoelectric focusing and capillary zone electrophoresis." J Chromatogr A 830: 453-63.

Ciucanu, I. and F. Kerek (1984). "A simple and rapid method for the permethylation of carbohydrates." Carbohydrate Research 131: 209-17.

Clarke, N. J., A. J. Tomlinson, et al. (1997). "Capillary isoelectric focusing of physiologically derived proteins with on-line desalting of isotonic salt concentrations." Anal Chem 69: 2786-92.

Colangelo, J. and R. Orlando (1999). "On-target exoglycosidase digestions, MALDI-MS for determining the primary structures of carbohydrate chains." Analytical Chemistry 71: 1479-82.

Conboy, J. J. and J. Henion (1992). "High-performance anion-exchange chromatography coupled with mass spectrometry for the determination of carbohydrates." Biol Mass Spectrom 21: 397-407.

Costa, J., D. A. Ashford, et al. (1997). "The glycosylation of the aspartic proteinases from barley (Hordeum vulgare L.) and cardoon (Cynara cardunculus L.)." European Journal of Biochemistry 243: 695-700.

De Beer, T., C. W. Van Zuylen, et al. (1996). "NMR studies of the free alpha subunit of human chorionic gonadotropin. Structural influences of N-glycosylation and the beta subunit on the conformation of the alpha subunit." Eur J Biochem 241: 229-42.

Dell, A. (1990). "Preparation and desorption mass spectrometry of permethyl and peracetyl derivatives of oligosaccharides." Methods in Enzymology 193: 647-60.

Dell, A., N. H. Carman, et al. (1987). "Fast atom bombardment mass spectrometric strategies for characterising carbohydrate-containing biopolymers." Biomedical and Environmental Mass Spectrometry 16: 19-24.

Dell, A., K.-H. Khoo, et al. (1993). FAB-MS and ES-MS of glycoproteins. Glycobiology: a Practical Approach. B. D. Hames. Oxford, IRL Press (Oxford University Press): 197-.

Domon, B. and C. E. Costello (1988). "A systematic nomenclature for carbohydrate fragmentations in FAB-MS/MS spectra of glycoconjugates." Glycoconjugate Journal 5: 397-409.

Doyle, P., L. de la Canal, et al. (1986). "Characterization of the mechanism of protein glycosylation and the structure of glycoconjugates in tissue culture trypomastigotes and intracellular amastigotes of Trypanosoma cruzi." Mol Biochem Parasitol 21: 93-101.

Duffin, K. L., J. K. Welply, et al. (1992). "Characterization of N-Linked Oligosaccharides by Electrospray and Tandem Mass Spectrometry." Analytical Chemistry 64: 1440-8.

Dwek, R. A. (1995). "Glycobiology: more functions for oligosaccharides." Science 269: 1234-5.

Dwek, R. A., C. J. Edge, et al. (1993). "Analysis of glycoprotein-associated oligosaccharides." Annu Rev Biochem 62: 65-100.

Edge, C. J., T. W. Rademacher, et al. (1992). "Fast sequencing of oligosaccharides: the reagent-array analysis method." Proc Natl Acad Sci U S A 89: 6338-42.

El Rassi, Z. and Y. Mechref (1996). "Recent advances in capillary electrophoresis of carbohydrates." Electrophoresis 17: 275-301.

Elbein, A. D. (1991). "The role of N-linked oligosaccharides in glycoprotein function." Trends Biotechnol 9: 346-52.

Endo, T., J. Amano, et al. (1986). "Structure identification of the complex-type, asparagine-linked sugar chains of beta-D-galactosyl-transferase purified from human milk." Carbohydr Res 150: 241-63.

Fergusson, M. A. J. (1993). GPI Membrane anchors: isolation and analysis. Glycobiology, A Practical Approach. A. Kobata. Oxford, IRL Press (Oxford University Press): 349-.

Ferranti, P., P. Pucci, et al. (1995). "Human alpha-fetoprotein produced from Hep G2 cell line: struc-ture and heterogeniety of the oligosaccharide moiety." Journal of Mass Spectrometry 30: 632-8.

Frears, E. R., A. H. Merry, et al. (1999). "Screening neutral and acidic IgG N-glycans by high density electrophoresis." Glycoconj J 16: 283-90.

Freeze, H. H. and D. Wolgast (1986). "Structural analysis of N-linked oligosaccharides from glycoproteins secreted by Dictyostelium discoideum. Identification of mannose 6-sulfate." J Biol Chem 261: 127-34.

Gawlitzek, M., U. Valley, et al. (1995). "Characterization of changes in the glycosylation pattern of recombinant proteins from BHK-21 cells due to different culture conditions." J Biotechnol 42: 117-31.

Geyer, H., S. Schmitt, et al. (1999). "Structural analysis of glycoconjugates by on-target enzymatic digestion and MALDI-TOF-MS." Analytical Chemistry 71: 476-82.

Gillece-Castro, B. L. and A. L. Burlingame (1990). "Oligosaccharide characterization with high-energy collision-induced dissociation mass spectrometry." Methods in Enzymology 193: 689-712.

Gonzalez, J., T. Takao, et al. (1992). "A method for determination of N-glycosylation sites in glycoproteins by collision-induced dissociation analysis in fast atom bombardment mass spectrometry: Identification of the positions of carbohydrate-linked asparagine in recombinant α-amylase by treatment with peptide-N-glycosidase F in ^{18}O-labelled water." Analytical Biochemistry 205: 151-8.

Goochee, C. F. (1992). "Bioprocess factors affecting glycoprotein oligosaccharide structure." Dev Biol Stand 76: 95-104.

Goodarzi, M. T. and G. A. Turner (1995). "Decreased branching, increased fucosylation and changed sialylation of alpha-1-proteinase inhibitor in breast and ovarian cancer." Clin Chim Acta 236: 161-71.

Görisch, H. (1988). "Drop dialysis: Time course of salt and protein exchange." Analytical Biochemistry 173: 393-8.

Gray, J. S. S. and R. Montgomery (1997). "The N-glycosylation sites of soybean seed coat peroxidase." Glycobiology 7: 679-85.

Greis, K. D., B. K. Hayes, et al. (1996). "Selective detection and site-analysis of O-GlcNAc-modified glycopeptides by beta-elimination and tandem electrospray mass spectrometry." Analytical Biochemistry 234: 38-49.

Greve, K. F., D. E. Hughes, et al. (1996). "Capillary electrophoretic examination of underivatized oligosaccharide mixtures released from immunoglobulin G antibodies and CTLA4Ig fusion protein." J Chromatogr A 749: 237-45.

Grossman, P. D., K. J. Wilson, et al. (1988). "Effect of buffer pH and peptide composition on the selectivity of peptide separations by capillary zone electrophoresis." Anal Biochem 173: 265-70.

Guile, G. R., D. J. Harvey, et al. (1998). "Identification of highly fucosylated N-linked oligosaccharides from the human parotid gland." Eur J Biochem 258(2): 623-56.

Guile, G. R., P. M. Rudd, et al. (1996). "A rapid high-resolution high-performance liquid chromatographic method for separating glycan mixtures and analyzing oligosaccharide profiles." Anal Biochem 240: 210-26.

Guile, G. R., S. Y. Wong, et al. (1994). "Analytical and preparative separation of anionic oligosaccharides by weak anion-exchange high-performance liquid chromatography on an inert polymer column." Anal Biochem 222: 231-5.

Guttman, A. (1997). "Multistructure sequencing of N-linked fetuin glycans by capillary gel electrophoresis and enzyme matrix digestion." Electrophoresis 18: 1136-41.

Guttman, A., F. T. Chen, et al. (1996). "High-resolution capillary gel electrophoresis of reducing oligosaccharides labeled with 1-aminopyrene-3,6,8-trisulfonate." Anal Biochem 233: 234-42.

Guttman, A. and C. Starr (1995). "Capillary and slab gel electrophoresis profiling of oligosaccharides." Electrophoresis 16: 993-7.

Hakomori, S. (1964). "A rapid permethylation of glycolipid, and polysaccharide catalysed by methylsulfinyl carbanion in dimethyl sulfoxide." Journal of Biochemistry 55: 205-8.

Hardy, M. R. and R. R. Townsend (1988). "Separation of positional isomers of oligosaccharides and glycopeptides by high-performance anion-exchange chromatography with pulsed amperometric detection." Proc Natl Acad Sci U S A 85: 3289-93.

Hardy, M. R. and R. R. Townsend (1994). "High-pH anion-exchange chromatography of glycoprotein-derived carbohydrates." Methods Enzymol 230: 208-25.

Hardy, M. R., R. R. Townsend, et al. (1988). "Monosaccharide analysis of glycoconjugates by anion exchange chromatography with pulsed amperometric detection." Anal Biochem 170: 54-62.

Harland, G. B., G. Okafo, et al. (1996). "Fingerprinting of glycans as their 2-aminoacridone deriv-atives by capillary electrophoresis and laser-induced fluorescence." Electrophoresis 17: 406-11.

Harvey, D. J. (1992). Mass spectrometry of picolinyl and other nitrogen-containing derivatives of lipids. Advances in Lipid Methodology - One. W. W. Christie. Ayr, The Oily Press. 1: 19-80.

Harvey, D. J. (1993). "Quantitative aspects of the matrix-assisted laser desorption mass spectrometry of complex oligosaccharides." Rapid Communications in Mass Spectrometry 7: 614-9.

Harvey, D. J. (1995). "Matrix-assisted laser desorption/ionization mass spectrometry of sphingo- and glycosphingo-lipids." Journal of Mass Spectrometry 30: 1311-24.

Harvey, D. J. (1997). Structural determination of protein-bound oligosaccharides by mass spectrometry. Glycopeptides and related compounds: Synthesis, Analysis and Applications. C. D. Warren. New York, Marcel Dekker: 593-629.

Harvey, D. J. (1999). "Matrix-assisted laser desorption/ionization mass spectrometry of carbohydrates." Mass Spectrometry Reviews 18: 349-451.

Harvey, D. J. (2000). "Collision-induced fragmentation of 2-aminobenzamide-labelled neutral N-linked glycans." The Analyst 125: 609-17.

Harvey, D. J. and M. G. Horning (1973). "Characterization of the trimethylsilyl derivatives of sugar phosphates and related compounds by gas chromatography and gas chromatography-mass spectrometry." Journal of Chromatography 75: 51-62.

Harvey, D. J. and A. P. Hunter (1998). "MALDI Mass spectrometry of carbohydrates on a magnetic sector instrument." Advances in Mass Spectrometry 14: BO6 TUP0083.

Harvey, D. J., B. Küster, et al. (1999). Matrix-assisted laser desorption/ionization mass spectrometry of N-linked carbohydrates and related compounds. Mass Spectrometry in Biology and Medicine. M. A. Baldwin. Totowa, Humana Press: 407-37.

Harvey, D. J., P. M. Rudd, et al. (1994). "Examination of complex oligosaccharides by matrix-assisted laser desorption/ionization mass spectrometry on time-of-flight and magnetic sector instruments." Organic Mass Spectrometry 29: 753-65.

Hellerqvist, C. G. (1990). "Linkage analysis using Lindberg method." Methods in Enzymology 193: 554-73.

Hellerqvist, C. G., B. Lindberg, et al. (1968). "Structural studies on the O-specific side-chains of the cell wall lipopolysaccharide from Salmonella typhimurium 395 MS." Carbohydrate Research 8: 43-55.

Hoffstetter-Kuhn, S., G. Alt, et al. (1996). "Profiling of oligosaccharide-mediated microheterogeneity of a monoclonal antibody by capillary electrophoresis." Electrophoresis 17: 418-22.

Hofmeister, G. E., Z. Zhou, et al. (1991). "Linkage position determination in lithium-catonised disaccharides: tandem mass spectrometry and semiempirical calculations." Journal of the American Chemical Society 113: 5964-70.

Honda, S., A. Makino, et al. (1990). "Analysis of the oligosaccharides in ovalbumin by high-performance capillary electrophoresis." Anal Biochem 191: 228-34.

Honda, S., M. Takahashi, et al. (1981). "Rapid, automated analysis of monosaccharides by high-performance anion-exchange chromatography of borate complexes with fluorimetric detection using 2-cyanoacetamide." Anal Biochem 113: 130-8.

Honda, S., K. Togashi, et al. (1997). "Unusual separation of 1-phenyl-3-methyl-5-pyrazolone deriv-atives of aldoses by capillary zone electrophoresis." Journal of Chromatography A 791: 307-11.

Huang, L. and R. M. Riggin (2000). "Analysis of nonderivatized neutral and sialylated oligosaccharides by electrospray mass spectrometry." Anal Chem 72: 3539-46.

Huddleston, M. J., M. F. Bean, et al. (1993). "Collisional fragmentation of glycopeptides by electrospray ionization LC/MS and LC/MS/MS - Methods for selective detection of glycopeptides in protein digests." Analytical Chemistry 65: 877-84.

Hunt, G., T. Hotaling, et al. (1998). "Validation of a capillary isoelectric focusing method for the recombinant monoclonal antibody C2B8." J Chromatogr A 800: 355-67.

Hunter, A. P. and D. E. Games (1995). "Evaluation of glycosylation site heterogeniety and selective differentiation of glycopeptides in proteolytic digests of bovine alpha-1-acid glycoprotein by mass spectrometry." Rapid Communications in Mass Spectrometry 9: 42-56.

Jackson, P. (1991). "Polyacrylamide gel electrophoresis of reducing saccharides labeled with the fluorophore 2-aminoacridone: subpicomolar detection using an imaging system based on a cooled charge-coupled device." Anal Biochem 196: 238-44.

Jackson, P. (1994). "The analysis of fluorophore-labeled glycans by high-resolution polyacrylamide gel electrophoresis." Anal Biochem 216: 243-52.

Jackson, P. (1994). "High-resolution polyacrylamide gel electrophoresis of fluorophore-labeled reducing saccharides." Methods Enzymol 230: 250-65.

Jackson, P. (1996). "The analysis of fluorophore-labeled carbohydrates by polyacrylamide gel electrophoresis." Mol Biotechnol 5: 101-23.

Jackson, P., M. G. Pluskal, et al. (1994). "The use of polyacrylamide gel electrophoresis for the analy-sis of acidic glycans labeled with the fluorophore 2-aminoacridone." Electrophoresis 15: 896-902.

Jilge, G., K. K. Unger, et al. (1989). "Evaluation of advanced silica packings for the separation of bio-polymers by high-performance liquid chromatography. VI. Design, chromatographic performance and application of non-porous silica-based anion exchangers." J Chromatogr 476: 37-48.

Karas, M., H. Ehring, et al. (1993). "Matrix-assisted laser desorption/ionization mass spectrometry with additives to 2,5-dihydroxybenzoic acid." Organic Mass Spectrometry 28: 1476-81.

Karas, M. and F. Hillenkamp (1988). "Laser desorption ionization of proteins with molecular masses exceeding 10,000 Daltons." Analytical Chemistry 60: 2299-301.

Karlsson, H., I. Carlstedt, et al. (1989). "The use of gas chromatography and gas chromatography - mass spectrometry for the characterisation of permethylated oligosaccharides with molecular mass up to 2300." Analytical Biochemistry 182: 438-46.

Khoo, K.-H. and A. Dell (1990). "Assignment of anomeric configurations of pyranose sugars on oligosaccharides using a sensitive FAB-MS strategy." Glycobiology 1: 83-91.

Kim, S. H., C. M. Shin, et al. (1998). "First application of thermal vapor deposition method to matrix-assisted laser desorption ionization mass spectrometry: determination of molecular mass of bis(p-methyl benzylidene) sorbitol." Rapid Communications in Mass Spectrometry 12: 701-4.

Klock, J. C. and C. M. Starr (1998). "Polyacrylamide gel electrophoresis of fluorophore-labeled carbohydrates from glycoproteins." Methods Mol Biol 76: 115-29.

Klyushnichenko, V. and M. R. Kula (1997). "Rapid SDS gel capillary electrophoretic analysis of proteins." J Capillary Electrophor 4: 61-4.

Knuver-Hopf, J. and H. Mohr (1995). "Differences between natural and recombinant interleukin-2 revealed by gel electrophoresis and capillary electrophoresis." J Chromatogr A 717: 71-4.

Kobata, A. (1992). "Structures and functions of the sugar chains of glycoproteins." Eur J Biochem 209: 483-501.

Kobata, A., K. Yamashita, et al. (1987). "BioGel P-4 column chromatography of oligosaccharides: effective size of oligosaccharides expressed in glucose units." Methods Enzymol 138: 84-94.

Kopp, K., M. Schluter, et al. (1996). "Monitoring the glycosylation pattern of recombinant interferon-omega with high-pH anion-exchange chromatography and capillary electrophoresis." Arzneimit-telforschung 46: 1191-6.

Kovácik, V., J. Hirsch, et al. (1995). "Oligosaccharide characterization using collision-induced dissociation fast atom bombardment mass spectrometry: Evidence for internal monosaccharide residue loss." Journal of Mass Spectrometry 30: 949-58.

Krause, J., M. Stoeckli, et al. (1996). "Studies on the selection of new matrices for ultraviolet matrix-assisted laser desorption/ionization time-of-flight mass spectrometry." Rapid Communications in Mass Spectrometry 10: 1927-33.

Krull, I.S. and J.R. Mazzeo (1992). "Capillary electrophoresis: the promise and the practice." Nature 357: 92-4.

Kunkel, J. P., D. C. Jan, et al. (1998). "Dissolved oxygen concentration in serum-free continuous culture affects N-linked glycosylation of a monoclonal antibody." J Biotechnol 62: 55-71.

Küster, B., A. P. Hunter, et al. (1998). "Structural determination of N-linked carbohydrates by matrix-assisted laser desorption/ionization mass spectrometry following enzymatic release within sodium dodecyl sulfate-polyacrylamide electrophoresis gels: application to species-specific glycosylation of a1-acid glycoprotein." Electrophoresis 19: 1950-9.

Küster, B., T. N. Krogh, et al. (2001). "Glycosylation of gel-separated proteins." Proteomics 1: 350-61.

Küster, B. and M. Mann (1999). "[18]O-labeling of N-glycosylation sites to improve the identification of gel-separated glycoproteins using peptide mass mapping and database searching." Analytical Chemistry 71: 1431-40.

Küster, B., T. J. P. Naven, et al. (1996). "Rapid approach for sequencing neutral oligosaccharides by exoglycosidase digestion and matrix-assisted laser desorption/ionization time-of-flight mass spectrometry." Journal of Mass Spectrometry 31: 1131-40.

Kuster, B., S. F. Wheeler, et al. (1997). "Sequencing of N-linked oligosaccharides directly from protein gels: in-gel deglycosylation followed by matrix-assisted laser desorption/ionization mass spectrometry and normal-phase high-performance liquid chromatography." Anal Biochem 250: 82-101.

Küster, B., S. F. Wheeler, et al. (1997). "Sequencing of N-linked oligosaccharides directly from protein gels: In-gel deglycosylation followed by matrix-assisted laser desorption/ionization mass spectrometry and normal-phase high performance liquid chromatography." Analytical Biochemistry 250: 82-101.

Lee, H. G. (1997). "Rapid high-performance isoelectric focusing of monoclonal antibodies in uncoated fused-silica capillaries." J Chromatogr A 790: 215-23.

Lee, K. B., D. Loganathan, et al. (1990). "Carbohydrate analysis of glycoproteins. A review." Appl Biochem Biotechnol 23: 53-80.

Lemoine, J., B. Fournet, et al. (1993). "Collision-induced dissociation of alkali metal cationized and permethylated oligosaccharides: influence of the collision energy and of the collision gas for the assignment of linkage position." Journal of the American Society for Mass Spectrometry 4: 197-203.

Lemoine, J., G. Strecker, et al. (1991). "Collisional-activation tandem mass spectrometry of sodium adduct ions of methylated oligosaccharides - sequence analysis and discrimination between alpha-NeuAc-(2->3) and alpha-NeuAc-(2->6) linkages." Carbohydrate Research 221: 209-17.

Liu, J. P., K. J. Volk, et al. (1993). "Structural characterization of glycoprotein digests by microcolumn liquid chromatography-ionspray tandem mass spectrometry." Journal of Chromatography 632: 45-56.

Liu, X., Z. Sosic, et al. (1996). "Capillary isoelectric focusing as a tool in the examination of antibodies, peptides and proteins of pharmaceutical interest." J Chromatogr A 735(1-2): 165-90.

Lowe, M. E. and B. Nilsson (1984). "A method of purification of partially methylated alditol acetates in the methylation analysis of glycoproteins and glycopeptides." Anal Biochem 136: 187-91.

Manzi, A. E., S. Diaz, et al. (1990). "High-pressure liquid chromatography of sialic acids on a pellicular resin anion-exchange column with pulsed amperometric detection: a comparison with six other systems." Anal Biochem 188: 20-32.

Marusyk, R. and A. Sergeant (1980). "A simple method for dialysis of small-volume samples." Analytical Biochemistry 105: 403-4.

Mattu, T. S., L. Royle, et al. (2000). "O-glycan analysis of natural human neutrophil gelatinase B using a combination of normal phase-HPLC and online tandem mass spectrometry: implications for the domain organization of the enzyme." Biochemistry 39: 15695-704.

Mawhinney, T. P. and D. L. Chance (1994). "Hydrolysis of sialic acids and O-acetylated sialic acids with propionic acid." Anal Biochem 223: 164-7.

McNeal, C. J., R. D. Macfarlane, et al. (1986). "A novel mass spectrometric procedure to rapidly determine the partial structure of heparin fragments." Biochemical and Biophysical Research Communications 139: 18-24.

McNerney, T. M., S. K. Watson, et al. (1996). "Separation of recombinant human growth hormone from Escherichia coli cell pellet extract by capillary zone electrophoresis." J Chromatogr A 744: 223-9.

Mechref, Y., A. G. Baker, et al. (1998). "Matrix-assisted laser desorption, ionization mass spectrometry of neutral and acidic oligosaccharides with collision-induced dissociation." Carbohydrate Research 313: 145-55.

Mechref, Y. and Z. el Rassi (1994). "Capillary zone electrophoresis of derivatized acidic monosaccharides." Electrophoresis 15: 627-34.

Merkle, R. K. and I. Poppe (1994). "Carbohydrate composition analysis of glycoconjugates by gas-liquid chromatography/mass spectrometry." Methods in Enzymology 230: 1-15.

Metzger, J. O., R. Woisch, et al. (1994). "New type of matrix for matrix-assisted laser desorption mass spectrometry of polysaccharides and proteins." Fresenius Journal of Analytical Chemistry 349: 473-4.

Mizuochi, T. (1993). "Microscale sequencing of N-linked oligosaccharides of glycoproteins using hydrazinolysis, Bio-Gel P-4, and sequential exoglycosidase digestion." Methods Mol Biol 14: 55-68.

Mo, W., H. Sakamoto, et al. (1999). "Structural characterisation of chemically derivatised oligosaccharides by nanoflow electrospray ionization mass spectrometry." Analytical Chemistry 71: 4100-6.

Mock, K. K., M. Davy, et al. (1991). "The analysis of underivatised oligosaccharides by matrix-assisted laser desorption mass spectrometry." Biochemical and Biophysical Research Communications 177: 644-51.

Mohr, M. D., K. O. Börnsen, et al. (1995). "Matrix-assisted laser desorption/ionization mass spectrometry: Improved matrix for oligosaccharides." Rapid Communications in Mass Spectrometry 9: 809-14.

Montreuil, J., S. Bouquelet, et al. (1986). Glycoproteins. Carbohydrate Analysis: A Practical Approach. J. F. Kennedy. Oxford, IRL Press: 143-204.

Moorhouse, K. G., C. A. Eusebio, et al. (1995). "Rapid one-step capillary isoelectric focusing method to monitor charged glycoforms of recombinant human tissue-type plasminogen activator." J Chromatogr A 717: 61-9.

Moorhouse, K. G., C. A. Rickel, et al. (1996). "Electrophoretic separation of recombinant tissue-type plasminogen activator glycoforms: validation issues for capillary isoelectric focusing methods." Electrophoresis 17: 423-30.

Müller, S., S. Goletz, et al. (1997). "Localization of O-glycosylation sites on glycopeptide fragments from lactation-associated MUC1. All putative sites within the tandem repeat are glycosylation targets in vivo." Journal of Biological Chemistry 272: 24780-93.

Naven, T. J. P. and D. J. Harvey (1996). "Effect of structure on the signal strength of oligosaccharides in matrix-assisted laser desorption/ionization mass spectrometry on time-of-flight and magnetic sector instruments." Rapid Communications in Mass Spectrometry 10: 1361-6.

Naven, T. J. P., D. J. Harvey, et al. (1997). "Fragmentation of complex carbohydrates following ionization by matrix-assisted laser desorption with an instrument fitted with time-lag focusing." Rapid Communications in Mass Spectrometry 11: 1681-6.

Neuburger, G. G. and D. C. Johnson (1987). "Comparison of the pulsed amperometric detection of carbohydrates at gold and platinum electrodes for flow injection and liquid chromatographic systems." Anal Chem 59: 203-4.

Nonami, H., S. Fukui, et al. (1997). "β-Carboline alkaloids as matrices for matrix-assisted ultraviolet laser desorption time-of-flight mass spectrometry of proteins and sulfated oligosaccharides: a comparative study using phenylcarbonyl compounds, carbazoles and classical matrices." Journal of Mass Spectrometry 32: 287-96.

Nonami, H., K. Tanaka, et al. (1998). "β-Carboline alkaloids as matrices for UV-Matrix-assisted laser desorption/ionization time-of-flight mass spectrometry in positive and negative ion modes. Analysis of proteins of high molecular mass, and of cyclic and acyclic oligosaccharides." Rapid Communications in Mass Spectrometry 12: 285-96.

Okafo, G., L. Burrow, et al. (1996). "A coordinated high-performance liquid chromatographic, capillary electrophoretic, and mass spectrometric approach for the analysis of oligosaccharide mixtures derivatized with 2-aminoacridone." Anal Chem 68: 4424-30.

Okafo, G., J. Langridge, et al. (1997). "High-performance liquid chromatographic analysis of complex N-linked glycans derivatized with 2-aminoacridone." Anal Chem 69: 4985-93.

Okamoto, M., K.-I. Takahashi, et al. (1995). "Sensitive detection and structural characterization of trimethyl(p-aminophenyl)-ammonium-derivatized oligosaccharides by electrospray ionization-mass spectrometry and tandem mass spectrometry." Rapid Communications in Mass Spectrometry 9: 641-3.

Orlando, R., C. A. Bush, et al. (1990). "Structural analysis of oligosaccharides by tandem mass spectrometry: Collisional activation of sodium adduct ions." Biomedical and Environmental Mass Spectrometry 19: 747-54.

Page, K. C., G. J. Killian, et al. (1990). "Sertoli cell glycosylation patterns as affected by culture age and extracellular matrix." Biol Reprod 43: 659-64.

Papac, D. I., J. B. Briggs, et al. (1998). "A high-throughput microscale method to release N-linked oligosaccharides from glycoproteins for matrix-assisted laser desorption/ionization time-of-flight mass spectrometric analysis." Glycobiology 8: 445-54.

Papac, D. I., A. Wong, et al. (1996). "Analysis of acidic oligosaccharides and glycopeptides by matrix-assisted laser desorption/ionization time-of-flight mass spectrometry." Anal Chem 68: 3215-23.

Parekh, R. B., R. A. Dwek, et al. (1989). "Cell-type-specific and site-specific N-glycosylation of type I and type II human tissue plasminogen activator." Biochemistry 28: 7644-62.

Paskach, T. J., H.-P. Lieker, et al. (1991). "High-performance anion-exchange chromatography of sugars and sugar alcohols on quaternary ammonium resins under alkaline conditions." Carbohydrate Research 215: 1-14.

Patel, A. B. and H. V. Bhatt (1993). "Effect of organic and inorganic mercury on serum total and perchloric acid-soluble sialic acids in rats." Indian J Physiol Pharmacol 37: 259-60.

Patel, T., J. Bruce, et al. (1993). "Use of hydrazine to release in intact and unreduced form both N- and O-linked oligosaccharides from glycoproteins." Biochemistry 32: 679-93.

Patel, T. P., R. B. Parekh, et al. (1992). "Different culture methods lead to differences in glycosylation of a murine IgG monoclonal antibody." Biochem J 285 (Pt 3): 839-45.

Pedersen, J., J. Andersen, et al. (1993). "Characterization of natural and recombinant nuclease isoforms by electrospray mass spectrometry." Biotechnol Appl Biochem 18 (Pt 3): 389-99.

Poulter, L. and A. L. Burlingame (1990). "Desorption mass spectrometry of oligosaccharides coupled with hydrophobic chromophores." Methods in Enzymology 193: 661-89.

Powell, A. K. and D. J. Harvey (1996). "Stabilisation of sialic acids in N-linked oligosaccharides and gangliosides for analysis by positive ion matrix-assisted laser desorption-ionization mass spectrometry." Rapid Communications in Mass Spectrometry 10: 1027-32.

Rabilloud, T. (1996). "Solubilization of proteins for electrophoretic analyses." Electrophoresis 17: 813-29.

Rademaker, G. J., J. Haverkamp, et al. (1993). "Determination of glycosylation sites in O-linked glycopeptides: a sensitive mass spectrometric protocol." Organic Mass Spectrometry 28: 1536-41.

Raju, T. S., J. B. Briggs, et al. (2000). "Species-specific variation in glycosylation of IgG: evidence for the species-specific sialylation and branch-specific galactosylation and importance for engineering recombinant glycoprotein therapeutics." Glycobiology 10: 477-86.

Reason, A. J., L. A. Ellis, et al. (1994). "Novel tyvelose-containing tri- and tetra-antennary N-glycans in the immunodominant antigens of the intracellular parasite Trichinella spiralis." Glycobiology 4: 593-603.

Reif, O. W. and R. Freitag (1995). "Studies of complexes between proteases, substrates and the protease inhibitor alpha 2-macroglobulin using capillary electrophoresis with laser-induced fluorescence detection." J Chromatogr A 716: 363-9.

Reinhold, V. N., B. B. Reinhold, et al. (1995). "Carbohydrate molecular weight profiling, sequence, linkage and branching data: ES-MS and CID." Analytical Chemistry 67: 1772-84.

Rice, K. G., M. H. Chiu, et al. (1995). "In vivo targeting function of N-linked oligosaccharides. Pharmacokinetic and biodistribution of N-linked oligosaccharides." Adv Exp Med Biol 376: 271-82.

Richardson, S., A. Cohen, et al. (2001). "High-performance anion-exchange chromatography-electrospray mass spectrometry for investigation of the substituent distribution in hydroxypropylated potato amylopectin starch." J Chromatogr A 917: 111-21.

Rodriguez-Diaz, R., T. Wehr, et al. (1997). "Capillary isoelectric focusing." Electrophoresis 18(12-13): 2134-44.

Rohrer, J. S. (1995). "Separation of asparagine-linked oligosaccharides by high-pH anion-exchange chromatography with pulsed amperometric detection: empirical relationships between oligosaccharide structure and chromatographic retention." Glycobiology 5: 359-60.

Rohrer, J. S. (2000). "Analyzing sialic acids using high-performance anion-exchange chromatography with pulsed amperometric detection." Anal Biochem 283: 3-9.

Rolf, D. and G. R. Gray (1982). "Reductive cleavage of glycosides." Journal of the American Chemical Society 104: 3539-.

Rouse, J. C. and J. E. Vath (1996). "On-the-probe sample cleanup strategies for glycoprotein-released carbohydrates prior to matrix-assisted laser desorption-ionization time-of-flight mass spectrometry." Analytical Biochemistry 238: 82-92.

Rudd, P. M. and R. A. Dwek (1997). "Rapid, sensitive sequencing of oligosaccharides from glycoproteins." Curr Opin Biotechnol 8(4): 488-97.

Rudd, P. M., G. R. Guile, et al. (1997). "Oligosaccharide sequencing technology." Nature 388: 205-7.

Rudd, P. M., H. C. Joao, et al. (1994). "Glycoforms modify the dynamic stability and functional activity of an enzyme." Biochemistry 33: 17-22.

Rudd, P. M., T. S. Mattu, et al. (1999). "Glycosylation of natural human neutrophil gelatinase B and neutrophil gelatinase B-associated lipocalin." Biochemistry 38(42): 13937-50.

Rudd, P. M., T. S. Mattu, et al. (1999). "Glycoproteins: rapid sequencing technology for N-linked and GPI anchor glycans." Biotechnol Genet Eng Rev 16: 1-21.

Rudd, P. M., M. R. Wormald, et al. (1999). "Oligosaccharide analysis and molecular modeling of soluble forms of glycoproteins belonging to the Ly-6, scavenger receptor, and immunoglobulin superfamilies expressed in Chinese hamster ovary cells." Glycobiology 9(5): 443-58.

Schwer, C. (1995). "Capillary isoelectric focusing: a routine method for protein analysis?" Electrophoresis 16: 2121-6.

Shimura, K., Z. Wang, et al. (2000). "Synthetic oligopeptides as isoelectric point markers for capillary isoelectric focusing with ultraviolet absorption detection." Electrophoresis 21: 603-10.

Somerville, L. E., A. J. Douglas, et al. (1999). "Discrimination of granulocyte colony-stimulating factor isoforms by high-performance capillary electrophoresis." J Chromatogr B Biomed Sci Appl 732: 81-9.

Spellman, M. W. (1990). "Carbohydrate characterization of recombinant glycoproteins of pharmaceutical interest." Anal Chem 62: 1714-22.

Spengler, B., D. Kirsch, et al. (1995). "Structure analysis of branched oligosaccharides using post-source decay in matrix-assisted laser desorption/ionization mass spectrometry." Journal of Mass Spectrometry 30: 782-7.

Stahl, B., A. Linos, et al. (1997). "Analysis of fructans from higher plants by matrix-assisted laser desorption/ionization mass spectrometry." Anal Biochem 246: 195-204.

Stahl, B., M. Steup, et al. (1991). "Analysis of neutral oligosaccharides by matrix-assisted laser desorption/ionization mass spectrometry." Analytical Chemistry 63: 1463-6.

Stahl, B., S. Thurl, et al. (1994). "Oligosaccharides from human milk as revealed by matrix-assisted laser desorption/ionization mass spectrometry." Analytical Biochemistry 223: 218-26.

Starr, C. M., R. I. Masada, et al. (1996). "Fluorophore-assisted carbohydrate electrophoresis in the separation, analysis, and sequencing of carbohydrates." J Chromatogr A 720: 295-321.

Strupat, K., M. Karas, et al. (1991). "2,5-Dihydroxybenzoic acid: a new matrix for laser desorption-ionization mass spectrometry." International Journal of Mass Spectrometry and Ion Processes 111: 89-102.

Sutton, C. W., J. A. O'Neill, et al. (1994). "Site-specific characterization of glycoprotein carbohydrates by exoglycosidase digestion and laser desorption mass spectrometry." Analytical Biochemistry 218: 34-46.

Sweeley, C. C., R. Bentley, et al. (1963). "Gas-liquid chromatography of trimethylsilyl derivatives of sugars and related substances." Journal of the American Chemical Society 85: 2497-507.

Takahashi, N. (1996). "Three-dimensional mapping of N-linked oligosaccharides using anion-exchange, hydrophobic and hydrophilic interaction modes of high-performance liquid chromatography." J Chromatogr A 720: 217-25.

Takasaki, S., T. Misuochi, et al. (1982). "Hydrazinolysis of asparagine-linked sugar chains to produce free oligosaccharides." Methods in Enzymology 83: 263-8.

Takasaki, S., T. Mizuochi, et al. (1982). "Hydrazinolysis of asparagine-linked sugar chains to produce free oligosaccharides." Methods Enzymol 83: 263-8.

Takeuchi, M., N. Inoue, et al. (1989). "Relationship between sugar chain structure and biological activity of recombinant human erythropoietin produced in Chinese hamster ovary cells." Proc Natl Acad Sci U S A 86: 7819-22.

Tang, S., D. P. Nesta, et al. (1999). "A method for routine analysis of recombinant immunoglobulins (rIgGs) by capillary isoelectric focusing (cIEF)." J Pharm Biomed Anal 19: 569-83.

Tarentino, A. L., C. M. Gómez, et al. (1985). "Deglycosylation of asparagine-linked glycans by peptide:N-glycosidase F." Biochemistry 24: 4665-5671.

Taverna, M., A. Baillet, et al. (1995). "N-glycosylation site mapping of recombinant tissue plasminogen activator by micellar electrokinetic capillary chromatography." Biomed Chromatogr 9: 59-67.

Taverna, M., N. T. Tran, et al. (1998). "Electrophoretic methods for process monitoring and the quality assessment of recombinant glycoproteins." Electrophoresis 19: 2572-94.

Thayer, J.R., J.S.Rohrer, et al. (1998). "Improvements to in-line desalting of oligosaccharides separated by high-pH anion exchange chromatography with pulsed amperometric detection." Anal Biochem 256: 207-16.

Thorne, J. M., W. K. Goetzinger, et al. (1996). "Examination of capillary zone electrophoresis, capillary isoelectric focusing and sodium dodecyl sulfate capillary electrophoresis for the analysis of recombinant tissue plasminogen activator." J Chromatogr A 744: 155-65.

Townsend, R. R., M. Hardy, et al. (1988). "Chromatography of carbohydrates." Nature 335: 379-80.

Townsend, R. R. and M. R. Hardy (1991). "Analysis of glycoprotein oligosaccharides using high-pH anion exchange chromatography." Glycobiology 1: 139-47.

Townsend, R. R., M. R. Hardy, et al. (1988). "High-performance anion-exchange chromatography of oligosaccharides using pellicular resins and pulsed amperometric detection." Anal Biochem 174: 459-70.

Tran, N.T., M.Taverna, et al. (2000). "One-step capillary isoelectric focusing for the separation of the recombinant human immunodeficiency virus envelope glycoprotein glycoforms." J Chromatogr A 866(1): 121-35.

Tretter, V., F. Altmann, et al. (1991). "Peptide-N4-(N-acetyl-?-glucosaminyl)asparagine amidase F cannot release glycans with fucose attached ?1(3 to the asparagine-linked N-acetylglucosamine residue." European Journal of Biochemistry 199: 647-52.

Tsarbopoulos, A., U. Bahr, et al. (1997). "Glycoprotein analysis by delayed extraction and post-source decay MALDI-TOF-MS." International Journal of Mass Spectrometry and Ion Processes 169/170: 251-61.

Tseng, K., J. L. Hedrick, et al. (1999). "Catalog-library approach for the rapid and sensitive structural elucidation of oligosaccharides." Analytical Chemistry 71: 3747-54.

Vestal, M. L., P. Juhasz, et al. (1995). "Delayed extraction matrix-assisted laser desorption time-of-flight mass spectrometry." Rapid Communications in Mass Spectrometry 9: 1044-1050.

Wada, Y., J. Gu, et al. (1994). "Diagnosis of carbohydrate-deficient glycoprotein syndrome by matrix assisted laser desorption time-of-flight mass spectrometry." Biological Mass Spectrometry 23: 108-9.

Wang, J., P. Sporns, et al. (1999). "Analysis of food oligosaccharides using MALDI-MS: quantification of fructooligosaccharides." J Agric Food Chem 47: 1549-57.

Wang, T. H., T. F. Chen, et al. (1987). "Mass spectrometry of L-b-aspartamido carbohydrates isolated from ovalbumin." Biomedical and Environmental Mass Spectrometry 16: 335-8.

Watson, E. and F. Yao (1993). "Capillary electrophoretic separation of human recombinant erythropoietin (r-HuEPO) glycoforms." Anal Biochem 210: 389-93.

Weiskopf, A. S., P. Vouros, et al. (1997). "Characterization of oligosaccharide composition and structure by quadrupole ion trap mass spectrometry." Rapid Communications in Mass Spectrometry 11: 1493-504.

Weiskopf, A. S., P. Vouros, et al. (1998). "Electrospray ionization-ion trap mass spectrometry for structural analysis of complex N-linked glycoprotein oligosaccharides." Analytical Chemistry 70: 4441-7.

Wheeler, S. F. and D. J. Harvey (2001). "Extension of the in-gel release method for structural analysis of neutral and sialylated n-linked glycans to the analysis of sulfated glycans: application to the glycans from bovine thyroid-stimulating hormone." Anal Biochem 296: 92-100.

Wu, J. and J. Pawliszyn (1995). "Protein analysis by isoelectric focusing in a capillary array with an absorption imaging detector." J Chromatogr B Biomed Appl 669(1): 39-43.

Xu, N., Z.-H. Huang, et al. (1997). "Structural characterization of peptidoglycan muropeptides by matrix-assisted laser desorption ionization mass spectrometry and postsource decay analysis." Analytical Biochemistry 248: 7-14.

Yamashita, K., A. Hitoi, et al. (1983). "Structural studies of the carbohydrate moieties of rat kidney gamma-glutamyltranspeptidase. An extremely heterogeneous pattern enriched with nonreducing terminal N-acetylglucosamine residues." J Biol Chem 258: 1098-107.

Yamashita, K., T. Mizuochi, et al. (1982). "Analysis of oligosaccharides by gel filtration." Methods Enzymol 83: 105-26.

Yamashita, K., T. Ohkura, et al. (1993). "Electrospray ionization-mass spectrometric analysis of serum transferrin isoforms in patients with carbohydrate-deficient glycoprotein syndrome." J Biochem (Tokyo) 114: 766-9.

Yang, M. and M. Butler (2000). "Effect of ammonia on the glycosylation of human recombinant erythropoietin in culture." Biotechnol Prog 16: 751-9.

Yao, Y. J., K. C. Loh, et al. (1995). "Analysis of recombinant human tumor necrosis factor beta by capillary electrophoresis." Electrophoresis 16: 647-53.

Yoshino, K., T. Takao, et al. (1995). "Use of the derivatizing agent 4-aminobenzoic acid 2-(diethylamino)ethyl ester for high-sensitivity detection of oligosaccharides by electrospray ionization mass spectrometry." Analytical Chemistry 67: 4028-31.

Zieske, L.R., D. Fu, et al. (1996). "Multi-dimensional mapping of pyridylamine-labeled N-linked oligosaccharides by capillary electrophoresis." J Chromatogr A 720: 395-407.

2. THE EFFECT OF CELL CULTURE PARAMETERS ON PROTEIN GLYCOSYLATION

V.RESTELLI and M.BUTLER*

*Department of Microbiology, University of Manitoba, Winnipeg, Manitoba, Canada. *Corresponding author, E-mail: butler@cc.umanitoba.ca*

1. Introduction

An understanding of the carbohydrate moieties of recombinant glycoprotein for therapeutic use is of importance for two main reasons. Firstly, the carbohydrate structures attached to a protein can affect many of its properties (Takeuchi et al., 1989; Narhi et al., 1991) including pharmacokinetics, bioactivity, secretion, *in vivo* clearance, solubility, recognition and antigenicity (Storring, 1992; Wasley et al, 1991), all of which influence the overall therapeutic profile of the glycoprotein. Secondly, quantitative and qualitative aspects of glycosylation can be affected by the production process in culture, including the host cell line (Goto et al., 1988; Gooche, 1992; Sheeley et al., 1997; Kagawa et al, 1988), method of culture (Jenkins & Curling, 1994 Gawlitzek et al., 1995; Schewikart et al., 1999), extracellular environment and the protein itself (Jenkins et al., 1996; Reuter and Gabius, 1999).

Glycosylation is a process that occurs in eukaryotic cells in which oligosaccharides are added to the protein during synthesis and processed through the endoplasmic reticulum (ER) and Golgi apparatus along the secretory pathway (Schachter, 1983). Glycoproteins occur as heterogeneous populations of molecules, called glycoforms (Rudd & Dwek, 1997). The potential variability of glycoforms presents a difficulty to industrial production and for regulatory approval of therapeutic glycoproteins. The challenge is to know how the glycoprotein heterogeneity is generated, and how to evaluate its significance with respect to the safety and efficacy of the product (Teh-Yung, 1992).

The glycoforms have characteristic profiles that can vary with the control parameters of bioprocessing. It is important to understand these culture control parameters to assure the reproducibility of a bioprocess to yield the same glycoform profile that went through clinical trials and to maximize the titre of the desired glycoforms.

Mammalian cells are widely used for the commercial production of therapeutic glycoproteins. It is clear that there are several advantages in using them as host-cells for recombinant glycoprotein production because of their ability to perform post translational modifications and achieve a product close to that produced *in vivo* (Kornfeld & Kornfeld, 1985). The use of other expression systems such as yeast, plant or insect cells is more limited because despite their potential economic advantages for culture and the

M. Al-Rubeai.(ed.), Cell Engineering, 61-92.

higher yields of these systems, their glycosylation capacities do not resemble those of mammalian cells (Jarvis et al., 1998; Hersecovics & Orlean, 1993; Matsumoto et al., 1995)

Glycosylation engineering is gaining importance as a tool to optimize desirable properties such as stability, antigenicity and bioactivity of glycosylated therapeutic pharmaceuticals. This is achieved by genetic engineering of the pathways of oligosaccharide synthesis in the mammalian host cells (Grabenhorst et al., 1999; Bailey, 1997). Incorporation of new glycosyltransferase activities can modify the product, compete with endogenous enzymes to produce novel glycoforms or maximize the proportion of beneficial ones (Jenkins & Curling, 1994; Umaña & Bailey, 1997).

The purpose of this chapter is to review the capabilities of various cell lines for recombinant glycoprotein production. Each cell line's capacity for glycosylation may be modified by parameters associated with growth of the cells in culture. These combined factors will lead to a heterogeneity of glycoforms that will be considered in relation to their acceptability for use as human therapeutic agents.

2.   Oligosaccharide Structures Present in Glycoproteins

There are three types of oligosaccharides attached to proteins – N-glycans, O-glycans and GPI anchors (Parekh, 1994):

2.1. N-GLYCANS

The oligosaccharide is bound via an N-glycosidic bond to an Asn residue within the consensus sequence (sequon) Asn-X-Ser/Thr. (See Figure 1 for a list of abbreviations.) However, the presence of this sequon in a protein does not guarantee glycosylation. The glycosylation of the sequon is variable and gives rise to a *macroheterogeneity* of glycoforms (variable site occupancy). Amongst other factors, the site occupancy will depend upon the tertiary structure of the protein. There may be a multiplicity of glycan structures at a particular site. Differences in these structures is referred to as *microheterogeneity* (Spellman, 1990).

The structure of N-linked oligosaccharides fall into three main categories: high-mannose, hybrid and complex-type. They all have the same core structure: $Man_3GlcNAc_2$-Asn but differ in their outer branches (See Figure 2):

   i) High-Mannose type: typically has two to six additional Man residues linked to the core

   ii) Complex type: contains two or more outer branches containing N-acetyl glucosamine (GlcNAc), galactose (Gal), and sialic acid (S.A)

   iii) Hybrid type: has features of both high-Man and complex type oligosaccharides (Meynial-Salles & Combes, 1996).

Common substituents to the N-glycan structures are Fucose (Fuc) linked to either the innermost core GlcNAc (proximal) or the outer arm GlcNAc (peripheral). Also, a "bisecting" GlcNAc may be linked $\beta 1,4$ to the central core Man residue (Kornfeld & Kornfeld, 1985).

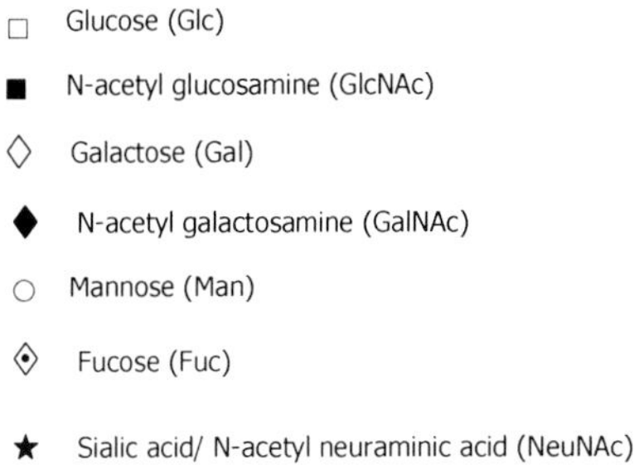

Glucose (Glc)
N-acetyl glucosamine (GlcNAc)
Galactose (Gal)
N-acetyl galactosamine (GalNAc)
Mannose (Man)
Fucose (Fuc)
Sialic acid/ N-acetyl neuraminic acid (NeuNAc)

Figure 1. Nomenclature and symbols used.

2.2. O-GLYCANS

The oligosaccharide is bound via an O-glycosidic bond to a Ser/Thr (O-glycosylation).

Eight core structures for O-glycosylation have been identified (See Figure 3). Any Ser or Thr residue is a potential site for O-glycosylation and no consensus sequence in the protein has been identified (Van den Steen et al, 1998).

2.3. GPI ANCHOR

The oligosaccharide is a component of the glycosyl phosphatidylinositol (GPI) membrane anchor. This becomes an integral part of the cell membrane. This modification is absent in the secreted form of any glycoprotein and will not be considered further in this review.

3. Assembly and Processing of Oligosaccharides on Proteins

3.1. ASSEMBLY OF ASN-LINKED OLIGOSACCHARIDES

This process is initiated in the endoplasmic reticulum as a protein is being synthesized from its constituent amino acids. The precursor oligosaccharide structure is added to the N-glycan sites and this is followed by a series of trimming reactions. High-mannose type structures are formed during the processing in the ER. However, the completion of the processing to form hybrid and complex glycans takes place in the Golgi. A

Sequon = Asn-X-Ser/Thr
= Fuc
= GlcNAc
= Man
= Gal
= NeuAc
Asn
High mannose
Asn
Hybrid
Asn
Complex

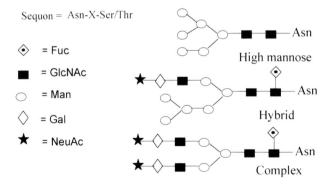

Figure 2. N-glycan structures of glycoproteins.

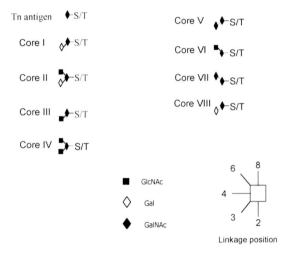

Figure 3. Core structures of mucin-type O-linked glycans.

simplified version showing the formation of a complex biantennary structure is shown in Figure 4. However, the full diversity of glycan structures occurs because of a series of competing processing enzymes associated with this pathway.

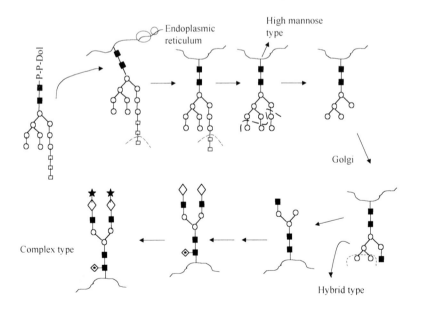

Figure 4. N-glycan processing.

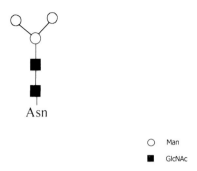

Asn

○ Man
■ GlcNAc

Figure 5. Core N-glycan structure.

All N-linked glycans on a glycoprotein share the same core (Figure 5) because they all come from the same precursor which is transferred to the nascent protein in the endoplasmic reticulum. The precursor consists of a lipid (dolichol) linked to an oligosaccharide by a py-rophosphate bond. During synthesis of this oligosaccharide, sugars are added in a step-wise fashion, the first seven sugars (two GlcNAc and five Man) derive from the nucleotide sugars UDP-GlcNAc and GDP-Man respectively (Kornfeld & Kornfeld, 1985 and Umaña et al., 1999).

The next seven sugars (four Man and three Glc) are derived from the lipid intermediates Dol-P-Man and Dol-P-Glc. The final product: $Glc_3Man_9GlcNAc_2$-P-P-Dol is the precursor for the N-linked glycans (Figure 6).

The glycosylation is initiated in the E.R where the N-glycan precursor is attached to the consensus sequence Asn-X-Ser/Thr by the enzyme, oligosaccharyltransferase. However, these sequences are not always glycosylated. Several factors can be identified to effect site-occupancy :

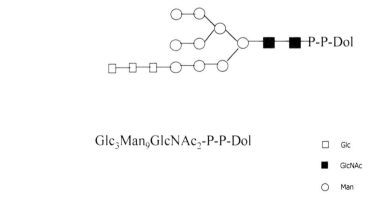

$Glc_3Man_9GlcNAc_2$-P-P-Dol

□ Glc
■ GlcNAc
○ Man

Figure 6. N-linked precursor.

i) The spatial arrangements of the peptide during the translation process can expose or hide the tripeptide signal

ii) The amino acid sequence around the attachment site (Asn-X-Ser/Thr) is an important determinant of glycosylation efficiency. X can be any amino acid except Pro or Asp. The occupancy level is high when X= Ser, Phe; intermediate for Leu, Glu and very low for Asp, Trp and Pro.

iii) The availability of precursors (lipid, nucleotide sugars and correctly assembled precursor) level of expression of the oligosaccharyltransferase enzyme disulphide bond formation, which can make the site inaccessible to the precursor.
(Rudd & Dwek, 1997).

3.2. OLIGOSACCHARIDE PROCESSING IN THE ENDOPLASMIC RETICULUM

Processing is initiated by glucosidases, which are located in the membrane of the ER. A specific α1,2 glucosidase I (Gluc I) removes the first terminal Glc. The next two Glc residues are removed by a single α1,3 glucosidase II (Gluc II). This product is the substrate for the α mannosidase I (Man I) that catalyses the removal of at least one Man residue (see Figure 7). The newly synthesised glycoproteins are then transported to the Golgi cisternae by means of vesicles (Rudd & Dwek, 1997).

3.3. OLIGOSACCHARIDE PROCESSING IN THE GOLGI

In the Golgi a series of glycosidases and glycosyltransferases act on the N-linked oligosaccharides and lead to a diversity of structures (*microheterogeneity*)

A mixture of Man9 (M9) and Man8 (M8) oligosaccharides are the substrates for the first enzyme in the Golgi: α1,2 Mannosidase I (Man I) that transforms them to Man5

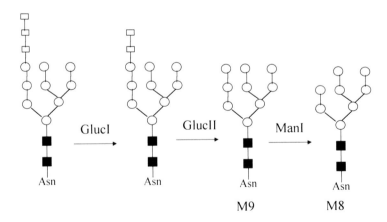

Figure 7. First stages of N-glycan processing in the endoplasmic reticulum.

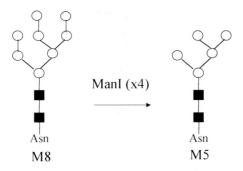

Figure 8. Production of high mannose structures. ManI trims M8 to M7, M6, and M5 structures.

(M5) products. M9 to M5 constitute the high mannose class of N-linked oligosac-charides (Figure 8).

Then, the enzyme β1,2N-acetylglucosaminyl transferase I (GnT I), transfers a GlcNAc residue creating the first hybrid oligosaccharide.

This is the substrate for the next enzyme: α1,2 Mannosidase II (ManII) that removes one or two Man residues, leaving the second branch available for extension by the second gluco-seminyltransferase (GnT II) to produce the first complex biantennary oligosaccharide (Figure 9).

Now this oligosaccharide can be processed by glucosaminyltransferase IV (GnT IV) or glucosaminyltransferase V (GnTV) that add a GlcNAc residue to the α1,3 mannose or α1,6 Mannose branch respectively creating two different types of complex-triantennary oligosaccharides. The tetraantennary complex compound can then be

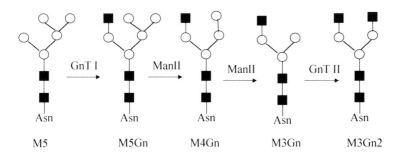

Figure 9. Production of hybrid and complex structures in the Golgi.

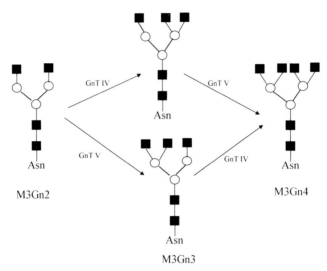

Figure 10. Production of tri- and tetra-antennary structures.

synthesised from both triantennary compounds by the action of GnT V and GnT IV respectively (Figure 10).

Competing with these enzymes, the β1,4 galactosyltransferase (Gal T) can extend any hybrid or complex oligosaccharide by adding a Gal to a non-reducing GlcNAc. Once a Gal residue is transferred, the modified oligosaccharide cannot be modified by the remaining GnTs or Man II enzymes.

Another enzyme competing for the same substrates is glucosaminyltransferase III (GnT III). This enzyme can modify any non galactosylated hybrid or complex glycan by transferring GlcNAc residue in a β1,4 linkage to the core Man. ("bisecting" GlcNAc [Gn^b]) (Figure 11). Gal T cannot extend this residue but it may modify the other non-reducing GlcNAc residues.

The final products can be modified further in the Golgi by the addition of:

i)   **sialic acid**: two main enzymes compete for the terminal Gal: α2,3 sialyltransferase (α2,3ST) and α2,6 sialyltransferase (α2,6ST)

ii)   **poly-N-acetyl lactosamine**: it is added by β N-Acetylglycosaminyltransferase (βGnT)

iii)   **Fucose**: by α1,6 fucosyltransferase, that adds fucose in an α1,6 linkage to the GlcNAc attached to Asn., or by α1,3 fucosyltransferase that adds fucose to C3 of the GlcNAc residue in the sequence Gal β 1-4 GlcNAc β 1-2 Man. The product is not a substrate for the α2,6ST. Fucose can be added at any time after M5 is synthesised but not after the action of Gal T or GnT III. Fucosylated oligosaccharides however, can be modified by these enzymes.

All the transferase-catalyzed reactions use sugar nucleotide co-substrates. It is evident that the key factor in determining the synthesis of particular N-linked oligosaccharides is the level of expression of the different glycosyltransferases.

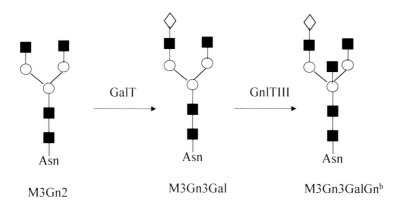

Figure 11. Synthesis of galactosylated and bisecting glycans.

The oligosaccharide profiles of glycoproteins are normally characteristic of the cell in which the protein is expressed and depends on cellular factors such as:

i) enzyme repertoire (Meynial-Salles & Combes, 1996)
ii) competition between different enzymes for one substrate (Umaña & Bailey, 1997)
iii) transit time of the glycoproteins (Hooker et al., 1999; Nabi & Dennis, 1998)
iv) levels of sugar nucleotide donors (Valley et al., 1999)
v) competition between different glycosylation sites on the protein for the same pool of enzymes (Schachter et al., 1983)

At any time, many glycoproteins may be trafficking through the glycosylation pathway, competing for the glycosylation enzymes. The oligosaccharides attached to the glycoprotein are processed by some enzymes and not by others.

Umaña and Bailey (1997), proposed a mathematical model based on the activities of a set of 8 enzymes and 32 reactions to determine the distribution of oligo-saccharides into the major structural classes: high Man, hybrid, bisected hybrid, bi-, tri, tri'- and tetraantennary complex and bisected complex oligosaccharides so the proportion of these structures could be calculated based on the kinetics of these enzymes (Figure 12).

3.4. ASSEMBLY OF O-LINKED OLIGOSACCHARIDES

O-linked glycans are added post-translationally to the fully folded protein. Glyco-sylation can occur on exposed Ser or Thr residues but no consensus sequence as been identified (Van den Steen et al., 1998).

The most commonly found O-glycans are the mucin-type, although other structures such as O-linked fucose or O-linked glucose do exist. The first step for the assembly of the mu-cin type O-glycans is the addition of N-acetylgalactosamine (GalNAc) residue to a Ser/Thr by a GalNAc transferase (Gal NacT) from UDP-GalNAc (Van den Steen et al., 1998).

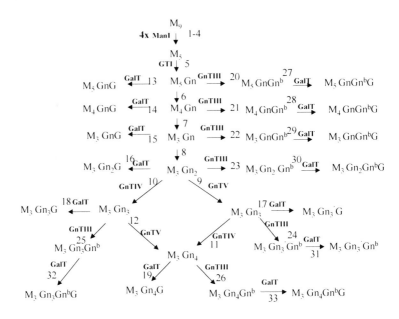

Figure 12. Reaction network for N-linked glycosylation. (from Umana and Bailey. 1997)

Although no consensus sequence has been identified for O-glycosylation, the glycosylated residue is often associated with regions of the peptide that contain a high proportion of Ser, Thr and Pro. It appears that this would make the polypeptide assume a favorable conformation making the Ser of Thr more accessible (Van den Steen et al., 1998). Thr residues appear to be glycosylated more efficiently than ser. Various GalNAcT have been identified, having a broad but different substrate specificity and are expressed in a tissue-specific manner. Further elongation leads to a large number of structures, synthesised by various glycosyltransferases, producing eight different core structures (see Figure 3). These core structures can be further modified by sialylation, fucosylation, sulfation, methylation or acetylation. A characteristic of transformed cells is that the initial GalNAc residue is not elongated and is known as the Tn antigen. Each of the core structures can give rise to a series of modified structures as illustrated for core 1 and core 2 glycans in Figure 13. An example of how metabolic engineering can lead to a modification of this metabolic network is discussed in section 6.2.

4.   Expression Systems and their Glycosylation Capabilities

The prokaryotes (mainly *E.coli*) were the first cells to be used for gene expression of recombinant proteins. These cells can be easily manipulated and grown in large scale but they lack the necessary glycosylation machinery and so the proteins produced are not glycosylated.

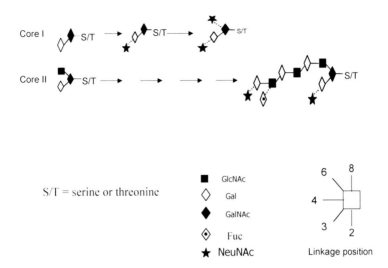

S/T = serine or threonine

■ GlcNAc
◇ Gal
◆ GalNAc
◉ Fuc
★ NeuNAc

Linkage position

Figure 13. Processing mucin-type O-linked glycans in CHO cells.

An alternative is offered by lower eukaryotes (yeast, insect and plant cells). However, the glycans produced in these cells differ significantly from those present in human glyco-proteins (Jenkins et al., 1996). Yeast, insect, plant and mammalian cells share the feature of N-linked oligosaccharide processing in the endoplasmic reticulum, including attachment of $Glc_3Man_9GlcNAc_2$-P-P-Dol and subsequent truncation to $Man_8GlcNAc_2$ structure. However, oligosaccharide processing by these different cell types diverges in the Golgi apparatus (Goochee, 1991). Although there is extensive heterogeneity of structures arising from any cell type, examples of predominant N-glycans that might occur from different systems is shown in Table 1.

4.1. INSECT CELLS

Lepidopteran insect cell lines (such as *Spodoptera frugiperda, Sf-9*) have been used extensively for expression of recombinant proteins using the baculovirus as a means of transfection (Jarvis et al, 1998). Alternative methods of protein expression are also available using efficient insect-associated promoter systems (Farrell et al, 1998). The advantage of the use of these cells is the high expression level and growth rate of the cells in culture. However, the glycosylation of proteins expressed by insect cells is limited. These cells can add $Glc_3Man_9GlcNAc_2$ precursors to appropriate N-glycan sites in a nascent polypeptide and convert them to $Man_9GlcNAc_2$. They also have the enzymes necessary to trim this oligosac-charide all the way down to $Man_3GlcNAc_2$. The formation of this product requires the action of the enzymes ManI, GnTI and ManII. However, there is little structural processing beyond the $Man_3GlcNAc_2$ core oligosac-charide apart from the possibility of fucosylation (Jarvis & Finn, 1996; Donaldson et al.,

Table 1. Typical predominant N-glycan structures from different cell types.

Cell-Type	N-Glycan	4.2. STRUCTURE
Bacteria, *E.coli*	None	-----------------
Yeast	High Mannose	
Insect	Fucosylated Core Structure	
Plant	Xylosylated and Fucosylated Core Structure	
Mammalian	Complex Biantennary	

Glycosyltransferase enzymes are either absent or at a low level of activity (Jarvis et al., 1998). Therefore, generally the insect expression system is incapable of synthesising sialylated lactosamine complex-type N-glycan or sialylated O-glycans.

However, some insect cells have been found to produce recombinant glycoproteins with elongated trimannosyl core structures containing terminal GlcNAc or Gal, and one recombinant glycoprotein acquired complex biantennary N-linked glycan containing sialic acid (Kulakosky et al., 1998; Davidson et al., 1990).

Jarvis et al., (1998) proposed a model to explain the N-linked oligosaccharide structures found in insect cell-produced glycoproteins (Figure 14). In this model the enzyme GlcNAc TII competes with N-acetylglucosaminidase (GlcNAdase) at a branch point. Depending up-on the relative activities of these competing processing enzymes, trimannosyl core or complex N-linked glycans may be produced. It would appear that insect cells have a high level of N-acetylglucosaminidase activity which can remove GlcNAc from a terminal position.

The production of complex sialylated N-linked structures could only occur when the recombinant protein is a very poor substrate for GlcNAdase or excellent substrate for the low level activities of the glycosyltransferases. The enzymic activity may vary considerably and this explains the differences in the N-glycan pattern of the same glycoprotein secreted by different insect cell lines (Kulakosky et al., 1998).

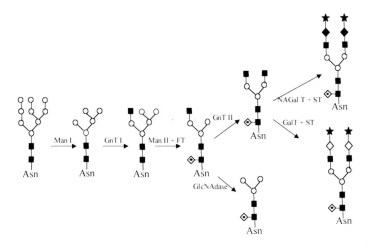

Figure 14. Proposed route of glycan processing in insect cells (from Jarvis et al. 1998).

The potential of insect cells for O-glycosylation was reported in a study in which 3 lepi-dopteran cell lines were shown to produce predominantly short O-glycan structures (Lopez et al, 1999). All 3 cell lines expressed GalNAcα1-O-Ser/Thr (Tn antigen), whereas the ability to synthesise Galβ1-3GalNAcα1-O-Ser/Thr (T-antigen) and Gal α1-4Galβ1-3GalN Acα1-O-Ser/Thr (PT- antigen) was more limited. This indicated low activity of β1-3-galac-tosyltransferase and α1-4-galactosyltransferase. There was no indication of sialylation of these structures suggesting the absence of sialyltransferase activity. The culture medium used to grow these cell lines had a major effect on the O-glycans expressed. The use of a semi-defined rich medium enhanced glycosyltransferase activity significantly compared to a minimal nutrient medium.

The limitations in the ability of insect cells for glycosylation have restricted their use for production of human therapeutics. Genetic engineering of these systems is under study in an attempt to improve the glycosylation machinery to produce more "humanized" glycoproteins (Jarvis et al., 1998). In particular the transformation of insect cells with the gene for mammalian β1,4-N-acetylgalactosyltransferase with a baculovirus expression vector leads to expression of the enzyme and results in more extensively processed N-glycans (Jarvis and Finn, 1996).

4.3. YEAST

The early steps in the addition of carbohydrate to proteins have been remarkably conserved during evolution. The synthesis of the $Glc_3Man_9GlcNAc_2$-P-P-Dol precursor, transfer to the polypeptide and early processing in the ER are common events shared by eukaryotic cells.

However, though relatively few N-linked sites have been characterized, it was noted that there is a trend in favour of the use of Asn-X-Thr sites over Asn-X-Ser in yeast glycoproteins. In contrast to mammalian cells, where several Man residues may be removed during processing, in S. *cerevisiae*, a single specific Man residue is cleaved to form $Man_8GlcNAc_2$. Most yeast and filamentous fungi synthesise carbohydrate chains of the high mannose type. Complex glycan structures are not observed among fungal glycoproteins.

Addition of Man residues to core oligosaccharides occurs very rapidly in the Golgi forming the characteristic high mannose structures (mannan) which can consist of more than 50 mannose residues and resulting in high molecular weight glycoproteins (Hersecovics & Orlean, 1993).

Proteins synthesised in yeast may also contain O-glycans consisting of linear poly-mannose structures attached to Ser or Thr. Similar to mammalian cells, O-glycosylation in yeasts has no obvious consensus sequence. However, unlike mammalian cells O-glycosylation in yeast is initiated with covalent attachment of mannose via a dolichol phosphate mannose precursor. Maras et al. (1997) showed that if the high mannose structures are trimmed *in vitro* by mannosidase, they can become acceptors for the re-combinant processing enzymes, N-acetylglucosaminyl-transferaseI, β1,4-galactosyl-transferase and α2,6-sialyl-transferase. Mutant strains of yeast also may synthesise truncated mannose structures.

4.4. PLANTS

Plant cells also conserve the early stages of N-glycosylation. However the processing of the oligosaccharide trimming and further modification of glycans in the Golgi differ from mammalian cells. Plant-derived oligosaccharides do not possess sialic acid and frequently contain xylose (Xyl), not normally present in mammalian N-linked oligosac-charides. Typically processed N-glycans in plants have a $Man_3GlcNAc_2$ structure with β1,2 Xylose and /or α1,3 fucose residues to the reducing terminal GlcNAc (Palacpac et al., 1999). The presence of these two residues makes plant recombinant glycoproteins less desirable as therapeutics because of the immunogenicity of these residues (Storring, 1992). Xylose is not present in mammalian glycan structures and fucose is attached to proximal (core) GlcNAc by α1,6 linkage in mammalian cells rather than α1,3. The absence of these determinants in mammals make them highly immunogenic if present in therapeutic glycoproteins (Parekh et al., 1989; Palacpac et al., 1999).

4.5. MAMMALIAN CELLS

Mammalian cells are the chosen host for the production of human glycoproteins because it has been recognised they meet the criteria for an appropriate glycosylation of recombinant human glycoproteins (Lamotte et al., 1997). They are capable of complex type N-glycan processing whereas the other systems are not. However, there are dif-ferent capabilities for glycosylation between mammalian cell lines as discussed in sec-tion 5.1. The most commonly used cell lines for recombinant protein production are the hamster-derived Chinese hamster ovary (CHO) and baby hamster kidney (BHK) cells. These cell lines have been chosen because of their favourable growth characteristics in culture as anchorage-dependent or suspension cells. They have been used as expression systems to produce proteins whose glycoforms are similar to the native human products.

5. Control of Oligosaccharide Processing in Mammalian Cell Culture

5.1. HOST CELL

An analysis of oligosaccharide structures on the same proteins from different species and even different tissues reveals that major variations frequently exist. It is evident that a key factor in determining the synthesis of particular N-linked oligosaccharides is the presence and/or level of expression of the various glycosyltransferases. Differences in the relative activity of these enzymes among species and tissues can account for many of the variations in oligosaccharide structures that are present (Kornfeld & Kornfeld, 1985; Goto et al, 1988).

An analysis of the glycan structures of IgG from 13 different species shows that there is significant variation in the proportion of terminal galactose, core fucose and bisecting GlcNAc (Raju et al, 2000). The predominant monosaccharide precursors may also differ structurally between species. For example the terminal sialic acid found in glycoproteins from goat, sheep and cows is predominantly N-glycolyl-neuraminic acid (NGNA) rather than N-acetyl-neuraminic acid (NANA) which is the sialic acid structure generally found in humans and rodents.

Chinese hamster ovary (CHO) and baby hamster kidney (BHK) are the most commonly used cell lines for the production of recombinant proteins with potential application as therapeutic agents. For such an application it would be desirable to obtain proteins with as near a human glycosylation profile as possible. However both CHO and BHK show differences in their potential for glycosylation compared to human cells. The sialyl transferase enzyme, $\alpha 2,6$ ST is not active in these cell lines, leading to exclusively $\alpha 2,3$ linked terminal SA residues. Furthermore, the absence of a functional $\alpha 1,3$ fucosyltransferase in CHO cells prevents the addition of peripheral Fuc residues and also the absence of N-acetylglucosamyltransferase III (GnTIII) prevents the addition of bisecting GlcNAc (Jenkins & Curling, 1994).

However, these differences in glycosylation potential between CHO and human cells do not appear to result in glycoproteins that are immunogenic. Natural human erythropoietin (EPO) consists of a mixture of sialylated forms – 60% being 2,3-linked and 40% being 2,6-linked. Because of the restricted sialylation capacity of CHO cells, the recombinant EPO is sialylated entirely via the $\alpha 2,3$ linkages. Nevertheless, recombinant EPO produced from CHO cells is currently employed as a highly effective therapeutic agent in the treatment of a variety of renal dysfunctions. There is no evidence of any adverse physiological effect due to the structural difference in the terminal sialic acid.

Mouse cells express the enzyme $\alpha 1,3$ galactosyltransferase, which generates Gal$\alpha 1,3$-Gal$\beta 1,4$-GlcNAc residues, not present in humans and is an epitope found to be highly immunogenic (Jenkins et al., 1996). The epitope is found in the glycan structures of glycoproteins from non-primate animals including rodents, pigs, sheep and cows as well as New World monkeys. Although there is evidence for the presence of the gene of this enzyme in CHO and BHK cells, there appears to be no activity of the $\alpha 1,3$galactosyl-transferase enzyme. The potential for immunogenicity associated with this epitope limits the use of murine cells for therapeutic glycoprotein production.

Sheeley et al., (1997) compared the glycosylation of CAMPATH (a recombinant humanized murine monoclonal immunoglobulin) expressed in two different cell lines: a murine hybridoma cell (NS0) and in CHO cells. The glycosylation expressed in CHO cells was consistent with the one found in native IgG while the antibody expressed in NS0 cells included potentially hypergalactosylated immunogenic glycoforms. These contain the α1,3-Gal-Gal terminal residues. Some of the characteristics of the glycosylation capacity of CHO cells and murine C127 cells are summarised in Table 2.

Kagawa et al., (1988), compared the oligosaccharides of natural human IFN-β1 produced in three different cell lines: CHO, a hamster derived cell line, C127 a mouse cell line and PC8 a cell line derived from human lung tumour. The CHO cells produced structures quite similar to those of the natural IFN-β1; C127 produced structures with α1,3-Gal-Gal sequences, completely missing in natural IFN-β1 although the occurrence of this type of substitution is common in mouse cell lines. Surprisingly, the human cell line PC8 produced the greatest variety of different structures, including α1,3-Gal-Gal terminal sequences.

Alterations of cell-type dependant glycosylation can also result from spontaneous or induced mutations affecting oligosaccharide synthesis. A series of CHO clones has been isolated possessing a variety of mutations affecting N- and O- glycosylation (Stanley, 1983). The mutants are characterized by the expression of aberrant lectins on the cell surface and are classified as a series of numbered LEC mutants. These mutations usually diminish the glycosylation capability. For example, Lec1 CHO mutant expresses no detectable GnTI and accumulates glycoproteins with Man$_5$GlcNAc$_2$ structures in the cell. A mutant affected in the same gene, Lec 1A was isolated from a sub-population of this mutant. This mutant produces a GnTI biochemically different than the one produced by the parental cell line, with new kinetic properties (Chaney & Stanley, 1986). Multiple enzymic defects may be an advantage in the production of glycoproteins with minimal carbohydrate heterogeneity (Stanley, 1989).

Table 2. Characteristics of the glycosylation capacity of two widely used mammalian cell lines compared to normal human cells.

1. CHO cells
- lacks α2.6-sialyltransferase activity. Sialylation by CHO cells only produce α2.3-linkages to terminal N-glycan structures
- unable to sulfate GalNAc residues. This is important in certain hormones.
- lacks α1.3-fucosyltransferase. This prevents the formation of peripheral fucose
- residues (fuc-α1.3-GlcNAc)
- lacks β-1.4-N-acetylglucosaminyltransferase III. This prevents the formation of GlcNAc attachment to the core mannose to form a bisected glycan.

2. Murine C127 cells
- lacks α2.3-sialyltransferase activity. Sialylation by C127 cells only produces α2.6-linkages to terminal N-glycan structures.
- expresses α1.3 galactosyltransferase activity. This results in gal-α1.3-gal which is antigenic in humans.

A mutation may also result in a mutant with a gain of function such as the CHO Lec 11 which expresses α1,3-fucosyltransferase (Zhang et al, 1999). CHO cells have a limited capacity for synthesis of elongated O-glycans because of the lack of core 2 GlcNAc-T activity, although this enzyme may be induced by butyrate treatment (Datti and Dennis, 1993).

5.2. CULTURE ENVIRONMENT

The control of the culture environment is important to maximize cell growth in order to attain a high cell density which is a prerequisite for producing cell products whether they be viruses, antibodies or recombinant proteins (Andersen & Goochee, 1994). However, it is also important to realise that the specific conditions of the culture can affect product glycosylation independently of the characteristics of the cell line. During the course of a batch culture nutrient consumption and product accumulation changes the cellular environment usually in a way that can gradually decrease the extent of protein glycosylation over time (Jenkins and Curling, 1994). Such changes are unacceptable in a cell culture bioprocess used for large-scale production of a protein that may be a therapeutic agent. It can lead to variable glycoform heterogeneity and significant batch to batch variation in the production processes. In order to maintain product consistency it is essential to understand the parameters of cell culture that can cause variations in glycosylation. As a more far-reaching objective it may be reasonable to control culture conditions in favour of reducing glycoform heterogeneity or producing a specific glycoform. For example, the maximization of product sialylation of therapeutic proteins could lead to higher specific biological activities *in vivo*.

Two lines of evidence suggest that the extracellular environment may affect glycosylation: One, significant *in vivo* changes in glycosylation are associated with the physiological state (e.g.: pregnancy) and disease (e.g.: diabetes) (Reuter and Gabius, 1999). Two, *in vitro* cell culture studies show direct effects of the extracellular environment on protein glycosylation. In some cases these have been reported from changes in the mode of culture. For example, the glycosylation of antibodies was found to be more consistent by *in vitro* culture than from ascites fluid (Maiorella et al, 1993) or from the adaptation of cells from serum to serum-free medium (Gawlitzek et al, 1995). In other reports, the specific culture parameters affecting an alteration in glycosylation has been analysed (Yang & Butler, 2000; Borys et al., 1993).

The choice of culture method, pH, nutrient concentration, dissolved oxygen, etc., are some of the parameters that have proved to affect the oligosaccharide structures of glycoproteins.

Among the potential mechanisms to explain such effects are:

i) depletion of the cellular energy state

ii) disruption of the local ER and Golgi environment

iii) modulation of glycosidase and glycosyltransferase activities

iv) modulation of the synthesis of nucleotides, nucleotide sugars and lipid precursors.

(Valley et al., 1999)

The awareness of such effects in the glycosylation of proteins makes the development of more defined culture media and conditions a very important issue for the development of a pharmaceutical product with defined oligosaccharide structures and batch consistency.

5.3. MODE OF CULTURE

CHO and BHK cells can be grown as anchorage-dependent cells in T-flasks or microcarriers. Alternatively, they can be adapted to suspension culture. This adaptation process leads to characteristic changes in the expression of cell surface proteins such as the integrins which affect viral susceptibility (Brown, 1998). Not surprisingly, the glycosylation process may also be affected by such changes. Watson et al (1994) reported that the sialylation of N-glycans of a secreted protein from CHO was reduced in microcarrier culture compared to suspension.. Gawlitzek et al (1995) also showed an affect on glycan antennarity of a BHK-produced protein related to the mode of culture.

The presence or absence of serum in the culture medium also has a significant effect on glycosylation. This is not surprising given the variable concentrations of hormones and growth factors in serum and even in different formulations of serum-free media. Cells grown in SFM (serum free medium) secrete a higher proportion of N-glycosylated and O-glycosylated protein with enhanced terminal sialylation and proximal fucosylation (Gawlitzek et al 1995). This result was attributed to the presence of high activities of sialidase and fucosidase in serum.

5.4. SPECIFIC GROWTH RATE AND PROTEIN PRODUCTIVITY

Schewikart et al., (1999) utilised three different bioreactor systems to evaluate the effect of different culture methods on glycosylation of a monoclonal IgA antibody. Although conditions such as nutrients, temperature, pH, oxygen, were kept constant, the environmental conditions in these systems were different, especially comparing immobilized versus suspension systems. Significant variations were detected in the pattern of N-linked oligosaccharide structures, especially in the degree of sialylation. These differences were attributed mainly to differences in growth rate, specific productivity and cell density among the bioreactors. The immobilized cells (in a fluidised bed bioreactor and hollow fibre bioreactor) showed higher growth rate and specific productivity compared to the cells in suspension (CSTR).

Glycosylation characteristics have been also related to the rate of glycoprotein production.

Nabi et al., (1998) slowed the transit time of a glycoprotein through the Golgi by controlling the polarization of MDCK cells in a culture or by reducing the temperature. The observed effect was an increase in polylactosamine glycosylation of LAMP-2 (lysosomal membrane glycoprotein). However, they found no differences in the activities of the glycosyltransferases. They demonstrated that the slower transit of the glycoprotein through the Golgi was responsible for the increase in polylactosamine glycosylation.

5.5. GLUCOSE

Low glucose concentrations have been reported to produce two distinct abnormalities in the synthesis of glycoproteins: attachment of aberrant precursors to the protein and absence of glycosylation at sites that are normally glycosylated. Both abnormalities would be related to a shortage of glucose- derived oligosaccharide precursors (Kornfeld & Kornfeld, 1985). Glucose starvation may result in an intracellular energy-depleted state or a shortage of glucose-derived oligosaccharide precursors (Rearick et al, 1981). Reduced site occupancy of Ig light chains was observed in mouse myeloma cells grown at a glucose concentration below 0.5 mM (Stark and Heath, 1979). Abnormal glycosylation of viral proteins is also observed at low glucose (Davidson and Hunt, 1985). In a chemostat culture of CHO cells Hayter et al (1993) showed an increase in non-glycosylated gamma-interferon under glucose limiting conditions. Pulsed additions of glucose restored normal glycosylation rapidly.

5.6. AMMONIA

Glutamine is normally added to culture medium at a concentration of 2-10 mM. The glutamine provides an energy source for cells as well as being an essential precursor for nucleotide synthesis. However, glutamine is a source of ammonia accumulation in culture medium which arises from either thermal decomposition of the glutamine or from metabolic deamination or deamidation. The accumulated ammonia is inhibitory to cell growth (Butler and Spier, 1984; Doyle and Butler. 1990) and also has a specific effect on protein glycosylation (Yang and Butler, 2000).

Castro et al., (1995) compared the effects of different concentrations of glutamine on the macroheterogeneity of IFN-γ produced in CHO cells. They observed an increase in the proportion of bi-glycosylated IFN-γ with increasing concentrations of glutamine.

Gawlitzek et al., (1998) examined the effect of different glutamine and NH_4^+ concentrations on the N-linked oligosaccharide structures of IL-Mu6 (recombinant human IL-2 N glycosylation mutant). In the absence of glutamine (low NH_4^+ production) the oligosaccharides revealed the most homogeneous pattern with the highest content of terminal sialic acid. Addition of glutamine and NH_4^+ both produced an increase in the complexity (antennarity) of oligosaccharides and a decrease in terminal sialylation.

A decrease in O-linked sialylation of G-CSF (granulocyte colony stimulating factor) produced in CHO cells was observed with increasing ammonia concentrations in the medium. This is consistent with the pH effect of ammonia in the Golgi compartments. At a concentration of 10mM, the expected pH change from 6.5 to 7.0 would result in approximately a two fold decrease in ST activity which correlates with the two fold decrease in sialylation found in G-CSF (Andersen & Gooche, 1995).

Glutamine and NH_4^+ increase the intracellular UDP-Glc/GalNAc pool, leading to the formation of more complex oligosaccharide structures. Valley et al., (1999) investigated the action of ammonium on the synthesis of the intracellular UDP-GNAc glycosylation precursors as well as on N-oligosaccharides of IL-Mu6. They used $^{15}NH_4^+$ to study the pathway leading to the increase of UDP-GNAc pool. They observed that the proportion of UDP-GNAc containing ^{15}N correlated with the increase of intracellular concentration of UDP-GNAc indicating that ammonium is channelled into the pathway of UDP-GNAc formation.

Ammonia caused a decrease in sialylation and antennarity of the glycan of recombinant erythropoietin (EPO) produced from CHO cells (Yang & Butler, 2000). The proportion of tetrasialylated-linked complex oligosaccharides in EPO expressed in CHO cells decreased from 49% in control cultures to 29% in ammonia exposed cultures. A reduction of the proportion of tetraantennary structures by 30% was also observed with a corresponding increase of bi- and triantennary structures.

The mechanism for the effect of ammonia on glycosylation has been well studied. The two major changes to the intracellular environment by ammonia are the increase in pH of the Golgi and the significant change in nucleotide pool concentrations, notably a sharp increase in the UDP-GNAc/ UTP ratio. The increase in UDP-GNAc is due to enhanced synthesis through incorporation of NH_4^+ via the glucosamine-6-phosphate isomerase reaction in which fructose 6-phosphate is converted to glucosamine 6-phosphate. Ammonium could act in two different modes on the glycosylation:

    i)   via induced elevation of the UDP-GNAc pool.

    ii)   via modifying the degree of sialylation due to altered pH conditions (Valley et al., 1999). Glycosylation enzymes have a pH optimum. Amines accumulate inside the cell in acidic intracellular compartments and raise the pH shifting it from the optimum pH of the ST enzymes. The elevated UDP-GNAc may also cause an impaired transport of CMP-NeuAc into the Golgi compartments (Gawlitzek et al., 1998; Anderson & Gooche, 1995). This is also likely to lead to feedback inhibition of the synthesis of CMP-NeuAc, which would reduce the availability of the sialic acid-nucleotide precursor even further (Figure 15). The availability and transport of these precursors into the Golgi has been recognised as a critical step of protein glycosylation (Hooker et al, 1999).

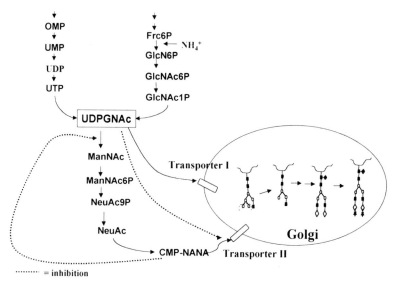

Figure 15. Ammonia decreases sialylation via an increase in intracellular UDP-GNAc.

5.7. pH

Under adverse external pH conditions the internal pH of the Golgi is likely to change resulting in a reduction of the activities of key glycosylating enzymes. The pH of the medium was shown to have some effect on the distribution of glycoforms of IgG secreted by a murine hybridoma (Rothman et al, 1989). Borys et al., (1993) related the extracellular pH to the specific expression rate and glycosylation pattern of recombinant mouse placental lactogen-I (mPL-I) by CHO cells. They observed that the maximum specific mPL-I expression rates occurred between pH 7.6 and 8.0. The level of site occupancy was maximum between these pH decreasing at lower (< 6.9) and higher (>8.2) pH values.

5.8. DISSOLVED OXYGEN CONCENTRATION

Oxygen plays a dominant role in the metabolism and viability of cells (Jan et al., 1997; Heidemann et al., 1998); it is a limiting nutrient in animal cell culture because of its low solubility in the medium.

Kunkel et al., (1998) studied the effect of dissolved oxygen concentrations on the glycosylation of a monoclonal antibody secreted by an hybridoma (CC9C10). They observed a decrease in galactosylation at reduced oxygen concentrations (10% DO). At this concentration, the glycans were mainly agalactosyl or monogalactosylated while at higher oxygen concentration (50 - 100% DO) there was a higher proportion of digalactosylated glycans (Figure 16).

The mechanism for the effect of DO on galactosylation is unclear. One explanation is that reduced DO causes a decline in the availability of the UDP-Gal. This might arise due to a sensitivity to reduced oxidative phosphorylation in the production of UDP-Gal

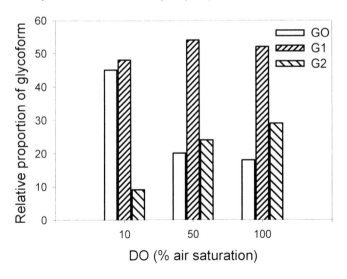

Figure 16. Effect of dissolved oxygen on the glycoform profile of IgG (Data from Kunkel et al, 1998).

or as a result of reduced UDP-Gal transport from the cytosol to the Golgi. A second explanation is based on evidence that the timing and rate of formation of the inter-heavy chain disulfide bonds in the hinge region of IgG determine the level of Fc galactosylation (Rademacher et al, 1996). Thus the addition of galactose may be impeded by the early formation of the inter-heavy chain disulfide bond. Low DO in the culture may cause a perturbation in the oxidating environment of the ER and/or the Golgi complex and the disturbance may result in a change in the pathway of inter-chain disulfide bond formation.

An effect of DO has also been observed in CHO cultures. Chotigeat et al., (1994) recorded a shift in the isoforms of human follicle stimulating hormone produced from CHO cells at different DO levels. An increase in the sialyltransferase activity was observed at higher oxygen concentrations that translated into an increase on sialylation of follicle-stimulating hormone (FSH) producing a shift of the isoforms to the lower pI fractions.

5.9. GROWTH FACTORS/CYTOKINES/ HORMONES

There are many reports of hormones involved in the regulation of protein glycosylation *in vivo*. Up and down regulation of specific glycosyltransferases has been observed frequently in conjunction with hormonal induction of cell differentiation. Presumably, transcriptional control of glycosylation enzymes concentration is responsible for many of the effects on oligosaccharide processing (Goochee & Monica, 1990). An example of glycosylation control *in vivo* is the cascade of events that occurs following the stimulation of the synthesis of thyrotropin by the tripeptide, thyrotropin-releasing hormone (TRH). This in turn promotes the synthesis and sialylation of thyroglobulin by thyroid cells (Ronin et al, 1986). The glycosylation of transferrin is regulated by prolactin in rabbit mammary glands (Bradshaw et al, 1985). In cell culture dexamethasone can affect glycan structures in rat hepatocytes (Pos et al, 1988).

Retinol and retinoic acid may play a role *in vivo* in epithelial cell differentiation and can be shown in culture to cause significant changes to protein glycosylation. This includes a shift from high mannose to complex glycans in chondrocytes (Bernard et al, 1984) and the extension of complex structures in mouse melanoma cells (Lotan et al, 1988).

Exogenous IL-6 induces changes in the activities of intracellular GnTs including a reduction in the activity of GnTIII and an increase in GnTIV and GnTV of a myeloma cell line that led to alterations in the glycan structure of the surface and secreted glycoproteins (Nakao et al, 1990).

5.10.   MEDIUM ADDITIVES FOR ENHANCED PRODUCTION

Butyrate affects glycosylation by inducing glycosyltransferases. Lamotte et al, (1999) demonstrated an increase in the sialylation of IFN-γ with the addition of 1mM Na-butyrate in to a culture of CHO cells. Na-butyrate treatment resulted in over-expression of mRNA coding for a variety of proteins. The mechanism for this is likely to be the increase of hyper acetylation of histones induced by Na-butyrate provoking the chromatin to loosen and allow increased access to RNA polymerase for mRNA synthesis. However butyrate caused a x4 increase in productivity of a chimeric antibody but without an affect on the glycoform distribution of the product (Mimura et al, 2001).

The availability of nucleotide sugar precursors may be a limiting factor for glyco-sylation. This is supported by the effect of the addition of precursors to cultures to en-hance glycosylation. Cystidine and uridine can alter protein glycosylation by increasing the availability of nucleotide sugars (Kornfeld & Kornfeld, 1985). The addition of N-acetyl mannosamine (ManNAc), a direct precursor of CMP-NeuAc, to CHO cultures increased significantly the sialylation of gamma-interferon (Gu and Wang, 1998).

5.11. EXTRACELLULAR DEGRADATION OF GLYCOPROTEIN OLIGOSACCHARIDES

Mammalian cells possess glycosidases that may be released extracellularly into the culture by cell secretion or upon cell lysis (Gramer & Goochee, 1993). Fucosidase, β galactosidase, β hexosaminidase and sialidase activities have been shown to accumulate in the extracellular medium of CHO cells (Warner, 1999). The action of these enzymes on secreted glycoproteins that have a variable residence time in the culture may result in significant heterogeneity of glycoforms.

Gramer & Goochee (1993) explored the presence of four glycosidases in CHO cells supernatant. They demonstrated that CHO cells possess a significant and stable sialidase activity that can accumulate in the extracellular medium and retains considerable activ-ity at pH 7.

The extent of glycan degradation depends on many factors, including the level of extracellular activity, pH, temperature and time of the glycoprotein exposure to the enzyme.

Bioprocesses that result in maintenance of high cell densities for long periods such as fed-batch or perfusion mode cultures may be particularly vulnerable to this type of glycan degradation. Early extraction of the product from the medium reduces the residence time of the glycoprotein in culture and may reduce glycoform heterogeneity.

6. Genetic Engineering of Mammalian Cells to Modify Glycosylation

Mammalian cell lines used for the production of glycoproteins may lack the enzymic profile to synthesise recombinant proteins that are glycosylated as authentic human proteins. This may be due to lack of processing enzymes, presence of alternative processing enzymes or through expression of glycosidases activities in the mammalian host cells (Warner, 1999).

Metabolic engineering provides a promising tool to modify the characteristics of the host mammalian cells by enhancing cell productivity, protein quality and bioactivity and by modifying the glycosylation pathway to obtain a final product with advantageous properties.

6.1. ENGINEERING OF HOST CELLS WITH NEW GLYCOSYLATION PROPERTIES

The two commonly used hamster cell lines, BHK-21 and CHO cells do not express $\alpha 2,6$ sialyltransferase, $\alpha 1,3$ fucosyltransferase or β-1,4-N-acetylglucosaminyltransferaseIII activities. As these enzymes are found in normal human cells, the products of the hamster cell lines may not possess some of the oligosaccharide structures found typically in human serum proteins. Transfection of the cells with the gene of the lacking glycosyltransferase may correct such deficiencies.

6.1.1. α2,6 Sialyltransferase (α2,6 ST)

Two different sialic acid linkages ($\alpha 2,3$ and $\alpha 2,6$) to the terminal Gal are found in N-linked oligosaccharides isolated from human glycoproteins. The enzymes responsible for these substitutions are $\alpha 2,3$ ST and $\alpha 2,6$ ST. Both enzymes compete for the same substrate and a mixture of both linkages is often found in native human glycoproteins. CHO and BHK-21 cells only produce $\alpha 2,3$ sialylated oligosaccharide structures (Takeuchi et al., (1988).

Grabenhorst et al., (1995) introduced the $\alpha 2,6$ ST gene into BHK cells expressing recombinant ATIII, EPO and β–TP (β trace protein). The modified cells produced glycoproteins with an increased level of sialylation which included a mixture of 2,3/6 sialylated oligosaccharide structures.

Lamotte et al., (1999) co-transfected CHO cells with genes for IFN-γ and $\alpha 2,6$ ST. The modified cells produced IFN-γ, 68% of which was sialylated with a $\alpha 2,6$ linkage. The over-all extent of sialylation was doubled compared to the product of the cells without the $\alpha 2,6$ ST gene. The addition of sodium butyrate enhanced the $\alpha 2,6$ ST activity and increased the extent of sialylation and the proportion of $\alpha 2,6$ linked SA to 82%. However, sodium butyrate had no effect on the sialylation of the product of the cells without the $\alpha 2,6$ ST gene insert

6.1.2. α 1,3 fucosyltransferase (α1,3FT)

This enzyme is required for the addition of a peripheral $\alpha 1,3$ fucose linkage to GlcNAc as found in certain human proteins. The co-expression of β-TP from recombinant BHK-21 cells with human $\alpha 1,3$FT successfully produced a glycoprotein, 50% of which had an $\alpha 1,3$ linked Fuc. (Grabenhorst et al., 1999). However, a significant decrease in the degree of sialylation of N-glycans was observed. It is suggested that this could be due to competition of $\alpha 1,3$FT with the endogenous $\alpha 2,3$ST for the same substrate.   The $\alpha 2,3$ sialyltransferase is unable to sialylate fucosylated structures.

6.1.3. β 1,4 N-acetyl glucosaminyltransferase (GnTIII)

The bisecting GlcNAc residue plays an important role in the branching and elongation of oligosaccharide structures by restricting the action of other enzymes. GnTIII is not expressed at significant levels in normal CHO cells so the synthesis of bisected oligo-saccharides in these cells depends on the over expression of the enzyme after transfec-tion with GnTII DNA.

Sburlati et al., (1998) created a CHO cell line capable of producing bisected oligosaccharides on the glycan structure of IFN-β. This structure has not been detected in native human IFN-β and the biological significance of this is unknown.

Immunoglobulin glycosylation is essential for complement fixation and antibody-dependent mediated cytotoxicity (ADCC). This is a lytic attack on antibody-targeted cells and is initiated after the binding of a lymphocyte receptor to the constant region (Fc) of the antibodies. This effector function may be essential for the therapeutic application of certain antibodies. Although human serum IgG contains low levels of bisecting GlcNAc, therapeutic antibodies containing bisected glycans may have an enhanced ADCC.

One example of this is a chimeric IgG1 (chCE7) anti neuroblastoma engineered in CHO cells which showed enhanced ADCC as a result of the presence of bisected glycoforms in the Fc region (Umaña et al., 1999). The antibody chCE7 was constructed by transfecting the CHO parental cell line with the GnTIII gene under tetracycline regulation. Over-expression of GnTIII led to a modified IgG containing a bisected gly-coform. The maximal activity of ADCC correlated with a high level of Fc-associated bisected complex oligosaccharides. Thus, enhanced ADCC activity of chCE7 together with the capacity of this antibody to recognise neuroblastoma cells, makes it a suitable candidate molecule for the treatment of these tumours.

The effect of GnTIII expression levels on glycan structures in CHO cells was pre-dicted by a mathematical model based upon enzyme kinetic constants and mass balances associated with the production of 33 different N-glycan structures (Umana and Bailey, 1997). GnTIII has the potential to act upon at least 7 independent glycan structures that result in either bisected complex and bisected hybrid glycans. The complex form is the required structure to maximize biological activity *in vivo*. Analysis of the competitive enzymic reactions in the central reaction network of the Golgi can be used to predict the activity of GnTIII required to maximize synthesis of the bisected complex glycoforms.

6.2. METABOLIC REGULATION OF O-GLYCOSYLATION

The predominant O-glycan structures formed in CHO cells are the core 1 type (Figure 3). However it has been shown by the simultaneous up-regulation and down-regulation of key enzymes this pathway can be altered. CHO cells that had been genetically engineered to express α 1,3 fucosyltransferase (α1,3FT) were selected for the co-expression of a CMP-sialic acid: Galβ1,3GalNAcα2,3-sialyltransferase (ST3Gal I) gene fragment set in the antisense orientation and the human UDP-GlcNAc:Galβ1,3GalNAc-R β1,6-N-acetylgluco-saminyltransferase (C2GnT) (Prati et al 2000).   This co-expression resulted in an increase in the activity of the C2GnT enzyme and a decrease in the activity of the ST3Gal I enzyme (Figure 17). The effect of this co-ordinated change was to divert the O-glycosylation pathway from the formation of core 1 glycans to core 2 glycans. The significance of this is the formation of proteins containing sialyl-Lewis X glycan structures that mediate interaction with selectins and cell-cell adhesion.

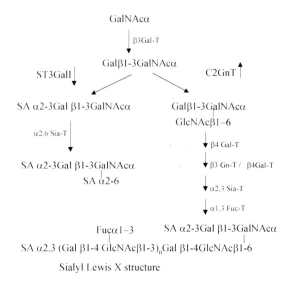

GalNAcα

|β3Gal-T

Galβ1-3GalNAcα

ST3Gall| C2GnT↑

SA α2-3Gal β1-3GalNAcα   Galβ1-3GalNAcα
GlcNAcβ1–6

α2.6 Sia-T| ↓ β4 Gal-T

SA α2-3Gal β1-3GalNAcα   ↓ β3 Gn-T / β4Gal-T
SA α2-6   ↓ α2.3 Sia-T

↓ α1.3 Fuc-T

Fucα1–3   SA α2-3Gal β1-3GalNAcα

SA α2.3 (Gal β1-4 GlcNAcβ1-3)ₙGal β1-4GlcNAcβ1-6

Sialyl Lewis X structure

Figure 17. Control of O-linked glycosylation pathways showing inhibition of the Core 1 pathway and inhibition of the Core 2 pathway (data from Prati et al. 2000).

6.3. ANTISENSE RNA AND GENE TARGETING

An alternative for metabolic engineering of producer mammalian cells focuses on the direct manipulation of the expression of endogenous proteins by the use of anti-sense RNA. This approach is suitable for removing an unwanted enzyme activity or to enhance the expression of endogenous proteins to improve the product quality or enhance cell productivity (Stout & Caskey, 1987; Nellen and Sczakiel, 1996).

One obvious target for this strategy in CHO cells is the soluble sialidase gene. A reduction of sialidase expression would improve the stability of secreted proteins in culture supernatant. The de-sialylation of therapeutic proteins reduces bioactivity because the resulting proteins with exposed terminal Gal residues are removed from the blood stream by hepatocyte asialo-glycoprotein receptors. Thus, the conservation of sialic acid in the oligosaccharide chains of glycoproteins is critical and must be maintained on the proteins during the production and purification processes (Rush et al., 1995).

Antisense expression of sialidase resulted in a 60% reduction of sialidase activity in the culture supernatant of CHO cells expressing DNAase (Ferrari et al., 1998). Although only an additional one mole of sialic acid per mole of protein was observed, this modest improvement in sialylation resulted in a dramatic effect on the serum clearance rate of the protein.

Antisense RNA targeting has proven to be a valuable means to revive silent genes or correct gene defects. More complete glycosylation of recombinant glycoproteins may be possible if the activities of endogenous glycosyltransferases are increased above normal levels (Warner, 1999).

Two strategies for the construction of anti-sense expressing cells are possible:

i) The creation of a universal host cell line which express the desired anti-sense RNA. Once the cell line is established, the product expression vector is introduced. The advantage of this approach is the availability of a universal cell line constitutively expressing anti-sense RNA.

ii) The introduction of the anti-sense expression vector into an existing recombinant host after growth and productivity parameters have been optimised. This reduces the possibility of modifying the anti-sense RNA.

7.   Genetic Engineering of Non-Mammalian Cells

7.1. ENGINEERING INSECT CELLS

Some success has been achieved in expanding the glycoprotein processing capabilities of the insect cell systems, which generally have low levels of specific glycosylating enzymes (Jarvis et al, 1998). In insect cells the endogenous N-acetylglucosamine-transferase activity is thought to be in competition with N-acetylglucosamidase activity. Lepidoptera (Sf9) cells co-transfected with genes for a human glycosyltransferase enzyme (GlcNAcTI) and a re-combinant influenza hemagglutinin produced a glyco-protein with an extended trimannosyl core with GlcNAc terminal residues (Wagner et al, 1996).   Further glycan processing can be promoted in these cells by the introduction of β-1,4-galactosyl transferase activity. A stable transfectant of Sf9 cells with multiple integrated copies of the β-1,4-galactosyl transferase gene supported expression of mammalian proteins such as the glycoprotein, gp64 and tissue-plasminogen activator (t-PA) with glycans containing terminal galactose residues (Jarvis & Finn, 1996; Hollister et al, 1998).

A future direction in the genetic manipulation of these cells would be the suppres-sion of the N-acetyl glucosaminidase and the induction of further endogenous processing enzymes.

7.2. ENGINEERING PLANT CELLS

Plant cells normally process N-glycans to trimannosyl core structures with or without attached xylose. Only rarely have complex type N-glycans been identified in plants.

However, transfection of tobacco BY2 cells with human GalT gene led to the ability to produce proteins containing glycans with Gal residues at the terminal non-reducing ends (Palacpac et al., 1999).

8.   Conclusion

The choice of the cell expression systems and the control of the production parameters at earlier stages of bioprocess development are key factors for ensuring the production of glycoproteins with consistent structures.

To optimise the glycoform distribution for a given glycoprotein produced by a given cell type it is important to understand the specific environmental factors affecting oligo-saccharide structures and how to control these factors at the cellular level. Given the biological complexities of cell growth and metabolism, the cellular and environmental parameters that can be potentially altered are enormous. An increased awareness of the importance of pharmaceutical protein glycosylation has lead to the increased importance of analysis of glycan structures.

Recent efforts in metabolic engineering are clearly justified in view of the demands for the production of proteins with a consistent glycoform profile and more cost effective, high productivity processes. Continuing efforts in metabolic engineering may lead to host cell lines capable of producing a restricted set of glycoforms for a specific glycoprotein, with enhanced bioactivity and reduced blood clearance rates.

9. References

Andersen D.C. and Goochee C.F. (1994) The effect of cell-culture conditions on the oligosaccharide structures of secreted glycoproteins. Curr. Op. Biotech. 5: 546-9.

Andersen D.C. and Goochee C.F. (1995) The Effect of Ammonia on the O-Linked Glycosylation of Granulocyte Colony-Stimulating Factor Produced by Chinese Hamster Ovary Cells. Biotechnol. Bioeng. 47: 96-105.

Bailey J.E., Umaña P., Minch S, Harrington M, Page M. and Sburlati-Guerini A. (1997) M.J.T. Carrondo et al. (Eds.) Animal Cell Technology p489-94. Metabolic engineering of N-linked glycoform synthesis systems in Chinese Hamster Ovary (CHO) cells.

Bernard B.A., De-Luca L.M., Hassell J.R., Yamada K.M. and Olden K. (1984) Retinoic acid alters the proportion of high mannose to complex type oligosaccharides on fibronectin secreted by cultured chondrocytes. J Biol Chem. 259: 5310-5.

Borys C, Linzer D.I.H. and Papoutsakis. E.T. (1993) Culture pH Affects Expression Rates and Glycosylation of Recombinant Mouse Placental Lactogen Proteins by Chinese Hamster Ovary (CHO) Cells. BioTechnology 11: 720-4.

Bradshaw J.P., Hatton J. and White D.A. (1985) The hormonal control of protein N-glycosylation in the developing rabbit mammary gland and its effect upon transferrin synthesis and secretion. Biochim Biophys Acta. 847: 344-51.

Brown F. (1998) Problems with BHK 21 cells. Dev Biol Stand 93: 85-8.

Butler M. and Spier R.E. (1984). The effects of glutamine utilisation and ammonia production on the growth of BHK cells in microcarrier cultures. J. Biotechnol. 1: 187-96.

Castro P.M., Ison A.P., Hayter P.M. and Bull A.T. (1995) The macroheterogeneity of recombinant human interferon-γ produced by Chinese-hamster ovary cells is affected by the protein and lipid content of the culture medium. Biotech. Appl. Biochem. 21: 87-100.

Chaney W. and Stanley P. (1986) Lec1A Chinese Hamster Ovary Cell Mutants Appear to Arise from a Structural Alteration in N-Acetylglycosaminyltransferase I. J.Bioch. Chem. 261: 10551-7.

Chotigeat W., Watanapokasin Y., Mahler S. and Gray P.P. (1994) Role of environmental conditions on the expression levels, glycoform pattern and levels of sialyltransferase for hFSH produced by recombinant CHO cells. Cytotechnology 15: 217-21.

Datti A. and Dennis J.W. (1993) Regulation of UDP-GlcNAc-Galβ1-3GalNAc-R β1-6-N-Acetyl-gluco-saminyl-transferase (GlcNAc to GalNAc) in Chinese hamster ovary cells. J. Biol. Chem. 268: 5409-16.

Davidson D.J., Fraser M.J. and Castellino F.J. (1990) Oligosaccharide Processing in the Expression of Human Plasminogen cDNA by Lepidopteran Insect (Spodoptera frugiperda) Cells. Biochemistry 29: 5584-90.

Davidson S.K. and Hunt L.A. (1985) Sindbis virus glycoproteins are abnormally glycosylated in Chinese hamster ovary cells deprived of glucose. J Gen Virol. 66: 1457-68.

Donaldson M., Wood H.A. and Kulakosky P.C. (1999) Glycosylation of a Recombinant Protein in the Tn5B1-4 Insect Cell Line: Influence of Ammonia, Time of Harvest, Temperature, and Dissolved Oxygen. Biotechnol. Bioeng. 63: 255-62.

Doyle C. and Butler, M. (1990). The effect of pH on the toxicity of ammonia to a murine hybridoma. J. Biotechnol. 15: 91-100.

Farrell P.J., Lu M., Prevost J., Brown C., Behie L. and Iatrou K. (1998) High-Level Expression of Secreted Glycoproteins in Transformed Lepidopteran Insect Cells Using a Novel Expression Vector. Biotechnol. Bioeng. 60: 656-63.

Ferrari J., Gunson J., Lofgren J., Krummen L. and Warner T.G. (1998) Chinese Hamster Ovary Cells with Constitutively Expressed Sialidase Antisense RNA Produce Recombinant DNase in Batch Culture with Increased Sialic Acid. Biotech. Bioeng. 60: 589-95.

Gawlitzek M., Valley U. and Wagner, R. (1998) Ammonium Ion and Glucosamine Dependent Increases of Oligosaccharide Complexity in Recombinant Glycoproteins Secreted from Cultivated BHK-21 Cells. Biotechnol. Bioeng. 57: 518-28.

Gawlitzek M., Valley U., Nimtz M., Wagner R. and Conradt H.S. (1995) Characterization of changes in the glycosylation pattern of recombinant proteins from BHK-21 cells due to different culture conditions. J. Biotech. 42: 117-31.

Goochee C.F., Gramer J., Andersen D.C., Bahr J.B. and Rasmussen, J.R. (1991) The oligosaccharides of glycoproteins: bioprocess factors affecting oligosaccharide structure and their effect on glycoprotein properties. Bio/Technology 9: 1347-55.

Goochee, C.F. and Monica, T. (1990) Environmental effects on protein glycosylation. Bio/Technology 8: 421-7.

Goto M, Akai K, Murakami A, Hashimoto C, Tsuda E, Ueda M, Kawanishi G, Takahashi N, Ishimoto A, Chiba H, Sasaki R. (1988) Production of recombinant human erythropoietin in mammalian cells: host-cell dependency of the biological activity of the cloned glycoprotein. BioTechnology 6: 67-71.

Grabenhorst E., Hoffmann A., Nimtz M., Zettlmeissl G. and Conradt H.S. (1995) Construction of stable BHK-21 cells coexpressing human secretory glycoprotein and human Gal (β1-4)GlcNAc-R α2,6-sialyltransferase. Eur. J. Biochem. 232: 718-25.

Grabenhorst E., Schlenke P., Pohl S., Nimtz M. and Conradt H.S. (1999) Genetic engineering of recombinant glycoproteins and the glycosylation pathway in mammalian host cells. Glycoconjugate J. 16: 81-97.

Gramer J. and Goochee, C.F. (1993) Glycosidases Activities in Chinese Hamster Ovary Cell Lysate and Cell Culture Supernatant. Biotechnol. Prog. 9: 366-73.

Gu X. and Wang D.I. (1998) Improvement of interferon-gamma sialylation in Chinese hamster ovary cell culture by feeding of N-acetylmannosamine. Biotechnol. Bioeng. 58: 642-8.

Hayter P.M., Curling E.M., Gould M.L., Baines A.J., Jenkins N., Salmon I., Strange P.G. and Bull, A.T. (1993) The effect of dilution rate on CHO cell physiology and recombinant interferon γ production in glucose-limited chemostat cultures. Biotech Bioeng 42: 1077-85.

Heidemann R., Lutkemeyer D., Buntemeyer H. and Lehmann, J. (1998) Effects of dissolved oxygen levels and the role of extra- and intracellular amino acid concentration upon the metabolism of mammalian cell lines during batch and continuous cultures. Cytotechnology 26: 185-97.

Hersecovics A. and Orlean P. (1993) Glycoprotein biosynthesis in yeast. FASEB J. 7: 540-50.

Hollister J.R., Shaper J.H. and Jarvis D.L. (1998) Stable expression of mammalian beta 1,4-galactosyltransferase extends the N-glycosylation pathway in insect cells. Glycobiology 8: 473-90.

Hooker A.D., Green N.H., Baines A.J., Bull A.T., Jenkins N., Strange P.G. and James D.C. (1999) Constraints on the Transport and Glycosylation of Recombinant IFN-γ in Chinese Hamster Ovary and Insect Cells. Biotechnol. Bioeng. 63: 559-72.

Jan D.C., Petch D.A., Huzel N. and Butler M. (1997) The effect of dissolved oxygen on the metabolic profile of a murine hybridoma grown in serum-free medium in continuous cultures. Biotech. Bioeng. 54: 153-64.

Jarvis D.L. and Finn E. (1996) Modifying the insect cell N-glycosylation pathway with immediate early baculovirus expression vectors. Nature Biotech. 14: 1288-92.

Jarvis D.L., Kawar Z.S. and Hollister J.R. (1998) Engineering N-glycosylation pathways in the baculovirus-insect cell system. Curr. Op. Biotech. 9: 528-33.

Jenkins N. and Curling E.M. (1994) Glycosylation of recombinant proteins: Problems and prospects. Enzyme Microb. Technol. 16: 354-64.

Jenkins N., Parekh R.B. and James D.C. (1996) Getting the glycosylation right: Implications for the Biotechnology industry. Nature BioTechnology 14: 975-81.

Kagawa Y., Takasaki S., Utsumi J., Hosoi K., Shimizu H., Kochibe N. and Kobata A. (1988) Comparative Study of the Asparagine-linked Sugar Chains of Natural Human Interferon -β1 Produced by Three Different Mammalian Cells. J. Biol. Chem. 263: 17508-15.

Kornfeld R. and Kornfeld S. (1985) Assembly of Asparagine-Linked Oligosaccharides. Ann. Rev. Biochem. 54: 631-64.

Kulakosky P.C., Shuler M.L. and Wood H.A. (1998) N-Glycosylation of a baculovirus-expressed recombinant glycoprotein in three insect cell lines. In Vitro Cell. Dev. Biol.-Animal 34: 101-8.

Kunkel J.P., Jan D.C.H., Jamieson J.C. and Butler M. (1998) Dissolved oxygen concentration in serum-free continuous culture affects N-linked glycosylation of a monoclonal antibody. J. Biotech. (1998) 62: 55-71.

Lamotte D., Buckberry L., Monaco L., Soria M., Jenkins N., Engasser J-M. and Marc A. (1999) Na-butyrate increases the production and α2,6-sialylation of recombinant interferon-γ expressed by α2,6-sialyltransferase engineered CHO cells. Cytotechnology 29: 55-64.

Lamotte D., Eon-Duval A., Acerbis G., Distefano G., Monaco L., Soria M., Jenkins N., Engasser J-M and Marc A. (1997) M.J.T. Carrondo et al. (eds.), Animal Cell Technology, p761-5. Controlling the glycosylation of recombinant proteins expressed in animal cells by genetic and physiological engineering.

Lopez, M., Tetaert D., Juliant S., Gazon M., Cerutte M., Verbert A. and Delannoy, P. (1999). O-glycosylation potential of lepidopteran insect cell lines. Biochim Biophys Acta 1427: 49-61.

Lotan R., Lotan D. and Amos B. (1988) Enhancement of sialyltransferase in two melanoma cell lines that are growth-inhibited by retinoic acid results in increased sialylation of different cell-surface glycoproteins. Exp Cell Res. 177: 284-94.

Maiorella B.L., Winkelhake J., Young J., Mayer B., Bauer R., Hora M., Andya J., Thomson J., Patel T. and Parekh R. (1993) Effect of culture conditions on IgM antibody structure, pharmokinetics and activity. Bio/Technology 11: 387-92.

Maras M., Saelens X., Laroy W., Piens K., Claeyssens M., Fiers W. and Contreras, R. (1997) In vitro conversion of the carbohydrate moiety of fungal glycoproteins to mammalian-type oligosaccharides. Evidence for N-acetylglucosaminyltransferase-I-accepting glycans from Trichoderma reesei. Eur. J. Biochem. 249: 701-7.

Matsumoto S., Ikura K., Masatsugu U. and Sasaki R. (1995) Characterization of a human glycoprotein (erythropoietin) produced in cultured tobacco cells. J. Biotech. 46: 1-14.

Meynial-Salles I, Combes D. (1996) In vitro glycosylation of proteins: An enzymatic approach. Plant Mol. Biol. 27: 1163-72.

Mimura Y., Lund J., Church S., Dong S., Li J., Goodall M. and Jefferis R. (2001) Butyrate increases production of human chimeric IgG in CHO-K1 cells whilst maintaining function and glycoform profile. J. Immunol Methods 247: 205-16.

Nabi I.R. and Dennis J.W. (1998) The extent of polylactosamine glycosylation of MDCK LAMP-2 is determined by its Golgi residence time. Glycobiology 8: 947-53.

Nakao H., Nishikawa A., Karasuno T., Nishiura T., Iida M., Kanayama Y., Yonezawa T., Tarui S. & Taniguchi N. (1990) Modulation of N-acetylglucosaminyltransferase III, IV & V activities and alteration of the surface oligo- saccharide structure of a myeloma cell line by interleukin 6. Biochem Biophys Res Commun. 172: 1260-6.

Narhi L.O., Arakawa T., Aoki K.H., Elmore R., Rohde M.F., Boone T. and Strickland T.W. (1991) The Effect of Carbohydrate on the Structure and Stability of Erythropoietin. J. Biol. Chem. 266: 23022-6.

Nellen W. and Sczakiel G. (1996) In Vitro and In Vivo Action of Antisense RNA. Mol. Biotech. 6: 7-14.

Palacpac N.Q., Yoshida S., Sakai H., Kimura Y. and Fujiyama K. (1999) Stable expression of human β 1,4-galactosyltransferase in plant cells modifies N-linked glycosylation patterns. Proc. Natl. Acad. Sci. 96, 4692-7.

Parekh R.B., Dwek R.A., Edge C.J. and Rademacher T.W. (1989) N-glycosylation and the production of recombinant glycoproteins. TIBTECH 7: 117-22.

Parekh R-B. (1994) Biologicals 22, 113-119 Gene Expression - Glycosylation. Biologicals 22, 113-9.

Pos O., van-Dijk W., Ladiges N., Linthorst C., Sala M., van-Tiel D. and Boers W. (1988) Glycosylation of four acute-phase glycoproteins secreted by rat liver cells in vivo and in vitro. Effects of inflammation and dexamethasone. Eur J Cell Biol. 46: 121-8.

Prati E.G.P., Matasci M., Suter T.B., Dinter A., Shuurlati, A.R. and Bailey J.E. (2000) Engineering of co-ordinated up- and down-regulation of two glycotransferases of the o-glycosylation pathway in Chinese hamster ovary (CHO) cells. Biotechnol. Bioeng. 68: 239-44.

Rademacher T.W., Jaques A. and Williams P.J. (1996) "The defining characteristics of immunoglobulin glycosylation" In: Isenberg, DA and Rademacher, TW (eds) publ. Wiley, NY pp1-44. Abnormalities of IgG Glycosylation and Immunological Disorders.

Raju T.S., Briggs J.B., Borge S.M. and Jones A.J. (2000) Species-specific variation in glycosylation of IgG: evidence for the species specific sialylation and branch-specific galactosylation and importance for engineering recombinant glycoprotein therapeutics. Glycobiology 10: 477-86.

Rearick J.I., Chapman A. and Kornfeld S. (1981) Glucose starvation alters lipid-linked oligosaccharide biosynthesis in Chinese hamster ovary cells. J Biol Chem 256: 6255-61.

Reuter G. and Gabius H.J. (1999) Eukaryotic glycosylation: whim of nature or multipurpose tool? Cell. Mol Life Sci. 55: 368-422.

Ronin C., Fenouillet E., Hovsepian S., Fayet G. and Fournet B. (1986) Regulation of thyroglobulin glycosylation. A comparative study of the thyroglobulins from porcine thyroid glands and follicles in serum-free culture. J.Biol Chem. 261: 7287-93.

Rothman R.J., Warren L., Vliegenthart J.F. and Hard K.J. (1989) Clonal analysis of the glycosylation of immunoglobulin G secreted by murine hybridomas. Biochemistry. 28: 1377-84.

Rudd P.M. and Dwek R.A. (1997) Glycosylation: Heterogeneity and the 3D Structure of Proteins. Crit. Rev. Biochem. Molec. Biol. 32: 1-100.

Rush R.S., Derby P.L., Smith D.M., Merry C., Rogers G., Rohde M.F. and Katta V. (1995) Microheterogeneity of erythropoietin carbohydrate structure.Anal Chem. 67: 1442-52.

Sburlati A.R., Umaña P. Prati E.G. and Bailey J.E. (1998) Synthesis of Bisected Glycoforms of Recombinant IFN-β by Over expression of B-1,4-N-Acetyl glucosaminyltransferase III in Chinese Hamster Ovary Cells. Biotechnol. Prog. 14: 189-92.

Schachter H., Narasimhan S., Gleeson P. and Vella G. (1983) Control of branching during the biosynthesis of asparagine-linked oligosaccharides. Can. J. Biochem. Cell Biol. 61: 1049-66.

Schewikart F., Jones R., Jaton J.C. and Hughes G.J. (1999) Rapid structural characterization of a murine monoclonal IgA α-chain: heterogeneity in the oligosaccharide structures at a specific site in samples produced in different bioreactor systems. J. Biotech. 69: 191-201.

Sheeley D.M., Merrill B.M. and Taylor L.C.E. (1997) Characterization of Monoclonal Antibody Glycosylation: Comparison of Expression Systems and Identification of Terminal α-Linked Galactose. An. Biochem. 247: 102-10.

Spellman M.W. (1990) Carbohydrate Characterization of Recombinant Glycoproteins of Pharmaceutical Interest. Anal. Chem. 62: 1714-22.

Stanley P. (1983). Lectin-resistant CHO cells: selection of new mutant phenotypes. Somatic Cell Genetics 9: 593-608.

Stanley P. (1989) Chinese Hamster Ovary Cell Mutants with Multiple Glycosylation Defects for Production of Glycoproteins with Minimal Carbohydrate Heterogeneity. Mol. Cell. Biol. 9: 377-83.

Stark N.J. and Heath E.C. (1979) Glucose-dependent glycosylation of secretory glycoprotein in mouse myeloma cells. Arch-Biochem-Biophys. 192: 599-609.

Storring P.L. (1992) Assaying glycoprotein hormones – the influence of glycosylation on immunoreactivity. TIBTECH 10: 427-32.

Stout J.T. and Caskey C.T. (1987) Antisense RNA Inhibition of Endogenous Genes. Meth. Enzym. 151: 519-30.

Takeuchi M., Inoue N., Strickland T.W., Kubota M., Wada M., Shimizu R., Hoshi S., Kozutsumi H., Takasaki S. and Kobata A. (1989) Relationship between sugar chain structure and biological activity of recombinant human erythropoietin produced in Chinese hamster ovary cells. Proc. Natl. Acad. Sci. USA 86: 7819-22.

Takeuchi M., Takasaki S., Miyazaki H., Kato T., Hoshi S., Kochibe N. and Kobata A. (1988) Comparative Study of the Asparagine-linked Sugar Chains of Human Erythropoietins Purified from Urine and the Culture Medium of Recombinant Chinese Hamster Ovary Cells. J. Biol. Chem. 263: 3657-63.

Teh-Yung Liu D. (1992) Glycoprotein pharmaceuticals: scientific and regulatory considerations, and the US Orphan Drug Act. TIBETCH 10: 114-9.

Umaña P. and Bailey J.E. (1997) A Mathematical Model of N-Linked Glycoform Biosynthesis. Biotechnol. Bioeng. 55: 890-908.

Umaña P., Jean-Mairet J., Moudry R., Amstuz H. and Bailey J.E. (1999). Engineered glycoforms of an antineuroblastoma IgG1 optimized antibody-dependent cellular cytotoxic activity. Nature Biotech. 17: 176-80.

Valley U., Nimtz M., Conradt H.S. and Wagner R. (1999) Incorporation of Ammonium into Intracellular UDP-Activated N-Acetylhexosamines and into Carbohydrate Structures in Glycoproteins. Biotechnol. Bioeng. 64: 401-17.

Van den Steen P., Rudd P.M., Dwek R.A. and Opdenakker G. (1998) Concepts and Principles of O-Linked Glycosylation. Crit. Rev.Biochem. Mol. Biol. 33: 151-208.

Wagner R., Liedtke S., Kretzschmar E., Geyer H., Geyer R., and Klenk H.D. (1996) Elongation of the N-glycans of fowl plague virus hemagglutinin expressed in Spodoptera frugiperda (Sf9) cells by coexpression of human beta 1,2-N-acetylglucosaminyltransferase I. Glycobiology 6: 165-75.

Warner T.G.. (1999) Enhancing therapeutic glycoprotein production in Chinese hamster ovary cells by metabolic engineering endogenous gene control with anti-sense DNA and gene targeting. Glycobiology 9: 841-50.

Wasley L.C., Timony G., Murtha P., Stoudemire J., Dorner A.J., Caro J., Krieger M. and Kaufman J. (1991) The Importance of N- and O-Linked Oligosaccharides for the Biosynthesis and *In Vitro* and *In Vivo* Biologic Activities of Erythropoietin. Blood 77: 2624-32.

Watson E., Shah B., Leiderman L., Hsu Y.R., Karkare S., Lu H.S. and Lin F.K. (1994) Comparison of N-linked oligosaccharides of recombinant human tissue kallikrein produced from Chinese hamster ovary cells on microcarrier beads and in serum-free suspension culture. Biotechnol Prog 10: 39-44.

Weikert S., Papac D., Briggs J., Cowfer D., Tom S., Gawlitzek M., Lofgren J., Mehta S., Chisholm V., Modi N.. Eppler S., Carroll K., Chamow S., Peers D., Berman P. and Krummen L. (1999) Engineering Chinese hamster ovary cells to maximize sialic acid content of recombinant glycoproteins. Nature-Biotechnol. 17: 1116-21

Yang M. and Butler M. (2000) Effects of Ammonia on CHO Cell Growth, Erythropoietin Production and Glycosylation. Biotechnol. Bioeng. 68: 370-80.

Zhang A., Potvin B., Zaiman A., Chen W., Kumar R., Phillips L. and Stanley P. (1999) The gain-of-function Chinese hamster ovary mutant LEC11B expresses one of two Chinese hamster FUT6 genes due to the loss of a negative regulatory factor. J-Biol-Chem. 274: 10439-50

3. GLYCOSYLATION OF RECOMBINANT IgG ANTIBODIES AND ITS RELEVANCE FOR THERAPEUTIC APPLICATIONS
(A Soupcon of Sugar Helps the Medicine Work)

R. JEFFERIS

Professor of Molecular Immunology. Division of Immunity and Infection, The Medical School, The University of Birmingham, Vincent Drive, Birmingham, UK., Email: R.Jefferis@bham.ac.uk

1. Introduction

The year 2000 saw the 25th anniversary of the publication of Kohler and Milstein's seminal paper "Continuous cultures of fused cells secreting antibody of predefined specificity" reporting the generation of monoclonal antibodies (Mabs) [Kohler & Milstein, 1975]. The techniques (not yet a technology) developed for the production of these monoclonal antibodies were a means to an end, understanding mechanisms for the generation of a virtually infinite repertoire of antibody specificities. However, the last two sentences of the paper read: "Such cells can be grown *in vitro* in massive cultures to provide specific antibody. Such cultures could be valuable for medicinal and industrial use". It is impossible to summarise and quantitate the impact that Mabs have had in the biosciences and beyond, however, the most visual impact must be the development, production and delivery of therapeutic Mabs. At the time of the anniversary article in Immunology Today sixty-six (66) monoclonal antibodies were in clinical developments [Glennie & Johnson, 2000]. That number has now risen to over one hundred and there are hundreds more being evaluated for future potential. The capital investment in this section of the biopharmaceutical industry and the market value is virtually inestimable since it is rising so rapidly. It is accepted, however, that currently and for the next several years there will be a worldwide shortage of production capacity. This situation creates opportunities and spurs innovation that is reflected in the development of less conventional means of production, viz. transgenic animals and plants.

The majority of therapeutic Mabs have been produced in either transfected Chinese hamster ovary (CHO), baby hamster kidney (BHK) murine NSO or Sp2/0 cell lines [see Krummen & Weikert - this volume]. Regulatory authorities put a premium on the quality and reproducibility of biopharmaceutical therapeutics and industry has responded by exhibiting a rapid learning curve in the characterisation of products and control of production processes. A movement to production in transgenic animals or plants will need a complete reappraisal of issues such as bio-efficacy, immunogenicity etc. As we enter the era of proteomics there is increasing emphasis on the importance of post-translational modifications, that are species and tissue specific, which make a

M. Al-Rubeai.(ed.). Cell Engineering. 93-107.
© 2002 *Kluwer Academic Publishers. Printed in the Netherlands.*

vital contribution to biological specificity and efficacy. The subject of this review is the post-translational modification: antibody glycosylation. Correct glycosylation of recombinant Mabs is crucial for expression of biological activities that are to be recruited *in vivo* for therapeutic efficacy – hence the subtitle (with apologies to Mary Poppins' fans!). It is an issue for Mabs, and other glycoproteins, produced in CHO, BHK, NSO or Sp2/0 cell lines since their products do not exhibit an oligosaccharide profile fully comparable to that observed for normal polyclonal human IgG. This deficit is being addressed by cell engineering with the application of knock-in (introduction of transgenes) and knockout protocols.

2. IgG Antibodies – Basic Structure

In humans five classes of antibody molecule have been defined, IgA, IgM, IgG, IgD and IgE. Within the IgG and IgA classes four and two subclasses are identified, respectively. The classes and subclasses together constitute the nine human antibody isotypes. Each antibody isotype expresses a unique profile of biological activities, Table 1. To date all recombinant antibodies developed for therapeutic application have been of the IgG isotype.

The choice of IgG subclass being determined by the profile of effector mechanisms thought to be optimal for therapeutic efficacy [Jefferis et al., 1998]. The IgG molecule is a multifunctional glycoprotein that binds antigen (pathogen) with the formation of immune complexes that activate effector mechanisms resulting in their clearance and destruction. The basic structure of an IgG molecule is of two light ($\sim$ 25kDa) and two heavy chains ($\sim$50 kDa) in covalent and non-covalent association to form three

Table 1. The effector ligand binding/activation properties of the human IgG subclasses.

Isotype	IgG1	IgG2	IgG3	IgG4
Binding and activation:				
C1	++	+	+++	–
$Fc_\gamma RI$	+++	–	+++	+
$Fc_\gamma RII*$	+	–	+	–
$Fc_\gamma RIII$	+	–	+	–
FcRn	+	+	+	+
MBL**	+	+	+	+
MR**	+	+	+	+
Binding:				
Staphylococcus protein A***	+	+	–	+
Streptococcus protein G	+	+	+	+
Catabolic half-life (days)	23	23	7	23

*   There are polymorphic variants of $Fc_\gamma RIIa$ that bind and are activated by IgG2
      (Jefferis et al., 1998)
**   Potential activities, not yet demonstrated
***   Proteins of allotype IgG3.G3m(s,t) do bind SpA   (Jefferis et al., 1998)

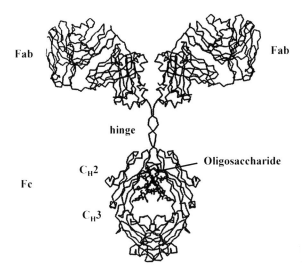

Figure 1. A computer generated model of a human IgG1 molecule.

independent protein moieties connected thorough a flexible linker (the hinge region), Figure 1. Two of these moieties are of identical structure and each expresses a specific antigen-binding site, referred to as Fab regions; the third, the Fc, expresses interaction sites for ligands that activate clearance mechanisms. The effector ligands include three structurally homologous cellular Fc receptor types (FcγRI, FcγRII, FcγRIII), the neo-natal Fc receptor (FcRn), the C1q component of complement, mannan binding lectin (MBL) and the mannose receptor (MR). All human IgG molecules are glycosylated through the N-linked glycosylation motif $-_{297}$Asn-Ser-Thr$_{299}$- within the Fc region [Diesenhofer, 1981]. The oligosaccharide accounts for only ~ 2% of the mass of the IgG molecule but is essential to biological function [Jefferis et al., 1998]. In addition ~ 10 – 20 % of polyclonal human IgG molecules bear N-linked oligosaccharides within the Fab regions. The latter oligosaccharides are expressed within the variable region of either the light or heavy chains and result from glycosylation at N-linked glycosylation motifs that are either germline encoded or result from antigen driven selection and somatic hypermutation [Jefferis 1993; Dunn-Walters et al., 2000].

Potential pathogens have evolved strategies to frustrate and/or evade the activation of protective effector mechanisms by elaborating Fc binding proteins that neutralise or compete for Fc effector sites; e.g. Staphylococcus protein A (SpA), Streptococcus protein G (SpG) and virus encoded FcγR.

The IgG-Fc region is a homodimer comprised of inter-chain disulphide bonded hinge regions, glycosylated C_H2 domains, bearing N-linked oligosaccharide at aspara-gine 297 (Asn-297) and non-covalently paired C_H3 domains [Diesenhofer, 1981]. Effector mechanisms mediated through FcγRI, FcγRII, FcγRIII and C1q are severely compromised or ablated for aglycosylated or deglycosylated forms of IgG [Lund et al. 1995; Mimura et al. 2001]. Multiple non-covalent interactions between the oligosaccharide and the protein result in a reciprocal influences of each on the conformation of the other. The interaction of IgG-Fc with FcγRI, FcγRII, FcγRIII and C1q is primarily through the protein moiety, however, the precise IgG-Fc protein conformation is determined through interactions with the oligosaccharide [Lund et al., 1991; Jefferis et al., 1998]. Interaction sites for FcRn [Burmeister et al., 1994], SpA [Diesenhofer, 1981], SpG [Sauer-Eriksson et al., 1995] and a rheumatoid factor autoantibody [Corper et al., 1997] are topographically overlapping protein structures at the C_H2/C_H3 interface region. MBL and MR also interact at this topographical region but directly with terminal N-acetylglucosamine residues of the oligosaccharide. The site of oligosaccharide attachment, Asn-297, is proximal to the N-terminal region of the C_H2 domain, whilst terminal sugar residues are exposed at the C_H2/C_H3 domain interface, Figure 2 [Diesenhofer, 1981].

The pharmokinetics of therapeutic Mabs is of great importance to both efficacy and cost. The catabolic half-life of human IgG it determined through interactions with the neonatal Fc receptor, FcRn; so named because it was first identified from study of the transport of IgG across the gut of the newborn rat. Interestingly, this FcR is a structural homologue of MHC Class I molecules, rather than the three classes of

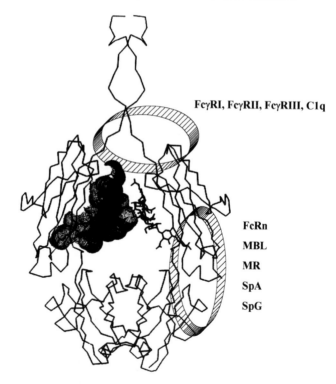

FcγRI, FcγRII, FcγRIII, C1q

FcRn

MBL

MR

SpA

SpG

Figure 2. The Fc Region of the Model in Figure 1, including the van der Waals surface of one oligosaccharide moiety (structure F, see Figure 5). The "bands" indicate topographical regions implicated in effector ligand binding – see text.

cellular FcγRs previously defined. FcRn is expressed on many tissues where it functions to regulate catabolism. It is proposed that pinocytotic vacuole formation by cells expressing FcRn results in uptake of IgG from the surrounding fluid. The pH within the vacuole is then lowered, to ~ 6.5, at which pH the FcRn become saturated with IgG. Enzymes that are active at the lowered pH degrade unbound IgG. When the membrane of the vacuole is recycled as the cellular membrane the pH reverts to that of the external milieu, 7.2, and the IgG is released into the surrounding tissue fluid. It is of interest to note that the catabolic half-life of human IgG3 is only ~7 days, in contrast to 21 days for IgG1, 2 & 4. Thus the efficacy of IgG3-FcRn interactions appears to differ when effecting catabolism or placental transport – see below. In contrast to placental transport the catabolic half-life does not appear to be dependent on IgG-Fc glycosylation [Ghetie & Ward, 2000]), however, it should be emphasised that only glycoforms of IgG bearing neutral oligosaccharides have been evaluated. It may be anticipated that sialylation could influence IgG-FcRn interactions since they would occupy space in the vicinity of the histidine residues involved in FcRn binding and could influence their pK values.

The neonatal receptor is also the receptor responsible for placental transport of human IgG [Ghetie & Ward, 2000]. The IgG class of antibodies only are transferred from mother to foetus to provide vital immune protection in the first months of life. Placental transport is an active process that results, at term, in higher IgG concentrations in cord blood than in the matched maternal blood. Each of the four human IgG subclasses are transferred across the placenta with facility, however, there is a slight positive gradient for IgG1 subclass proteins. It is also observed that during pregnancy the level of galactosylation of maternal IgG increases and that there is preferential transport of galactosylated IgG across the placenta [Williams et al., 1995; Kibe et al., 1996]. This provides circumstantial evidence to suggest that the affinity of IgG for FcRn may differ between glycoforms, under conditions operative at the interface between the mother and the placenta.

3. Oligosaccharide Heterogeneity and IgG Glycoforms

3.1. NORMAL POLYCLONAL HUMAN IgG

Analysis of oligosaccharides released from normal polyclonal human IgG allows definition of a core heptasaccharide: $GlcNac_2Man_3GlcNac_2$ [Jefferis et al., 1990; Jefferis et al., 1998] with variable attachment of outer arm sugars (sialic acid, galactose, fucose and bisecting N-acetylglucosamine) and the generation of heterogeneous array of IgG glycoforms Figure 3. The total number of variant oligosaccharides attached to IgG heavy chains is approximately 30. Random association of differently glycosylated heavy chains can, potentially, generate hundreds of different IgG glycoforms. Since glycosylation is essential to the expression of vital biological activities different IgG glycoforms can be anticipated to exhibit variation in their efficacy of effector function activation. It should be emphasised, however, that most of our knowledge of IgG glycosylation has been obtained by analysis oligosaccharides or

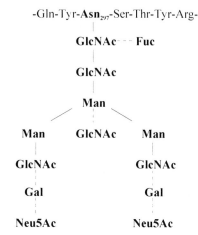

Figure 3. The Biantennary Oligosaccharide of IgG-Fc. The full lines define the "core" oligosaccharide structure; the dashed lines show additional sugar residues that may be attached: GlcNAc, N-acetylglucosamine; Man, mannose; Fuc, fucose; Gal, galactose; Neu5Ac, 5-N-acetyl sialic acid.

glycopeptides released from the intact molecule. This provides information on the relative proportions of the oligosaccharides present but does not give information on how those oligosaccharides were paired in the parent molecule. The latter information is required in order to determine the glycoforms present in IgG. Glycoform profiles can be determined with the new generation of mass spectrometers [Mimura et al., 2000; Matsuda et al., 2000].

Analysis of released oligosaccharides shows that only a minority bears sialic acid; 15–20% monosialylated and 3–10% disialylated. Although a majority of IgG molecules

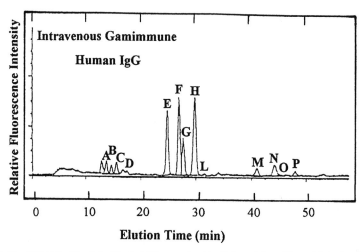

Figure 4. The HPLC Profile of the Neutral Oligosaccharides Released from Normal Polyclonal IgG [1].

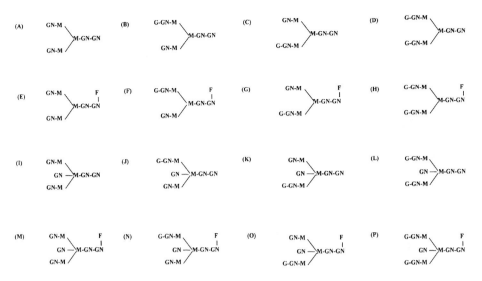

Figure 5.   Structures of the Neutral Oligosaccharides Identified as Peaks in Figure 4.: GN, N-acetylglucosamine; M, mannose; F, fucose; G, galactose.

are asialylated this does not influence catabolism since it is effected through the unique neonatal Fc receptor, FcRn [Burmeister et al., 1994; Ghetie et al., 2000] and not the asialyl-glycoprotein receptor. A typical HPLC profile of the neutral oligosaccharides released from desialylated IgG is shown in Figure 4, and the component oligosaccharides identified in Figure 5 [Jefferis et al., 1990; Takahashi et al., 1995]. The minimal structure observed is the heptasaccharide having terminal N-

fucose on the primary N-acetylglucosamine residue is observed (E & M). The variable addition of galactose generates further heterogeneity, e.g. B, C and D. Oligosaccharides bearing bisecting N-acetylglucosamine have been shown to constitute < 10 % the total oligosaccharide content; they are not fully resolved under the HPLC condition applied in generation of Figure 4.

3.2. RECOMBINANT IgG

3.2.1. Cell Culture

The cell line most widely used for the production of recombinant proteins is CHO [Stanley, 1998]. However, in its basic form it has numerous deficiencies and it is the subject of repeated cell engineering in attempts to improve its productivity and the structural and biological fidelity of the product. The oligosaccharide profile of antibody produced by CHO cells shows that biantennary structures dominate, similar to those released from polyclonal human IgG, however, minor populations of abnormal glycoforms, such as aglycosyl and high mannose forms are invariably present. Alternative cell lines and technologies are under development, including the generation of transgenic animals producing antibody secreted into milk [Pollock et al., 1999], transgenic plants [Gidding et al., 2001], yeasts [Horwitz et al., 1988], bacteria etc. A major obstacle for these vehicles will be the loss of glycosylation fidelity, with subsequent reduction/loss of biological activity, and/or the generation of immunogenicity [Adair, 2000].

The antibody product of CHO cells is predominantly comprised of fucosylated agalactosylated oligosaccharides that, potentially, could activate the lectin pathway of complement through MBL [Malhotra et al., 1995; Vorup-Jensen et al., 2000] and/or the MR expressed on antigen presenting cells [Dong et al., 1999]. Whilst CHO cells do add sialic acid to terminal galactose residues it does so through an $\alpha(2 - 3)$ linkage whereas in normal human IgG the linkage is $\alpha(2 - 6)$. We have engineered CHO cells with the introduction of the rat $\alpha(2 - 6)$ sialyltransferase and shown that the product, although only bearing 35 % of $\alpha(2 - 6)$ sialic acid structures, has improved FcγRI and complement activating properties [Jassal et al. 2001]. CHO cells do not have the capacity for addition of bisecting N-acetylglucosamine; cell engineering and the introduction of the GTIII transferase has been shown to yield an antibody product that promoted killing of target cells at approximately 10- to 20-fold lower concentrations than the parent molecule lacking bisecting N-acetylglucosamine [Umana et al., 1999; Davies et al., 2001]. Fortunately, there is only one reported instance of CHO cells producing a product bearing the galactose $\alpha(1 - 3)$ galactose- epitope; see below. The productivity of CHO cells and the oligosaccharide profile of Mab product are influenced by conditions of cell culture, which need to be optimised for each individual Mab producing cell line.

NSO and Sp2/0 cells are derived from mouse antibody secreting plasma cell lines, not surprisingly, therefore, they are more productive than CHO cells, for Mabs. As for CHO cells the predominant product bears fucosylated, agalactosyl oligosaccharides. Again, conditions may be optimised to increase galactosylation, however, in addition to a capacity to add galactose in $\beta(1-4)$ linkage to N-acetylglucosamine these cells may add a further galactose linked $\alpha(1-3)$ to the primary galactose. All mammals with the exception of higher primates and humans express the transferase for the addition of the galactose $\alpha(1-3)$ galactose- structure. To humans this structure represents an immunogenic epitope that they encounter in their environment and consequently it is estimated that 1% of circulating IgG is specific antibody directed against the galactose $\alpha(1-3)$ galactose- epitope [Jefferis et al., 1998; Chen et al., 2000]. Infusion of recombinant antibody bearing this epitope would, therefore, result in the formation of immune complexes and could provoke a systemic inflammatory response. In addition to adding sialic acid at the $\alpha(2-3)$ position NSO cells also adds a variant N-glycolyl form of sialic acid at the $\alpha(2-3)$ position. NSO cells do not have the capacity to add bisecting N-acetylglucosamine. These cells are also targets for cell engineering in attempts to develop optimal performance [Tey et al., 2000].

Salutary experiences were reported at the IBC "Well Characterised Biologicals" meeting (held in Seattle 2001); L. Durant (Protein Design Lab.) reported a case study in which a cell line selected to manufacture Mab was found not to deliver acceptable productivity. A new production line was selected that delivered acceptable product yield, however, the cell line was shown to have a single amino acid substitution that introduced a light chain N-linked glycosylation site that was fully glycosylated, with a biantennary structure. The quality of the established physico-chemical and biological characterisation allowed the originators of the Mab to demonstrate product comparability, to the satisfaction of the regulatory authorities. A further experience obtained when production of a therapeutic antibody was moved to a new facility, constructed and operated to the same protocol as the parent. The product was found to have an unacceptably reduced bio-efficacy, due to under galactosylation. Fortunately, relatively minor adjustments to operating conditions restored galactosylation levels and bio-efficacy (J. O'Connor [Genentech]. Given the massive number of cell divisions effected within commercial scale production there exits a significant potential for spontaneous mutations to occur; analytical protocols to detect such an event must be in place – see below.

3.2.2. Transgenic Animals and Plants

Recombinant proteins have been expressed in transgenic mammals under tissue specific promoters directing expression within the mammary gland and secretion into milk [Pollock et al., 1999]. This is, potentially, a very efficient means for the production of recombinant human antibodies and levels of 10 grams/litre can be achieved. Production can be expanded rapidly, through a breeding programme, in significantly less time than is required to build and commission a new cell culture facility. Sheep, goats and cattle are the main species under development; however, analyses of the glycosylation profiles of recombinant antibodies isolated from milk are not available. Analysis of the sheep and goat IgG isolated from blood shows that they are relatively highly galactosylated and sialylated, however, only with the N-glycolyl form of sialic acid. An additional issue for Mab secreted into the milk of transgenic animals is that the Mab is exposed to lactose, over an extended period, both in vivo and through storage and processing stages. There results a potential for random non-enzymatic glycation of protein molecules. Glycation through glucose, in patients with diabetes, has been shown to compromise antibody effector functions [Dolhofer et al., 1985].

A recent development has been the generation of transgenic hens producing recombinant proteins that can be recovered from their eggs [www.AviGenics.com]. It has been shown to be viable, within a laboratory setting, and commercial scale activities are in progress. Antibodies have also been produced in a variety of plants [Giddings et al., 2001]. These routes could provide cheap antibody product for diverse commercial applications but, due to significant differences in post-translational modifications, there will be a need to demonstrate that they can mediate native effector functions and are not immunogenic before they can be considered for routine therapeutic application [Adair 2000].

4. Glycosylation Engineering

It is evident that there could be advantage to being able to produce selected homogenous glycoforms of recombinant antibody molecules. Manipulation of culture conditions appears to have limited potential contributing only to minor changes in the proportions of the glycoforms produced. Increased cell productivity, achieved through culture of a hybridoma cell line in the presence of butyrate, was shown to increase Mab productivity but did not compromise either the glycoform profile or biological activity [Mimura et al., 2000].

As indicated, above, subtle structural parameters may also influence the glycosylation profile. Replacement of single amino acid side chains making non-covalent interactions with sugar residues resulted in down modulation of biological activities and gross changes in the glycoform profile of product [Lund et al, 1996]. In most cases an increase in galactosylation and sialylation was observed. Similarly, radical differences in glycoform profile were observed for a series of truncated IgG molecules generated in order to evaluate the contribution of structural components to glycosylation and function [Lund et al., 2000]. The series included IgG1 lacking a C_H3 domain pair, single-chain Fv fusion proteins with Fc or a hinge-C_H2 domain, Fc with/out a hinge, and a single C_H2 domain. Analysis of released oligosaccharides indicates that intact IgG1 and scFvFc antibodies are galactosylated and sialylated to levels similar to those observed previously for normal human IgG1. The truncated forms expressed increased levels of digalactosylated (30 - 83%) or sialylated (9 - 21%) oligosaccharide chains; the highest levels being observed for the single C_H2 domain. These data show which architectural components influence IgG glycosylation processing and that the $(C_H3)_2$ pair is particularly influential.

It is also possible to modify the glycoform profile, *in vitro*, using appropriate glycosidases or glycosyltransferases and activated sugar precursors. We employed serial glycosidase digestion to determine the minimal oligosaccharide structure that could provide both structural stability and biological function for IgG-Fc [Mimura et al., 2000]. The study showed that the initial GlcNAc-GlcNAc-Man trisaccharide conferred significant stability and activity in comparison with the aglycosylated form. Similarly, we have generated fully galactosylated IgG-Fc to further probe structure/function relationships [Gilhespy-Muskett, 1994]. There is promise that similar *in vitro* modifications to the glycoform profile could be economic on a commercial scale (www.NEOSE.com).

5. Product Characterisation

The hallmark of a pharmaceutical is product consistency. This is more demanding for biopharmaceuticals than for small molecule drugs, and particularly so for Mabs due to their relatively large size and multiple physico-chemical and biological characteristics. The early promise of a Mab will have been established in pre-clinical studies with material produced on a, relatively, small scale. It is essential that the Mab should be fully characterised at this stage so that if it progresses through Phase I, II and III trials to approval and marketing product comparability can be established. Comparability refers to a product made within a validated manufacturer's process. The concept does not mean identical but implies that physico-chemical, functional biological, pharmacological and toxicological results are indistinguishable between pre- and post-production changes. These criteria must be re-established following any change that is introduced into the manufacturing process, e.g. source of chemicals, separation media, newly commission plant etc. Virtually every available physico-chemical technique may be used to fully characterise a Mab and subsequently to demonstrate comparability; e.g. amino acid sequencing, peptide mapping, mass spectrometry, nuclear magnetic spectroscopy, iso-electric focusing etc, etc. It is of prime importance to demonstrate that the product exhibits a consistent oligosaccharide profile.

6. IgG Glycoforms and Fc Effector Functions

It is established that glycosylation of the IgG-Fc is essential for optimal expression of biological activities mediated through FcγRI, FcγRII, FcγRIII and the C1q component of complement [Jefferis et al., 1998]. Present evidence suggests that it does not influence interactions with FcRn and consequently, presumably, the catabolic half-life or transport across the placenta. Bacterial IgG-Fc binding proteins, e.g. SpA, SpG are also unaffected. The association constant of aglycosylated IgG1 or IgG3 binding for FcγRI is reduced by two orders of magnitude, relative to that observed for the normally glycosylated form, however, aglycosylated IgG3 antibody can mediate ADCC if a high level of target cell sensitisation is achieved. Activation through FcγRII or FcγRIII appears to be completely ablated [Lund et al., 1991]. The association constant for C1q binding is reduced by an order of magnitude for aglycosylated IgG, however, this results in a complete loss of complement mediated cellular cytotoxicity (CDCC) [Lund et al., 1996; Mimura et al., 2000]. Protein engineering, employing alanine scanning, has been used to "map" amino acid residues deemed to be critical for FcγR and C1q binding. These studies "map" the binding site for all four of these ligands to the hinge proximal or lower hinge region of the C_H2 domain [Lund et al., 1996; Idusogie et al., 2000; Shields et al., 2001]. Formal proof for one of these receptors has been obtained recently through x-ray crystallographic analysis of IgG-Fc in complex with soluble recombinant forms of FcγRIII [Sondermann et al., 2000; Radaev et al., 2001]. Interestingly, the structure for human FcγRIII/IgG-Fc reveals a possible contribution of the primary N-acetylglucosamine residue to binding [Sondermann et al., 2000] whilst the structure for mouse FcγRIII/IgG-Fc holds that there is no direct contact. Clearly, any contribution is minimal thus the oligosaccharide contributes indirectly to the binding of these ligands.

The x-ray crystallographic analysis of the IgG-Fc fragment, residues 216–446, reveals electron density for residues 238–443 only [Deisenhofer, 1981]. Thus the lower hinge region is assumed to be mobile and not to have defined structure. This might appear not to be compatible with the suggestion that the lower hinge region provides for distinct interaction sites for multiple ligands. We have proposed that this region of the molecule is comprised of multiple conformers, in dynamic equilibrium, that are influenced through reciprocal interactions between the oligosaccharide and the protein moiety [Mimura et al., 2000]. Certain of these conformers are compatible with specific ligand recognition. In the absence of the oligosaccharide a different set of conformers is generated that is not compatible with ligand binding. The complex of IgG-Fc with FcγRIII supports this thesis since the interaction site is seen to include asymmetric binding to discrete conformations of each heavy chain within the lower hinge residues. Other residues of the hinge proximal region of the C_H2 domain also form contacts with the FcγRIII.

Another critical requirement is explained by this structure – that the IgG-Fc is univalent for the FcγR. This is essential since if monomeric IgG were divalent it could cross link cellular receptors and hence constantly activate inflammatory reactions. X-ray crystal structures of IgG-Fc in complex with SpA, FcRn, SpG and the autoantibody rheumatoid factor show that each of these ligands interacts with sites embracing residues of both the C_H2 and C_H3 domain, at their junction [Deisenhofer, 1981; Burmeister et al., 1994; Sauer-Eriksson et al., 1995; Corper et al., 1997]. The IgG-Fc is divalent for these ligands. It is evident, therefore, that the distribution of effector ligand binding and activation sites results from selection for valency, amongst other properties.

Currently there are two functional activities only for which direct binding to the oligosaccharide is established – the mannan binding ligand (MBL) [Vorup-Jensen et al., 2000] and the cellular mannose receptor (MR) [Dong et al., 1999]. Each of these lectins recognise arrays of sugar residues that may be presented on the surface of micro organisms and they form a link between the innate and adaptive immune response. The sugar residues recognised include N-acetylglucosamine and there is evidence that immune complexes of G0 IgG can be activating for these lectins. There is evidence that immune complexes of G0 IgG can promote the adaptive immune response by uptake through the mannose receptor expressed on dendritic cells. There is a possible downside to this activity since most recombinant antibody molecules are potentially immunogenic; therefore G0 IgG glyco-forms could potentiate for the presentation of idiotypic determinants to the immune system. This property could be turned to advantage in situations where an anti-idiotypic response is the therapeutic agent.

7. Glycosylation and Disease

Changes in the profile of oligosaccharides released from IgG have been observed for a number of inflammatory diseases, e.g. rheumatoid arthritis, systemic lupus erythematosis, tuberculosis, Crohn's disease, [Routier et al., 1998; Axford 1999], Wegener's granulomatosis and microscopic polyangiitis [Holland et al. 2001]. A deficit in galactosylation results in a high proportion of IgG molecules bearing oligosaccharides terminating in N-acetylglucosamine – referred to as the G0 glycoform, in contrast to the G1 and G2 glycoforms, which bear oligosaccharides with one or two galactose residues, respectively. Immune complexes formed with G0 IgG have been shown to potentiate the induction of autoimmunity [Rademacher, 1994] and the processing of immune complexes for presentation of epitopes to T cells [Dong et al., 1999]. Further, G0 IgG has been shown to be capable of activating MBL and the MR, each of which has potential to effect immuno-regulatory processes [Malhotra et al., 1995; Abadeh et al., 1997; Dong et al., 1999]. These finding for clinical IgG samples may provide guidance for the choice of recombinant IgG antibody glycoforms developed as therapeutics. Similarly, analysis of the mechanism of action of recombinant IgG therapeutics will further inform us of the biological activity of the IgG isotypes *in vivo*. Intriguingly, the level of galactosylation of normal polyclonal IgG increases during pregnancy but declines again following parturition. There is evidence that this might facilitate placental transport [Williams et al., 1995; Kibe et al., 1996].

Since the oligosaccharide profile of polyclonal IgG is so complex it is of interest to determine whether the glycosylation profile of the IgG product of individual plasma cells is unique or complex. If the former then presumably the profile observed for poly-clonal IgG is the sum of all of the unique contributions; if the latter, is it the same as that observed for polyclonal IgG. Analysis of monoclonal IgG, isolated from the sera of patients with multiple myeloma, showed each monoclonal protein to exhibit a unique (clonotypic) glycoform profile [Jefferis et al., 1990]. However, there was also evidence that the protein structure, in the form of the IgG subclass, also influenced the profile, particularly the arm preference for the addition of galactose, by the enzyme β-galactosidase.

In free solution β-galactosidase has been shown to preferentially add galactose to the α(1-3) arm, however, for polyclonal IgG there is an apparent preference for addition to the α(1-6) arm. Analysis of monoclonal IgG proteins shows that preference for addition to the α(1-6) arm is characteristic for IgG1 and IgG4 proteins whilst for IgG2 proteins addition to the α(1-3) arm is preferred. These data demonstrate that subtle structural characteristics can influence glycoprotein biosynthesis. The apparent α(1-6) arm preference observed for polyclonal IgG is due to the preponderance of the IgG1 (60-65%) subclass over IgG2 (20-25%), IgG3 (~10%) and IgG4 (~5 %) [Jefferis et al., 1998].

8. Conclusions

It is evident that glycosylation profoundly influences the biology of antibody mole-cules. This presents both a challenge and an opportunity for the generation and produc-tion of antibody therapeutics. New protocols are being developed for the modification of glycoforms *in vitro* which will allow us to exploit the oligosaccharide as a structural *"rheostat"* through which effector functions can be modulated. The realisation of these advances will require continued integration and collaboration between the basic and applied sciences.

9. References

Abadeh S., Church S., Dong S., Lund J., Goodall M. and Jefferis R. (1997) Remodelling the oligosaccharide of human IgG antibodies: effects on biological activities. Biochem.Soc.Trans. 25: S661

Adair F. (2000) Immunogenicity: The last hurdle for clinically successful therapeutic antibodies BioPharm 13: 42-6.

Axford J.S. (1999) Glycosylation and rheumatic disease Biochimica et Biophysica Acta.1455: 219-29.

Burmeister W.P., Huber A.H. and Bjorkman P.J. (1994) Crystal structure of the complex of rat neonatal Fc receptor with Fc. Nature 372: 379-83.

Chen, Z.C., Radic M.Z. and Galili U. (2000) Genes coding evolutionary novel anti-carbohydrate antibodies: studies on anti-Galproduction in alpha 1.3 galactosyltransferase knock out mice. Molecular Immunology. 37: 455-66.

Corper, A.I., Sohi MK., Bonagura VR., Steinitz M., Jefferis R., Feinstein A., Beale D., Taussig MJ. and Sutton BJ. (1997) Structure of human IgM rheumatoid factor Fab bound to its autoantigen IgG Fc reveals a novel topology of antibody-antigen interaction. Nature Structural Biology 4: 374-81.

Davies J., Jiang L., Pen L-Z., LaBarre M.J., Anderson D and Reff M. (2001) Expression of GTIII in a recombinant anti-CD20 CHO production cell line: Expression of antibodies of altered glycoforms leads to an increase in ADCC thro' higher affinity for FcRIII. Biotech.Bioeng. 74: 288-94.

Deisenhofer, J. (1981) Crystallographic refinement and atomic models of a human Fc fragment and its complex with fragment B of protein A from Staphylococcus aureus at 2.9- and 2.8-Å resolution. Biochem. 20: 2361-70.

Dolhofer R., Siess E.A. and Wieland O.H. (1985) Nonenzymatic glycation of immunoglobulins leads to an impairment of immunoreactivity. Biological Chemistry Hoppe-Seyler. 366: 361-6.

Dong, X., Storkus W.J. and Salter R.D. (1999) Binding and uptake of agalactosyl IgG by mannose receptor on macrophages and dendritic cells". J.Immunol. 163: 5427-34

Ghetie V. and Ward E.S. (2000) Multiple roles for the major histocompatibility complex class I- related receptor FcRn. Annual Review of Immunology. 18: 39-66

Giddings G., Allison G., Brooks D. and Carter A. (2001) Transgenic plants as factories for biopharmaceuticals. Nature Biotechnology. 18: 1151-6.

Gilhespy-Muskett A.M., Partridge L., Jefferis R. and Homans S.W. (1994) A novel ^{13}C isotopic labelling strategy for probing the structure and dynamics of glycan chains in situ on glycoproteins. J Glycobiology 4: 485-90.

Glennie, M.J. and Johnson, P.M. (2000) Clinical trials of antibody therapy. Immunology Today. 21: 403-10.

Holland M., Takada K., Takahashi N., Kato K., Adu D., Harper L., Savage C.O.S. and Jefferis R. (2001) Hypogalactosylation of IgG isolated from the sera of patients with anti-neutrophil cytoplasmic antibodies (ANCA) -associated vasculitis. Submitted for publication.

Horwitz A.H., Chang C.P., Better M., Hellstrom K.E., Robinson R.R. (1988) Secretion of functional antibody and Fab from yeast cells. Proc.Natl.Acad.Sci.USA. 85: 8678-82

Idusogie. E.E., Presta L.G., Gazzano-Santoro H., Totpal K., Wong P.Y., Ultsch M., Meng Y.G. and Mulkerrin M.G. (2000), Mapping of the C1q binding site on rituxan, a chimeric antibody with a human IgG1 Fc. J.Immunol. 164: 4178-84.

Jassal, R., Jenkins, N., Charlwood, J., Camilleri, P., Jefferis, R. and Lund, J. (2001) Sialylation of human IgG-Fc carbohydrate by transfected rat $\alpha(2-6)$ sialyltransferase. Biochem.Biophys.Res.Comm. In press.

Jefferis R., Lund J., Mizitani H., Nakagawa H., Kawazoe Y., Arata Y. and Takahashi N. (1990) A comparative study of the N-linked Oligosaccharide structures of human IgG subclass proteins. Biochem.J. 268: 529-37.

Jefferis, R. (1993) Glycosylation of antibody molecules: functional significance. Glycoconjugate J. 10: 357-61.

Jefferis, R., Lund J. and Pound J.P. (1998) IgG-Fc mediated effector functions: molecular definition of interaction sites for effector ligands and the role of glycosylation. Immol.Rev. 163: 50-76.

Kibe T., Fujimoto S., Ishida C., Togari H., Wada Y., Okada S., Nakagawa H., Tsukamoto Y. and Takahashi N. (1996) Glycosylation and placental transport of immunoglobulin G J.Clin.Biochem.Nutr. 21: 57-63

Kohler G., and Milstein C. (1975) Continuous culture of fused cells secreting antibody of predefined specificity. Nature 256: 495-7.

Lund J., Mizitani H., Nakagawa H., Kawazoe Y., Arata Y. and Takahashi N. (1991) Human FcγRI and Fcγ RII interact with distinct but overlapping sites on human IgG. J.Immunol. 147: 2657-62.

Lund J., Takahashi N., Pound J., Goodall M., Nakagawa H. and Jefferis, R. (1995) Oligosaccharide-protein interactions in IgG can modulate recognition by Fcγ receptors. FASEB J. 9: 115-9.

Lund J., Takahashi N., Pound J., Goodall M. and Jefferis R. (1996) Multiple interactions of IgG with its core oligosaccharide can modulate recognition by complement and human FcγRI and influence the synthesis of its oligosaccharide chains. J.Immunol. 157: 4963-9.

Lund J., Takahashi N., Popplewell A., Goodall M., Pound J., Tyler R., King D and Jefferis R. (2000) Expression and characterisation of truncated glycoforms of humanised L243 IgG1: architectural features can influence synthesis of its oligosaccharide chains and affect superoxide production triggered through human FcγRI". Eur.J.Biochem. 267: 7246-57

Malhotra R., Wormald M.R., Rudd P.M., Fischer P.B., Dwek R.A. and Sim R.B. (1995) Glycosylation changes of IgG associated with rheumatoid arthritis can activate complement via the mannose-binding protein. Nature Medicine. 1: 237-43.

Masuda. K., Yamaguchi Y., Kato K., Takahashi N., Shimada I., and Arata Y. (2000) Pairing of oligosaccharides in the Fc region of immunoglobulin G. FEBS Letters. 473: 349-57.

Mimura Y., Church S., Ghirlando R., Dong S., Goodall M., Lund J. and Jefferis R. (2000) The influence of glycosylation on the thermal stability and effector function expression of human IgG1-Fc: properties of a series of truncated glycoforms. Molec. Immunol. 37: 697-706

Mimura Y., Lund J., Church S., Goodall M., Dong S. and Jefferis R. (2000) Butyrate increases production of human chimeric IgG in CHO-K1 cells whilst maintaining function and glycoform profile. J.Immunol.Meth. 247: 205-16.

Pollock D.P., Kutzko J.P., Birck-Wilson E., Williams J.L., Echelard Y., Meade H.M. (1999) Transgenic milk as a method for the production of recombinant antibodies. Journal of Immunological Methods. 231: 147-57.

Radaev S., Motyka S., Fridman W.H., Sautes-Fridman C. and Sun P.D. (2001) The structure of human type III Fcγ receptor in complex with Fc. J.Biol.Chem. 276: 16469-77

Rademacher TW. Williams P. Dwek RA. (1994) Agalactosyl glycoforms of IgG autoantibodies are pathogenic. Proc.Natl.Acad.Sci.USA. 91: 6123-7.

Routier FH., Hounsell, EF., Rudd. PM., Takahashi, N., Bond A., Axford, J., Hay, F. and Jefferis R. (1998) Quantitation of human IgG glycoforms isolated from rheumatoid sera: A critical evaluation of chromatographic methods. J.Immunol.Meth. 213: 113-30.

Sauer-Eriksson et al. (1995) Crystal structure of the C2 fragment of streptococcal protein G in complex with the Fc domain of human IgG". Structure 3: 265-78.

Shields. R.L., et al. (2001) High Resolution Mapping of the Binding Site on Human IgG1 for FcRI, FcRII, FcRIII, and FcRn and Design of IgG1 Variants with Improved Binding to the FcR". J.Biol.Chem. 276: 6591-604.

Sondermann P., Huber R. and Jacob U. (2000) The 3.2-A crystal structure of the human IgG1 Fc fragment-FcγRIII complex. Nature. 406: 267-73.

Stanley. P. (1998) In the land of chocolate, excitement about sugars. Trends in Cell Biology. 8: 128-30.

Takahashi N., Nakagawa H., Fujikawa K., Kawamura Y. and Tomiya N. (1995) Three-dimensional elution mapping of pyridylaminated N-linked neutral and sialyl oligosaccharides. Annals of Biochemistry. 226: 139-46.

Tey BT. Singh RP. Piredda L. Piacentini M. Al-Rubeai M. (2000) Bcl-2 mediated suppression of apoptosis in myeloma NS0 cultures. Journal of Biotechnology. 79: 147-59.

Tey B.T., Singh R.P., Piredda L., Piacentini M. and Al-Rubeai M. (2000) Influence of bcl-2 on cell death during the cultivation of a Chinese hamster ovary cell line expressing a chimeric antibody. Biotechnology & Bioengineering. 68: 31-43.

Umana P., Jean-Mairet J., Moudry R., Amstutz H., and Bailey J.E. (1999) Engineered glycoforms of an antineuroblastoma IgG1 with optimized antibody-dependent cellular cytotoxic activity. Nature Biotechnology. 17: 176-80.

Vorup-Jensen T., Petersen S.V., Hansen A.G., Poulsen K., Schwaeble W., Sim R.B., Reid K.B., Davis S.J., Thiel S. and Jensenius J.C. (2000) Distinct pathways of mannan-binding lectin (MBL)- and C1-complex autoactivation revealed by reconstitution of MBL with recombinant MBL-associated serine protease-2. J.Immunol. 165: 2093-100.

Williams P.J. Arkwright P.D. Rudd P. Scragg I.G. Edge C.J. Wormald M.R. Rademacher T.W. (1995) Short communication: selective placental transport of maternal IgG to the fetus. Placenta. 16: 749-56.

4. CELLULAR MODIFICATION FOR THE IMPROVEMENT OF THE GLYCOSYLATION PATHWAY

L. KRUMMEN and S. WEIKERT
Department of Cell Culture & Fermentation R&D, Genentech, Inc., South San Francisco, California, USA

1. Introduction

Recombinant glycoproteins and monoclonal antibodies produced by mammalian cell lines are currently being developed as therapeutics for a spectrum of diseases. Consistent and appropriate control of the quality of N-linked glycans synthesized during post-translational modification of recombinant proteins is a major issue in successful commercial protein manufacture. This chapter will overview factors that can influence the glycosylation of recombinant proteins and review methods for engineering of cells to produce more highly galactosylated and sialylated oligosaccharides and/or introduce specialized glycan structures through overexpression of glycosyltrans-ferases.

2. Overview of N-linked Glycosylation in Mammalian Cells

N-linked glycosylation in mammalian cells is a complex process of successive addition and modification of a growing oligosaccharide structure linked to an asparagine residue (Asn-Xaa-Thr/Ser) on the nascent polypeptide molecule. The process requires precise step-wise interaction of the oligosaccharide acceptors with specific glycosyltransferases and donor nucleotide sugars within the compartments of rough endoplasmic reticulum (RER) and Golgi.

The process begins with en bloc transfer of a 14-mer oligosaccharide $Glc_3Man_9GlcNAc_2$ preassembled on dolichol phosphate to an Asn residue of the newly synthesized protein in the RER (see Figure 1); a process known as core-glycosylation. α-glucosidases I and II and $\alpha1,2$ mannosidase then remove the terminal glucose residues and a single mannose residue. The protein containing the resulting high mannose structure ($Man_8GlcNAc_3$) is then shuttled to the Golgi. In the Golgi a series of specific mannosidases act further to reduce the oligosaccharide structure to the $Man_3GlcNAc$ core structure which is characteristic of mammalian complex-type N-linked carbohydrates. The stepwise action of several separate enzymes, including N-acetylglucosaminyltransferases I - V, $\beta1,4$ galactosyltransferase ($\beta1,4$ GT), fucosidases and $\alpha2,3$ and/or $\alpha2,6$ sialyltransferases ($\alpha2,3$ ST and a2,6 ST, respectively), then results in the addition

M. Al-Rubeai.(ed.). Cell Engineering. 109-130.
© 2002 *Kluwer Academic Publishers. Printed in the Netherlands.*

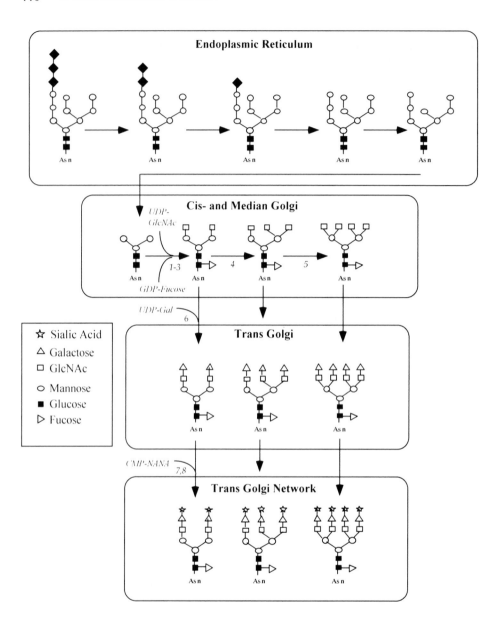

Figure 1. The N-linked Glycosylation Pathway. Enzymes identified are 1-3) N-acetylglucosaminyltransferase I and II and α1.6 fucosyltransferase. 4) N-acetylglucosaminyltransferase IV. 5) N-acetyl glucosaminyltransferase V. 6) β1.4 galactosyltransferase and 7.8) α2.3 sialyltransferase and α2.6 sialyltransferase, respectively.

and sequential maturation of up to 4 complex branch structures to the growing oligosaccharide. The N-linked branches found on fully processed human glycoproteins with complex-type N-glycans typically consist predominately of the lactosamine-containing trisaccharide GlcNAc-β1,4Gal-NANA, where sialic acid (NANA) is linked to the underlying galactose with either an α2,3 or an α2,6 linkage. A bisecting GlcNAc branch, synthesized through the activity of N-acetylglucosaminyltransferase III (GnT-III) may also be present in some cases.

3. Glycan Microheterogeneity

In vivo, N-linked glycans often display a range of structural heterogeneity. First, the site occupancy and antennarity of specific structures can be influenced by the surrounding poly-peptide sequence, resulting in preferential synthesis of bi- tri- or tetra-antennary structures at specific glycosylation sites (Jeffries et al., 1998; Grabenhorst et al., 1999, Spellman et al., 1989). For example the glycosylation sites of interferon β or of the Fc region of IgG-type antibodies contain predominately biantennary, complex type oligosaccharide sidechains. In contrast, among the three potential glycosylation sites on tissue plasminogen activator (tPA), the glycan added in vivo to sites 117 and 448 contain mixtures of bi- tri- and tetra- antennary structures (Spellman et al., 1989) in predictable ratios whereas the glycan at Asn 184 contains exclusively high mannose structures. Further, the degree to which individual antennae of complex carbohydrates are completely decorated with terminal galactose and sialic acid appears to be dependent on 1) the context of the glycan within the polypeptide background, 2) the availability of appropriate donor sugars and glycosyltransferases within the synthesizing cell or tissue type and 3) the physiological state of the organism (Varki, 1993). In nature this microheterogeneity may provide an important mechanism for regulating the solubility, secretion, stability, specific activity or bioavailability of key physiological glycoproteins.

Glycan microheterogeneity is also observed on proteins produced by mammalian cells in vitro. The primary protein sequence and structure of the recombinant protein may specify glycan site occupancy, antennarity and whether or not a particular structure is sialylated (Goochee et al., 1992, Grabenhorst et al., 1999). In addition, the diversity of terminal structures produced appears also to be influenced by factors in the cell culture environment (Goochee et al., 1992, Andersen & Goochee, 1994). Such factors (changes in media composition, process, by-product accumulation, specific productivity) may affect the availability or activity of glycosyltransferases in the secretory pathway or availability of required donor nucleotide sugar pools leading to inconsistency in the glycosylation of products as commercial cell culture processes are developed. Such variability can present a particular challenge for the manufacture of recombinant proteins whose pharmacokinetic properties or specific activity are strongly dependent on the structure of their complex N-linked carbohydrates. For example, insufficient or inconsistent sialylation and galactosylation can result in variable clearance of these proteins through the asialoglycoprotein or mannose/GlcNAc receptor-mediated pathways

(Ashwell & Hartford, 1982, Boyd et al., 1995), potentially posing a significant problem for adequate, reproducible dosing of the drug. Similarly, the extent of galactosylation of monoclonal antibodies has been shown to alter C1Q binding and compliment-dependent cellular cytotoxicity (CDCC) in vitro (Boyd et al., 1995), illustrating that in some instances, variability of oligosaccharide synthesis in culture can undermine the specific activity

vance of carbohydrate, variation in protein glycosylation as processes are developed or improved can restrain the process scientist from full process optimization.

4. Glycosylation Capabilities of Mammalian Cell Lines Used for Manufacturing

Among the various host cell lines used for pharmaceutical protein production, Chinese hamster ovary (CHO), baby hamster kidney (BHK), murine myeloma NSO and murine fibroblasts (C127) are among the most widely used for this purpose. Data collected on glycoforms produced by all of these cells indicate that the complex N-linked carbohydrate structures they synthesize exhibit many features characteristic of glycans naturally occurring on human proteins (Goochee et al., 1992). However certain cell-type specific dissimilarities exist. Murine cell lines, unlike hamster lines, may introduce significant quantities of terminal N-glycolylneuramic acid (NGNA) in place of terminal NANA (Bhatia and Mukhopadhay, 1998), or display an $\alpha 1,3$galactosyltransferase activity, that may lead to synthesis of glycans that contain structures potentially immunogenic in humans. However CHO and BHK cells lack the ability to provide certain structures normally found on some human glycoproteins, including terminal sialic acid residues attached in an $\alpha 2,6$ linkage to galactose or a bisecting GlcNAc residue (Jenkins and Curling, 1994).

It has long been suggested that the inability of many mammalian cell lines to produce glycans with completely human qualities is a limitation in the use of these cell-types for expression of recombinant molecules (Lee et al, 1989; Monaco et al., 1996). Particularly researchers have speculated that $\alpha 2,6$ ST activity may be required for proper biologic activ-ity or targeting of some glycoproteins (Monaco et al., 1996). Indeed, an important class of proteins has been identified which use sialic acid as a cellular homing signal (Crocker et al, 1999). $\alpha 2,6$ as well as $\alpha 2,3$ linked sialic acids may be important in providing binding specificity to these molecules. As mentioned above, many mammalian cell lines, particularly CHO and BHK do not normally express the $\alpha 2,6$ ST activity. In addition, data are also available that indicate that bisecting GlcNAc structures residues, which are not found on the oligosaccharides produced by recombinant CHO cells, may provide new or enhanced functionality to certain antibody molecules (Lifely et al., 1995; Umana et al., 1999). For example, addition of bisecting GlcNAc residues to an antibody directed against neuroblastoma has been suggested to be associated with acquisition of ADCC activity for this antibody (Umana et al., 1999).

5. The Role of Glycosylation Engineering

A mammalian host expression cell line with properties allowing consistant and maximal galactosylation and sialylation of N-linked carbohydrate structures, in spite of modest process modifications, would be a considerable advantage for the manufacture of glycoprotein therapeutics. In addition, a recombinant host cell line capable of elaborating all of the glycan structures and linkages represented on human proteins would be of value in ensuring that the most "human-like" glycoforms could be reproducibly synthesized in highly-productive, recombinant cell cultures. Therefore there has been considerable efforts over the past decade to explore whether engineering of production cell lines to overexpress the glycosyltransferases responsible for terminal glycosylation of N-linked glycans could be used as a strategy to promote specialized or more consistent glycosylation patterns. Initial efforts in this field focused on modifying the functionality of the post-translational processing pathway in hamster cells (CHO and BHK) to include functional α2,6 ST activity (Lee et al., 1989; Monaco et al., 1996; Grabenhorst et al., 1995; Minch et al., 1995). Early work by the Paulsen laboratory demonstrated that α2,6 ST gene expression was feasible and resulted in production of cellular protein containing α2,6 sialic acid. This work was extended by the laboratories of Jen-kins, Bailey and Grabenhorst who also demonstrated specific incorporation of α2,6 sialic acid into the glycans of recombinant proteins.

In our laboratory we have engineered DHFR⁻ CHO K1 host cells (DUX-B11) which have been previously transfected and optimized to carry on high titer product expression to overexpress several individual or combinations of exogenous glycosyltransferases (Weikert et al., 1999). Using this experimental approach we have successfully increased the sialic acid and galactose content of recombinant proteins to >90% of available branches and introduced significant quantities of both terminal α2,6 sialic acid and bisecting GlcNAc residues into the N-linked glycan of the recombinant protein product.

The remainder of this chapter will summarize this work, provide methods for the successful implementation of glycosylation engineering, and share observations made during the course of this work that should be considered when deciding upon the context in which this technology might be implemented.

6. Methods

6.1. GENERAL CONSIDERATIONS

6.1.1. Level of Transferase Gene Expression

Our experiences indicate that the level of transferase gene expression, measured by either mRNA levels or enzyme activity assay needs to exceed that in control CHO cells by several orders of magnitude (100-1,000 X) in order to achieve maximum alteration of glycoprotein product quality. Achievement of lesser levels of gene expression can be accompanied by a dose-related decrease in effect. The explanation for this is likely that only a fraction of the enzyme overexpressed is localized to the region of the Golgi where appropriate donor sugar pools and acceptor substrate are present. Maximal impact on glycan quality by co-expression with glycosyltransferases therefore appears to require efficient transfection & expression technology to achieve high-level gene expression.

6.1.2. Assays for Transferase Overexpression

It is also key to the success of these experiments to have a quantitative mRNA assay and/or activity assay available to evaluate the outcome of transfections as well as to screen clones for high glycosyltransferase activity levels. Quantitative RT-PCR assays can be set up fairly easily for high throughput screening of expression of any glycosyltransferase gene. We have also used modified versions of the protocols of Keusch et al. (1995) and Gross et al. (1990) very successfully for measurement of sialyltransferase and galactosyltransferase enzyme activity.

6.1.3. Appropriate Analytical Tools

Perhaps the most important factor in establishing the successfulness of the genetic engineering approach is the availability of a panel of analytical tools which allow detailed quantitative and structural analysis of the recombinant glycans. Methods such as MALDI-TOF MS (Papac et al., 1998; Papac et al., 1996; James, 1996) and HPAEC-PAD (Basa and Spellman, 1990; Hardy and Townsend, 1994) allow a much fuller appreciation of the impact of glycosylation engineering on oligosaccharide antennarity and monosaccharide composition than is afforded by strict compositional analysis.

6.2. MODEL SYSTEMS

Over the last several years we have studied the effects of overexpression of various combinations of glycosyltransferases in established cell lines making 5 separate model proteins with various primary and glycan structures. To date we have seen no consistant effect of overexpression of any of the transferase genes on cell growth, viability or specific productivity of the product gene when compared to the non-engineered cell line. Initial effects of overexpression can be readily monitored in smaller-scale non-instrumented cell cultures. However, bioreactor cultures afford the best opportunity to evaluate the net impact of glycosyltransferase overexpression on process and product quality.

6.3. GLYCOSYLTRANSFERASE SEQUENCES

Sequences for many of the important glycosyltransferases involved in the key terminal reactions in N-linked glycosylation have been published. cDNAs encoding the genes chosen for overexpression can be obtained by standard hybridization screening methods or by PCR of human, bovine or rodent cDNA libraries or mRNA samples. In our studies, we've used full length coding sequences for the human α2,3-ST (1130 bp fragment, upper primer 5'-ATGGGACTCTTGGTATTTGT-3', lower primer 5'-ATCTAAGCAG TGGCATCTGA-3'), human β1,4 GT gene (1390 bp fragment, upper primer 5'-CTTCTTAAAGCGGCGGCG-GGAA-3', lower primer 5'-TCACATGCCGAGCCAAG TTGGG-3')], human α2,6 ST (1,221bp fragment, upper primer 5'-TTCATTAT GATTCACACCAAC-3' and lower primer 5'-AATGGTCCGGA-3') isolated from a human placenta γ library by PCR. Primers were designed based on published sequences (Kitagawa and Paulson, 1993; Weinstein et al., 1987; Masri et al., 1988). The rodent GnT-III gene was isolated by RT-PCR from mRNA extracted from the YB2/0 cell line (Kilmartin et al., 1982;), which is known to express this activity. Primers were based on the sequence of Ihara et al. (1993) (1610 bp fragment, upper primer 5'ATGAG ACGCTACAAGCTTTTTCTCA-3', lower primer 5'CTAGCCCTCCGTTGTATCCA ACTTG-3').

6.4. CONSTRUCTION OF UNIQUE EXPRESSION VECTORS

The vectors used in our lab for overexpression of glycosyltransferases were modified from those previously described by Lucas et al (1996). These plasmids contain a puro-mycin selectable marker between 5'-splice donor and 3'-splice acceptor sites which define an intronic sequence separating the myeloproliferative sarcoma virus (MPSV) promoter/enhancer elements from the transcriptional start site of the glycosyltransferase gene. Imper-fect splicing of the primary dicistronic transcript produced from this vector yields small quantities of mRNA encoding puromycin-resistance along with the majority of the correctly spliced glycosyltransferase transcript. Levels of drug resistance sufficient to impart resistance to the concentrations of puromycin added to the media usually result in high levels of expression of the glycosyltransferase gene. Other proven strategies used to achieve high-level gene expression in mammalian cells have been described elsewhere (Werner et al., 1998; Fussenegger et al., 1999).

6.5. TRANSFECTION AND SELECTION OF GLYCOSYLTRANSFERASE OVEREXPRESSING CELL LINES

3×10^6 cells/100 mm plate from established serum-free, suspension adapted CHO cell lines producing a recombinant product were transfected with plasmids containing the glycosyltransferase sequences using lipofectamine according to the manufacturer's instructions (Gibco Lifetechnologies, Gaithersburg, MA). Briefly, 60 µl of lipofectamine were mixed with 15 µg plasmid DNA in 300 µl of a DMEM:F-12 based serum free growth media, and incubated for 30 min at room temperature before addition to the cells. Following overnight incubation at 37°C, media were replaced with fresh media con-taining 5% diafiltered FBS and 10 µg/ml puromycin. Adherent colonies appearing on the plates were pooled and readapted to serum-free suspension growth. Enzyme overex-pression was verified by mRNA assay and enzyme activity assays where possible. Meth-otrexate was added to all to maintain expression of the DHFR/product gene construct.

To expedite the proof of concept experiments described in this chapter, we have evaluated product quality from pools of cells resulting from the transfection. This approach is more rapid than selecting single clones for study and, in this case, has not appreciably hampered our ability to see robust effects. However, if enzyme expression among the cells in a given population is highly variable, there is a danger that the effects of overexpression will be underestimated, due to contamination of the product with normally-glycosylated molecules produced from cells with low glycosyltransferase levels. Therefore, screening of clones and selection for high levels of enzyme expression may be helpful in clearly documenting the effects of overexpression. Screening & selection of clones with high-level enzyme expression is of course required for establishment of a true production cell line or new host cell line.

6.6. CELL CULTURE AND RECOVERY OF RECOMBINANT PROTEINS

Product quality experiments to confirm an effect of enzyme expression can be performed in stirred tank bioreactors or in non-instrumented spinner flasks. For the TNFR-IgG and TNK-tPA studies described, 3 L bioreactors (Applikon, Foster City, CA) equipped with calibrated dissolved oxygen, pH and temperature probes were used. Dissolved oxygen was controlled on-line through sparging with air and/or oxygen at 60 $\pm$ 5% air saturation. pH was maintained at 7.2 $\pm$ 0.1 and controlled on-line through the addition of CO_2 or base (Na_2CO_3). Temperature was maintained using an electric heating jacket.

Monoclonal antibody experiments were carried out in 3L spinner vessels in temperature controlled CO_2 incubators (5.0%).

Control and re-engineered cell lines were cultivated under identical conditions in serum-free, peptone-supplemented DMEM-F:12 based medium. For bioreactor experiments, the cell culture process used was identical to that optimized for large-scale production of each product. Samples were taken daily for determination of cell concentration, viability and titer. Bioreactor cultures were harvested by centrifugation after 7-9 days when viability remained > 70%. Spinner cultures were harvested between days 4 and 7 (> ~70% viability). Proteins were purified by affinity chromatography.

6.7. ANALYTICAL METHODS

6.7.1. MALDI-TOF/MS

In the studies shown, oligosaccharides released from the recombinant proteins using PNGase F were analyzed by MALDI-TOF as described in detail previously (Papac et al., 1998). The Voyager Elite mass spectrometer (Perseptive Biosystems, Framingham, MA) was used to acquire the spectra as described by Papac et al (1996). Using this method, these workers have shown that individual oligosaccharide species can be quantitatively detected when mixtures of known quantities of acidic or neutral oligosaccharides are ionized. Sialic acid-containing acidic species are resolved in the negative ion mode, whereas uncharged oligosaccharides are resolved in the positive ion mode. A typical negative ion mode spectrum of charged glycans is shown in Figure 2. The notation used to identify peaks consists of 4 digits. The first digit represents the number

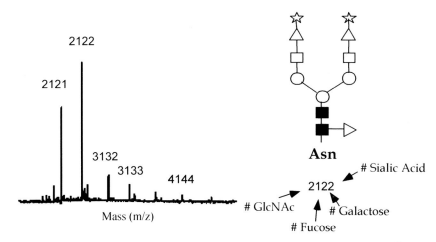

Figure 2. Nomenclature used to denote glycans identified in MALDI-TOF MS negative ion mode spectra.

of GlcNAc residues (or antennae), the second number represents the number of fucose, the third galactose and the fourth, sialic acid.

6.7.2. HPAEC-PAD

To resolve structures containing sialic acid in an α2,3 Vs α2,6 linkage, high-pH anion-exchange chromatography with pulsed amperometric detection (HPAEC-PAD) can be used. In our studies the analysis was performed using a DX-500 chroma-tography system (Dionex) consisting of a GP40 gradient pump, an ED40 electrochemical detector and an electrochemical cell outfitted with a gold electrode. A 4 x 250 mm CarboPac PA 100 column (Dionex) was used for the analysis. Solvent A consisted of 0.1 M NaOH and solvent B consisted of 0.1 M NaOH containing 0.5 M sodium acetate. Following a three minute isocratic period at 2.5% solvent B, oligosaccharides were eluted using a linear gradi-ent from 2.5% to 40% solvent B over 50 minutes. The flow rate was 1 ml/min. The spectra produced by HPEAC-PAD groups released oligosaccharides into clusters of charged species. The identity of bi- tri and tetra- antennary species can be assigned from the known retention time of standard oligosaccharide structures as well as through cor-relation with predominant structures identified in the MALDI spectrum (see Figure 3).

7.   Overexpression of α2,6 and/ or α2,3 ST Improves the Sialic Acid Content of Recombinant Proteins

TNK-tPA is a single chain glycoprotein variant (T103N, N117Q, KHRR(296-299)-AAAA-tPA) of wild-type tPA (Keyt et al., 1994). It contains three N-linked oligosac-charide side chains that exhibit variable degrees of undersialylation when the molecule is

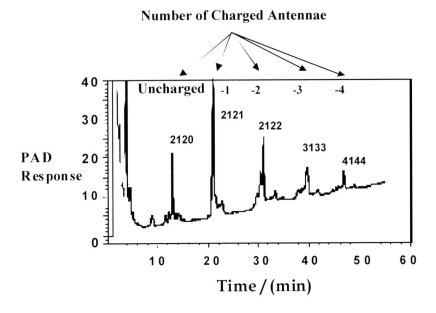

Figure 3. Identification of glycan structures by HPAEC-PAD

produced in fed-batch culture. This molecule was chosen for study because the pharmacokinetic properties of this molecule are known to be dependent upon sialic acid content. TNK expressing cells were co-transfected with genes coding for $\alpha 2,3$-ST, $\alpha 2,6$-ST or a combination of both enzymes.

MALDI-TOF MS of sialylated TNK-tPA oligosaccharides (negative ion mode) from the control cell line (Figure 4A) revealed a heterogeneous mixture of different oligosaccharides consisting of fully sialylated N-glycans (2122, 3133, 4144) as well as significant amount of bi-, tri- and tetraantennary structures with missing sialic acid. 74% of the N-linked carbohydrates showed terminal sialic acid, 22% exposed galactose and 3% exposed GlcNAc. In contrast, mass spectra obtained from N-glycans synthesized by cell lines overexpressing one or both sialyltransferases (Figure 4B, 4C, 4D) revealed a significant increase of branches capped with sialic acid (81-88%). Simultaneously, the amount of structures with exposed galactose declined.

Biantennary structures with two exposed galactose (2120) were the predominant neutral species present in the positive mode ion spectra for control tPA-TNK material (Figure 5). After overexpression of $\alpha 2,3$- and/or $\alpha 2,6$-sialyltransferase (Figure 5B, 5C, 5D), asialyl, galactose-containing N-glycans (2120, 2110) were significantly reduced from the spectrum presumably because they were sialylated and now appear in the negative mode as 2122, 2121 or 2111 structures. Collectively these data demonstrate that overexpression of either sialyl-transferase is very effective in increasing the sialic acid

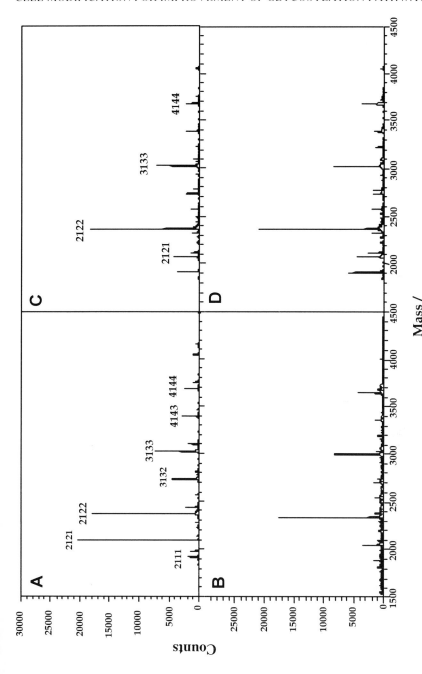

Figure 4. Negative ion mode spectrum of TNK expressed from control cells (A) or cells overexpressing α2.3 ST (B), α 2.6 ST(C) or both enzymes (D).

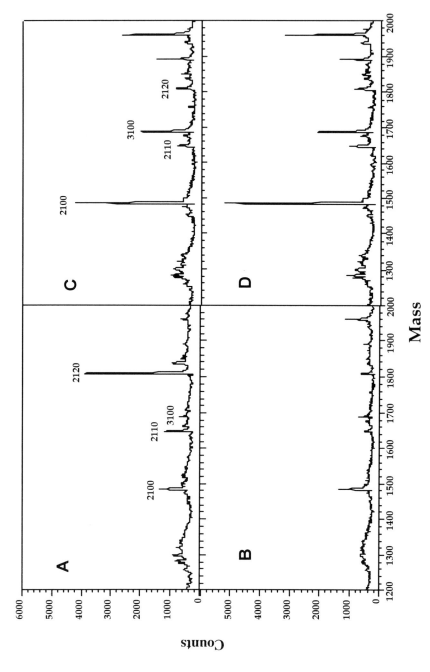

Figure 5. Positive ion mode spectrum of TNK expressed from control cells (A) or cells overexpressing α2,3 ST (B), α2,6 ST (C) or both enzymes (D).

content of recombinant proteins to a near maximum level. The impact of enhanced sialylation on the pharmacokinetics of TNK was studied in rabbits. As expected, the α2,3-ST, α2,6-ST or α2,3-ST/α2,6-ST modification resulted in significantly slower TNK-tPA plasma clearance in rabbits compared to control TNK-tPA. Clearance was reduced by approximately 34% for α2,3-ST and 20% for both the α2,6-ST and com-bination groups relative to the control group, respectively. These data suggest that alterations in the sialic acid content of recombinant proteins such as those achieved through glycosylation engineering can impact the clearance properties of glycoprotein therapeutics.

Interestingly, protein isolated from cell lines transfected with α2,6ST showed unexpectedly higher levels of undergalactosylated (2110 and 2111) structures compared with cells transfected with α2,3ST (see Figure 5). The reason for this phenomenon is uncertain, but may be related to inhibition of/or interference with galactosyltransferase activity when α2,6 ST is overexpressed. It is unclear whether the slight increase in exposed GlcNAc residues present in the material produced from α2,6ST cell lines was related to the slightly faster clearance of this material in the pharmacokinetic study. However, this observation illustrates that it is important to consider overall glycan quality in evaluating the outcome of a glycosylation engineering effort.

8.   Overexpression of α2,6 or α2,3 ST Can Impact the Quality of Sialic Acid in Recombinant Proteins

HPAEC-PAD analysis of major peaks found in the chromatogram of control and ST engineered TNK-tPA are consistent with the relative quantities of structures identified in the mass spectra generated by MALDI-TOF/MS. Purified control material showed a mixture of uncharged and charged N-glycans containing 0 – 4 sialic acids (Figure 6A).

Due to the absence of α2,6-ST in CHO cells, sialic acid on all oligosaccharides from control and α2,3-ST transfected material are in the α2,3-conformation. In contrast, signal intensities for α2,3-linked sialylated N-glycans are dramatically reduced for TNK-tPA produced by α2,6 ST overexpressing cells and new peaks eluting only a few minutes earlier than the α2,3 sialylated species can be identified that represent material with an α2,6-linked sialic acid. Overall, α2,6-conformation is the dominant linkage of sialic acid on biantennary N-glycans. However, several additional small signals in the area of carbohydrates with 3 or 4 negative charges also suggest that some of the tri- and tetraantennary structures have not only α2,6- or α2,3-linked sialic acid, but also a mixture of both linkages.

TNK-tPA isolated from cells transfected with equal quantities of cDNA for the α2,3-ST/α2,6-ST genes displayed only a small amount of oligosaccharides with sialic acid attached with only an α2,3-linkage. In this case, a significant percentage of biantennary structures had N-glycans with only α2,6-linked sialic acid. However co-expression of α2,3 ST resulted in approximately 50% of 2122 structure, as well as most of 3133 and 4144 structures, having both α2,6- and α2,3- linked sialic acid. These data agree with those previously reported by Grabenhorst et al (1995) and indicate that both STs

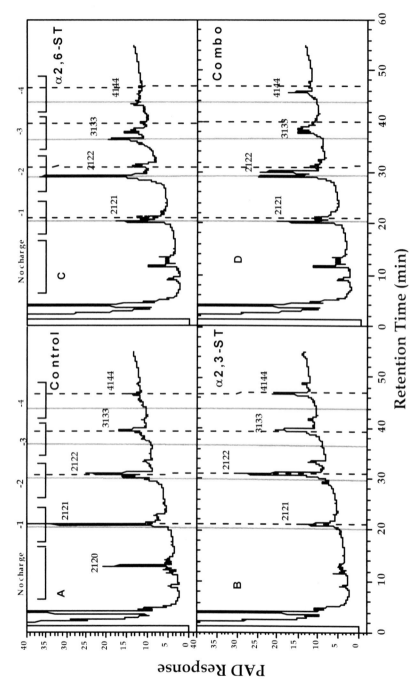

Figure 6. HPAEC analysis of glycans released from TNK-tPA produced from control cells (A), cells overexpressing α2,3 ST (B), α2,6 ST (C) or a combination of both enzymes (D). Solid verticle lines align α2,3 sialic acid containing glycans. dotted verticle lines align α2,6 containing glycans.

compete effectively for the Gal(β1-4)-GlcNAc acceptor on N-linked oligosaccharides. Based on our data it appears that for this molecule, α2,6 ST may be more effective in sialylating biantennary structures than α2,3 ST. However, α2,3 ST may compete more effectively with α2,6 ST for more highly branched structures. Thus, the precise composition of sialic linkages on a recombinant protein may be determined by the underlying structure of the oligosaccharide and the relative expression level of the α2,3 and α2,6ST enzymes.

9.   Overexpression of Galactosyltransferase Increases the Content of Galactose on Recombinant Proteins and Monoclonal Antibodies

We have confirmed overexpression of ST can alter the structure of N-linked glycans on other recombinant proteins in addition to TNK-tPA. TNFR-IgG is a receptor-IgG chimera that contains complex-type N-linked glycans in the receptor binding domain (Ashkenazi et al., 1991). These oligosaccharides can be undersialylated as well as undergalactosylated. Like TNK-tPA, the clearance of TNFR-IgG is dependent upon its carbohydrate structure.

However, in this case, the amounts of both exposed galactose and GlcNAc present on the molecule are known to be important (Jones ASJ, in preparation). Experiments performed with TNFR-IgG indicated that co-expression of both ST and GT may be beneficial in promoting maximal sialylation in cases where heterogeneity results from both undergalactosylation as well as under-sialylation. When both enzymes were co-expressed, branches terminating with sialic acid represented 90% of total oligosaccharides present in the negative ion spectra of GT/ST engineered TNFR-IgG whereas structures terminating with exposed GlcNAc (2111) were decreased to < 1% (Figure 7).

The IgG1 portion of the TNFR-IgG molecule contains a single biantennary glycan. Like those found on the Fc portion of monoclonal antibodies, Fc glycans of TNFR-IgG are not readily sialylated. The positive-ion spectra of TNFR-IgG glycans purified from control cul-tures are shown in Figure 8A. These structures terminate with galactose on one or two bran-ches (2110 or 2120) or with GlcNAc alone (2100). The majority of oligosaccharides found in the positive mode for control TNFR-IgG are non-galactosylated (65%). Figure 8B shows that the uncharged oligosaccharides present on TNFR-IgG co-expressed with GT were dramatically shifted towards more highly galactosylated, 2120 forms (68% 2120 (GT) vs. 7% 2120 (Control)). Neutral oligosaccharides containing one or two galactose residues were also significantly more abundant in the positive ion spectra of TNFR-IgG oligosaccharides produced by GT/ST cultures than in control cultures (Figure 8A vs. 8D). However in this case, there was a relative enrichment in the percentage of 2100 observed (Figure 8B Vs 8D). Although it is not possible to reliably quantitate movement of glycoforms between the positive and negative ion modes, it is likely that the relative enrichment of 2100 glycoforms in the GT/ST case is the result of sialylation of the increased levels of 2110 and 2120 structures, removing them from the positive mode spectra.

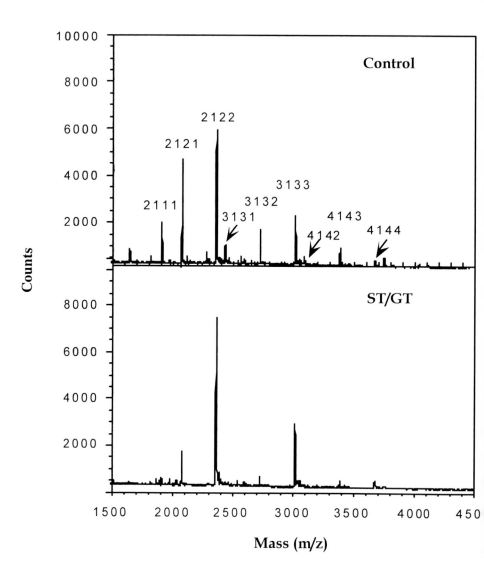

Figure 7. Negative mode spectrum (MALI-TOF MS) of TNFR-IgG glycans synthesized in control cells (top) and cells overexpressing both ST and GT (bottom).

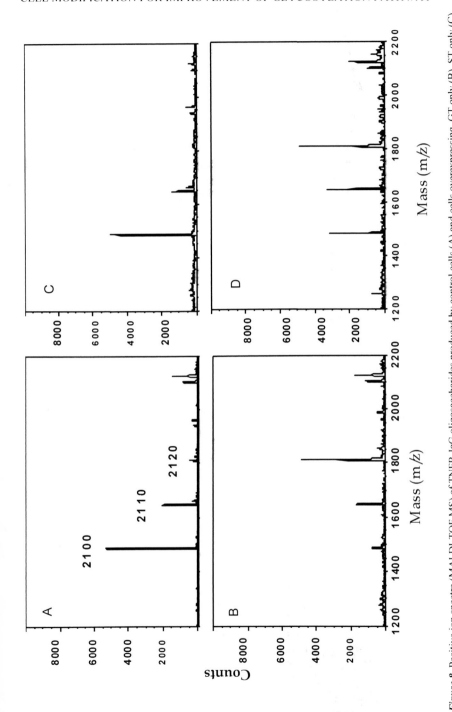

Figure 8. Positive ion spectra (MALDI-TOF MS) of TNFR-IgG oligosaccharides produced by control cells (A) and cells overexpressing GT only (B), ST only (C) or both enzymes (D).

These results suggest that engineering of the galactose content of monoclonal antibodies or introduction of a bisecting GlcNAc structure on these molecules would also be possible through cell engineering with glycosyltransferases. Indeed we have demonstrated that this approach could be used to increase the galactose content of several monoclonal antibodies. Figure 9A and C show the control MALDI-TOF spectra for two different non-engineered monoclonal antibody cell lines. It is evident that GT overexpression enhances the galactose content of oligosaccharides on both antibodies, albeit to different degrees (Figure 9B and D).

The differential effect of GT overexpression between these antibodies does not appear to be due to inherent limitations imposed by the antibodies' structure. Incubation of purified antibody in the presence of galactosyltransferase and UDP-galactose in vitro results in production of materials with > 90% 2120 structures after overnight incubation in both cases. The effect also does not appear to be due to insufficient or differential expression of GT in the two cell lines. Overexpression of increasing amounts of enzyme, documented by mRNA analysis did not result in quantitatively greater transfer of galactose to the N-glycans. Rather we hypothesize that there may be a limited exposure time of antibodies to the appropriate Golgi microenvironments as they transit the post-translational machinery in vivo. This may define the maximal galactosylation profile for a given protein that can be achieved in an engineered cell line. However, we do not believe this is related to the specific productivity of the recombinant protein *per se*. The cells shown in Figure 9A/B and Figure 9C/D have similar specific productivities. In addition, we have shown that two-fold increase in specific productivity of the TNFR-IgG process does not undermine the effectiveness of GT or ST overexpression.

10.   Overexpression of GlcNAc T-III in Cell Lines Producing Monoclonal Antibodies

Recently, there has been increased interest in the possibility that engineering antibody glycans to contain a bisecting GlcNAc residue may enhance antibody ADCC activity. This idea was stimulated by the observation that the same antibody expressed in CHO cells compared to a rat myeloma cell line YB2/0 displayed dramatically different ADCC activity (Lifely et al., 1995). One, but by no means the only difference observed in analytical quality between the materials produces by CHO Vs YB2/0 cells was the presence of bisecting GlcNAc on the N-linked oligosaccharides. Subsequently, the GnT-III gene has been overexpressed in CHO cells, which do not normally express the activity, by Umana et al (1999) as well as others. It has been demonstrated that antibody produced from these cells contains the bisecting Glc NAc residue and exhibits variably enhanced ADCC activity. However, unlike the relationship observed between increased sialylation and decreased clearance, studies in our lab have indicated that the magnitude of the enhancement of ADCC activity does not strictly correlate with relative change in content of bisecting structures. For example, in our hands a model antibody secreted from YB2/0 cells was found to contain < 20% structures with a bisecting GlcNAc residue whereas the same molecule secreted by CHO cells engineered to overexpress

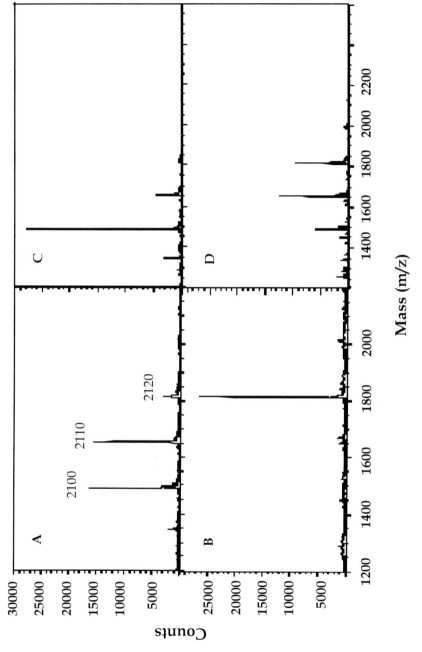

Figure 9. Effects of overexpression of GT on two different monoclonal antibodies.

GnT-III contain a bisecting GlcNAc on > 75% branches. Nonetheless, the magnitude of enhancement of ADCC in an in vitro assay appeared to be similar regardless of which material was compared to control. In this regard it is important to realize that over-expression of GnT-III results in other modifications of the glycan structure in addition to the addition of bisecting GlcNAc residue. For example in CHO cells the percentage of non-fucosylated and high mannose glycans also change modestly. The relative enhancement in ADCC achieved may also be highly protein dependent. Clearly further study of other analytical changes of molecules co-expressed with GnT-III, in addition to Fc receptor binding studies will be of importance to clarify mechanism of enhanced ADCC with GnT-III overexpression.

11. The Concept of the Universal Host Cell for Expression of Recombinant Glycoproteins

The success of the pilot studies described in this chapter and by others make it appealing to suggest that a new universal host cell line might be developed that has superior glycosylation properties for the expression of recombinant proteins. Indeed, one could imagine production of a GT/ST host that would be used to maximize sialylation and galactosylation of expressed molecules. In practice such a host would likely be beneficial for non-antibody molecules, but introduction of enhanced amounts of sialic acid in addition to galactose might not be optimal for antibodies. We have observed that for at least one of the antibodies we co-expressed with GT and ST that introduction of enhanced amounts of sialic acid increased aggregation of the product by approximately two-fold over that observed for the control antibody or antibody co-expressed with GT only. Thus for antibodies, it may be appropriate to enhance and stabilize molecule galactose content through use of a GT only host.

In terms of developing hosts with new or specialized glycosylation properties, studies certainly support that cell lines can be engineered to allow $\alpha 2,6$ sialic acid linkages or addition of bisecting GlcNAc. However where introduction of these residues is intended to modify biological activity, it will likely be critical to have fairly precise knowledge of the quality of the desired product and to design the cell line to achieve this target product quality on a case by case basis. Particularly, for the GnT-III activity, enhanced ADCC may be beneficial for some antibodies, but introduce an added risk of toxicity for others. Cells certainly can be fine-tuned to facilitate production of product with a predefined quality by regulating the relative expression level of the various glycosyltransferases. However, at this point it may remain preferable to assess the efficacy, safety and manufacturing consistency needs of each protein individually and use engineering approaches in a directed rather than universal fashion.

12. References

Andersen D, Goochee CF (1994) The effect of cell culture conditions on the oligosaccharide structures of recombinant glycoproteins. Curr Op Biotech 5(5): 546-9.

Ashkenazi A, Marsters S, Capon DJ, Chamow SM, Figari IS, Pennica D, Goeddel DV, Palladino MA, Smith DH (1991) Protection against endotoxic shock by a tumor necrosis factor receptor immunoadhesin. PNAS (USA) 88: 10535-9.

Ashwell G, Hartford J (1982) Carbohydrate specific receptors of the liver. Ann Rev Biochem 51:531-554.

Basa LJ, Spellman MW (1990) Analysis of glycoprotein-derived oligosaccharides by high-pH anion-exchange chromatography. J Chromatography 499: 205-20.

Battersby JE, Vanderlaan M, Jones AJS (1999) Purification and quantitation of tumor necrosis factor receptor immunoadhesin using a combination of immunoaffinity and reversed-phase chromatography. Journal of Chromatography 728(1): 21-33.

Bhatia PK, Mukhopadhay A. (1998) Protein glycosylation: implications for in vivo functions and therapeutic applications. Adv Biochem Eng Biotech 64: 157-201.

Boyd PN, Lines AC, Patel AK (1995) The effect of the removal of sialic acid, galactose and total carbohydrate on the functional activity of CAMPATH-1. Molecular Immunol 32(17/18): 1311-8.

Crocker PR, Vinson M, Kelm S, Drickamer (1999) Molecular analysis of sialoside binding to sialoadhesin by MNR and site-directed mutagenesis. Biochem J 341(2): 355-61.

Fussenegger M., Bailey JE, Hauser H, Mueller PP (1999) Genetic optimization of recombinant glycoprotein production by mammalian cells. TIBTECH 17: 35-42.

Goochee CF, Gramer MJ, Andersen DC, Bahr JC, Rasmussen JR (1992) The oligosaccharides of glycoproteins: Factors affecting their synthesis and their influence on glycoprotein properties. In: Todd P, Subhas K. Sikdar and Bier M (eds) Frontiers in Bioprocessing II. American Chemical Society, Washington DC. 198-240.

Grabenhorst E, Hoffmann A, Nimtz M, Zettmeissl G, Conradt HS (1995) Construction of stable BHK-21 cells co-expressing human secretory glycoproteins and human Gal(β1-4) GlcNAc-r α2,6-sialyltransferase. α2,6-linked NeuAc is preferentially attached to the Gal(β-1-4)GlcNAc(α-1-2)Man(α-1,3)-branch of diantennary oligosaccharides from secreted recombinant β trace protein. Eur J Biochem 232: 718-25.

GrabenhorstE., Schlenke P., Pohl S, Nimtz M., Conradt HS. (1999) Genetic engineering of recombinant glycoproteins and the glycosylation pathway in mammalian host cells. Glycoconjugate J. 16: 81-97.

Gross HJ, Sticher U, Brossmer R (1990) A highly sensitive fluorometric assay for sialyltransferase activity using CMP-9-fluorescinyl-NeuAc as donor. Anal Biochem 186: 127-34.

Hardy MR, and Townsend RR (1994) HPEAC-PAD, Methods Enzymol, 230: 208-25.

Ihara Y, Nishikawa A, Tohma T, Soejima H, Niikawa N, Taniguchi N (1993) cDNA cloning, expression and chromasomal localization of human N-acetylglucosaminyltransferase III (GnT-III) J. Biochem 113:692-8.

James DC (1996) Analysis of recombinant glycoproteins by mass spectrometry. Cytotechnology 22: 17-24.

Jeffries R, Lund J, Pound JD (1998) IgG-Fc mediated effector functions: molecular definition of interaction sites for effector ligands and the role of glycosylation. Immunol Rev 163: 59-76

Jenkins N, Curling EMA (1994) Glycosylation of recombinant proteins: problems and prospects. Enzyme Microb Technol 16: 354-65.

Keusch J, Lydyard PM, Isenberg DA, Delves PJ (1995) β1,4-galactosyltransferase activity in B cells detected using a simple ELISA-based assay. Glycobiology 5(4): 365-70.

Keyt BA, Paoni NF, Refino CJ, Berleau L, Nguyen H, Chow A, Lai J, Pena L, Pater C, Ogez J, Etcheverry T, Botstein D, Bennett WF (1994) A faster and more potent form of tissue plasminogen activator. PNAS (USA) 91: 3670-4.

Kilmartin JV, Wright B, Milstein C (1982) Rat monoclonal antitubulin antibodies derived from by a new non-secreting rat cell line. J Cell Biol 93: 576-82.

Kitagawa H and Paulson JC (1993) Cloning and expression of human Gal beta 1,3(4)GlcNAc alpha 2,3-sialyltransferase Biochem. Biophys. Res. Commun. 194, 375-82.

Lee UE, Roth J, Paulsen JC (1989) Alteration of terminal glycosylation sequences on N-linked oligosaccharides of Chinese Hamster Ovary cells by expression of β galactoside α2,6-sialyltransferase. J Biol Chem 164: 13848-55.

Lifely MR, Hale C, Boyce S, Keen MJ, Phillips J (1995) Glycosylation and biological activity of CAMPATH-1H expressed in different cell lines grown under different culture conditions. Glycobiology 5(8): 813-22.

Lucas. BK, Giere-L-M. DeMarco RA, Shen A, Chisholm V, Crowley C (1996) High level production of recombinant proteins in CHO cells using a dicistronic DHFR intron expression vector. Nucleic Acids Res 24(9): 1774-9.

Masri KA, Appert HE, Fukuda M (1988) Identification of the full-length coding sequence for human galactosyltransferase (N-acetyl-glucosaminide: β-1,4-galactosyltransferase) Biochem. Biophys. Res. Comm.:157: 657-63.

Minch SL, Kallio PT, Bailey JE (1995) Tissue plasminogen activator co-expressed in Chinese Hamster ovary cells with α-2,6-sialyltransferase contains NeuAc-α (2,6)Gal-β (1,4)GlcNAc linkages. Biotechnol Prog 11: 348-51.

Monaco L, Marc A, Eo-Duval A, Acerbis G, Distefano G, Lamotte D, Engasser Jean-Marc, Soria M, Jenkins N (1996) Genetic engineering of α2,6 sialyltransferase in recombinant CHO cells and its effect on the sialylation of recombinant interferon-γ. Cytotechnology 22: 197-203.

Papac DI, Wong A, Jones AJS (1996) Analysis of acidic oligosaccharides and glycopeptides by matrix-assisted laser desporption/ionization time -of-flight mass spectrometry. Anal. Chem 68: 3215-23.

Papac DI, Briggs JB, Chin ET, Jones AJS. (1998) A high-throughput microscale method to release N-linked oligosaccharides from glycoproteins for matrix-assisted laser desorption/ionization time-of-flight mass spectrometric analysis. Glycobiology 8(5): 445-54.

Spellman M, Basa LJ, Leonard, CK, Chakel JA, O'Connor JV (1989) Carbohydrate Structures of Human Tissue Plasminogen Activator Expressed in Chinese Hamster Ovary Cells. J Biol Chem 264(24): 14100-11.

Umana P, Jean-Mairet J, Moudry R, Amstutz H, Bailey JE (1999) Engineered form of an antineuroblastoma IgG1 with optimized antibody-dependent cellular toxicity activity. Nature Biotech 17: 176-80.

Varki A. (1993) Biological roles of oligosaccharides: all of the theories are correct. Glycobiology 3(2): 97-130.

Weikert S, Papac D, Briggs J, Cowfer D, Tom S, Gawlitzek M, Lofgren J, Mehta S, Chisholm V, Modi N, Eppler N, Carroll K, Chamow S, Peers D, Berman P, Krummen L. (1999) Nature Bio 17(11): 1116-21.

Weinstein J, Lee UE, McEntee K, Paulson JC (1987) Primary structure of β galactoside 2,6 sialyltransferase: conversion of membrane –bound enzyme to soluble forms by cleavage of the NH2-termnal signal anchor. J. Biol. Chem., 262: 17735-43.

Werner RG, Noe W, Kopp K, Schluter M (1998) Appropriate mammalian expression systems for biopharmaceuticals . Drug Res 48(II), Nr 8: 870-9.

5. CONTROLLING CARBOHYDRATES ON RECOMBINANT GLYCOPROTEINS

N. JENKINS*[1], J. LUND[2] and L. MONACO[3]

[1] Bioprocess Research & Development, Eli Lilly and Company, Indianapolis IN 46285, USA
[2] Division of Immunity and Infection, The Medical School, University of Birmingham, Vincent Drive, Edgbaston, Birmingham B15 2TT UK
[3] DIBIT, Department of Biological and Technological Research, San Raffaele Scientific Institute, via Olgettina 58, 20132 Milano, Italy
*Corresponding author, fax (317) 276-5499, email: njenkins@lilly.com

1. Introduction

Mammalian cell culture has progressed in recent years from a lab scale enterprise, mainly used for in vitro testing of chemical agents, to a major source of therapeutic and prophylactic agents. Industrial scale cell culture had its origins in vaccine production from adherent fibroblasts in roller bottles, however the dominant technology employed today uses suspension-adapted CHO or hybridoma cells in large-scale (up to 12,000L) tanks.

Since microbial cells are easier to manipulate and have lower production costs than mammalian cells, they are the method of choice for peptides and simple proteins such as insulin and growth hormone. However, many therapeutic proteins require complex post-translational modifications such as glycosylation, gamma-carboxylation, and site-specific proteolysis, and mammalian cells are uniquely equipped to perform these operations.

This chapter will explore the methods and interpretation used to optimize the most extensive post-translational protein modification: glycosylation.

Many animal cell proteins are modified by the covalent additions of oligosaccharides. These additions often result in significant mass increases and can cover a large portion of their surfaces. There are two common classes of oligosaccharides N- and O-linked. The oligosaccharides (also called glycans) usually contain high levels of sialic acids, mannose, galactose, N-acetylglucosamine (GalNAc) and fucose residues. N-linked oligosaccharides are linked through an N-glycosidic bond between a GlcNAc residue and an asparagine residue (Kornfeld and Kornfeld, 1985) and fall into two broad categories: high mannose and complex types. In most cases the core unit is composed of a pentasaccharide core consisting of three branched mannose residues sequentially attached to two GlcNAc residues via an amide linkage to asparagine. This common core structure is formed from the same precursor (dolichol-linked oligosaccharide) which is responsible for the glycans transfer to nascent peptides, and is then further processed to form a large variety of structures. In high and oligomannose glycans, found in yeast and insect cells, no further glycan processing is observed. However, in proteins containing

M. Al-Rubeai.(ed.), Cell Engineering, 131-148.

and the complex glycans typically found in mammals and higher plants, further process-sing takes place in cis, medial and trans components of the Golgi apparatus, resulting in addition of GlcNAc, fucose, galactose, and sialic acids. Besides bi-antennary structures containing two arms for each glycan, complex triantennary and tetra-antennary struc-tures are also found on some glycoproteins such as EPO and blood clotting factors. The O-linked oligosaccharide is linked through an O-glycosidic bond between the GlcNAc residue and either serine or threonine, and is not processed by the same enzymes as N-glycans. Multiple O-glycosylation sites are found in mucin-type proteins. Individual glycosylation sites may contain a mixture of oligosaccharides (microheterogeneity). This is thought to be due to competition between glycosyltransferase enzymes that add nucleotide-monosaccharide units to the glycoprotein as it travels through the Golgi compartments.

Translation of both cytoplasmic and membrane proteins is initiated on the processed mRNA in the cytoplasm. During and following protein translation in the endoplasmic reticulum (ER) a number of modifications can occur such as proteolytic cleavage, pro-tein folding and glycosylation. There is evidence that certain chaperone proteins recog-nize different glycosylation modifications and act as a quality control mechanism for ER-Golgi and intra-Golgi translocations (Fagioli & Sitia, 2001). The presence of cellu-lar proteases and exoglycosidases in the culture media, particularly those released by dead and dying cells, can often result in product degradation towards the end of culture.

Glycosylation is thought to have several biological roles including protein stability (resistance to proteolysis and aggregation, improved protein folding), clearance in vivo, antigenicity and in some cases biological activity. Both mannose and asialoglycoprotein (ASG) receptors in the liver remove glycoproteins with terminal mannose or galactose residues from the bloodstream (Bianucci & Chiellini, 2000) and influence the plasma half-life of a glycoprotein in vivo. In the case of immunoglobulins, effector functions such as Fc receptor binding and complement activation can be influenced by the glyco-sylation state of the Fc region (Jefferis et. al., 1998)

2. Glycoprotein Analysis

Increased availability and reliability of carbohydrate isolation and detection methods have lead to major advances in our understanding of protein glycosylation and how it is controlled. In general, the level of carbohydrate analysis expected by the regulatory authorities increases as the glycoprotein drug passes from early (IND) to late phase (BLA) clinical trials. A common theme is a requirement to demonstrate the influences of oligosaccharide content on the glycoprotein's biological efficacy in humans.

Basic information on the presence or absence of oligosaccharide (macrohetero-geneity) can be obtained by using polyacrylamide gel electrophoresis, Western blotting using carbohydrate-specific lectins, or the FACE technique (Jackson, 1996). Another technique used for the rapid analysis of glycoproteins is capillary electrophoresis (CE), involving the separation of glycoforms by electrophoretic flow through a narrow bore capillary based on net charge, molecular weight and micellar association. Recent improvements on the basic CE technique include microbore capillary iso-electric focusing (cIEF).

Common methods for the analysis of microheterogeneity include lectin affinity chromatography, exoglycosidase digestion, mass spectrometry (MS), high performance anion exchange chromatography, and nuclear magnetic resonance (NMR). Lectins are carbohydrate binding proteins, which are specific for oligosaccharide structures, and when these are covalently immobilized in silica individual oligosaccharides or glycopeptides can be separated from a heterogeneous population. Lectin affinity HPLC analysis can provide a basic fingerprint for most glycoproteins although several lectin columns would be necessary to complete the structural analysis (Lee et al., 1990).

The availability of larger quantities (>1g) of bio-pharmaceuticals allows analytical meth-ods to reach higher accuracy but with low sensitivity. NMR is a powerful technique allowing complete structural information to be recovered with the disadvantage of requiring large amounts of protein. Interpretation of 1H NMR spectra requires knowledge of uniquely known structures such as the H-2 and H-3 of mannose, H-3 of sialic acid, the methyl protons of N-acetyl groups and specific glycan linkages (Bush et. al. 1999). In contrast, the most sensitive approach to glycosylation analysis involves various forms of mass spectrometry (James et. al., 1996). Fast-atom bombardment (FAB-MS) has the highest mass accuracy of the spectrometric methods, however is also expensive and requires a large amount of sample. Two other methods of ionization developed recently are particularly suitable for glycoprotein and glycan analysis: matrix assisted laser desorption/ionization (MALDI-MS) and Electrospray ionization (ES-MS). Both are relatively rapid techniques used to study the both the primary (peptide) structure and post-translation modifications of recombinant proteins For MALDI-MS the sample is mixed with excess matrix which strongly absorbs the laser light (e. g. sinapinic acid). When the mixture is subjected to laser light it is vaporized into ions which can be measured by a time-of-flight (TOF) analyzer, with an effective mass range of approximately 0.5-200kDa. Several laboratories have successfully used this technique to elucidate molecular weight and structural details of oligosaccharides in combination with an exoglycosidase array sequencing (Sutton et. al. 1994, James et. al. 1996). ES-MS detectors are often coupled to HPLC, allowing in-line analysis of glycopeptides generated by controlled digest of glycoproteins. This method allows complete structural analysis of glycoproteins with a very high mass accuracy.

3. Cell Type, Metabolism and Environment

The cell line or tissue in which a glycoprotein is produced has dramatic effects on the oligosaccharide attached to the glycoprotein. This is due to the differences in relative activities and groups of glycosyltransferases and nucleotide-sugar donors that differ between species and tissue type. The N-linked carbohydrate populations associated with both Asn25 and Asn97 glycosylation sites of IFN-γ have been characterized by MALDI-MS in combination with exoglycosidase array sequencing (Hooker et al., 1999; James et. al. 1995, 1996; Ogonah et al 1996). Recombinant IFN-γ produced by Chinese hamster ovary cells (CHO) showed that N-glycans were predominantly of the complex bi- and triantennary type at both N-glycosylation sites. In the same recombinant glycoprotein produced by baculovirus-infected Sf9 insect cells glycans were mainly trimannosyl core structures, with no direct evidence of post-ER glycan-processing events other than core fucosylation and de-mannosylation. Transgenic mouse derived IFN-gamma exhibited considerable site-specific variation in N-glycan structures with Asn25-linked carbohydrates of the complex, core fucosylated type and Asn97-linked carbohydrates) were mainly of the oligomannose type, with smaller proportions of hybrid and complex N-glycans (Kemp et al. 1996). These data indicate that endoplasmic reticulum to cis-Golgi transport is a predominant rate-limiting step in both expression systems. However both sialyltransferase activity and CMP-NeuAc substrate were found to be absent in uninfected or baculovirus-infected Sf9 Sf21 and Ea4 insect cells. These data demonstrate the profound influence of host cell type and protein structure on the N-glycosylation of recombinant proteins, and the considerable challenges faced in re-engineering non-vertebrate cells to reproduce mammalian type glycosylation.

4. Effects of Chemical Supplements on Protein Glycosylation

Changes in the cell culture environment can also lead to changes in the oligosaccharide structures observed (Jenkins, 1996). Some of the conditions studied include the age of the batch culture, glucose limitation (Hayter et. al;. 1991), oxygen starvation (Regoeczi et. al. 1991), intracellular ammonium ion accumulation (Yang & Butler, 2000) and pH excursions (Borys et. al. 1993). In general, these batch and fed-batch conditions lead to poorer glycosylation profiles of recombinant glycoproteins compared to those produced in perfusion or chemostat cultures (Grabenhorst et al., 1999). This is most probably due to a combination of nutrient depletion and waste metabolite accumulation in batch and fed-batch systems.

The control of protein glycosylation is a major goal of the biotechnology industry, since glycosylation is the most extensive post translation modification made by animal cells. Protein structure, cell type and cell culture environment all effect the N-glyco-sylation machinery and contribute to glycoprotein heterogeneity. Exogenous chemicals that are not essential to maintain cell growth in defined media are sometimes added to improve recombinant protein production. Sodium butyrate has been shown to unbind the chromatin structure thus making it more accessible to RNA polymerase, promoting mRNA replication (Sowa & Sakai, 2000). As transfected genetic material will integrate at the most accessible loci, it has been shown that it is those genes that are most likely to be transcribed during sodium butyrate. It has also been shown that sodium butyrate can increase expression of certain chaperone-like proteins (i. e. GRP78 and 94) by preventing histone phosphorylation and increasing hyperacetylation. Sodium butyrate can also cause a dilation of the endoplasmic reticulum and influence cell cycle kinetics by stabilizing a population in the G1 phase. Lamotte et. al. (1997) also studied the effects of butyrate on the glycosylation pathway of CHO cells, and found a positive correlation between both IFN-γ production and N-glycan complexity with increasing doses of sodium butyrate. Dicloroacetate (DCA) is known to inhibit cell growth, glutamine catabolism and production of pyruvate and alanine. Little is known about the effect of this chemical on glycosylation but literature shows that recombinant antibody yields can be increased by up to 60% (Murray & Dickson, 1997). N-acetyl manno-samine, a sialic acid precursor able to enter the cell, has been shown to be effective in increasing the intracellular sialic acid pool and sialic acid content of recombinant glycoproteins produced by CHO cells, but its high cost may prohibit its utility in large-scale culture (Gu et al., 1998).

Lipids play an essential role in the glycosylation of many proteins, they act as sugar carriers and the dolichol donor is used for the block transfer of oligosaccharide units from the cytosol to the ER. Both complex lipid supplements and dolichol derivatives have been shown to improve glycosylation at Asn 97 of γ-IFN (Jenkins et. al. 1996, Green et. al. 1997). Cholesterol can also be an important growth supplement for CHO cells and is required for optimal NS0 cell growth. Consequently cholesterol, phospholipids (i.e. phosphatidyl choline) and fatty acids (linoleic and oleic acids) are usually included in serum-free formulations.

There follows a detailed description of two examples from our laboratories using transfection of the 2,6-sialyltransferase (2,6-ST) gene to correct an inherent defect in CHO cells and produce recombinant glycoproteins with modified properties.

5. Transfection of 2,6-ST to Produce a Universal Host CHO Line and Its Effects on Recombinant IFN-γ

Chinese hamster ovary cells (CHO) are widely employed to produce glycosylated recombinant proteins and are equipped with a glycosylation machinery very similar to the human (Monaco et al., 1996). A notable difference concerns sialylation: N-linked glycans of human origin carry terminal sialic acid residues in both $\alpha 2,3$- and $\alpha 2,6$-linkages, whereas only $\alpha 2,3$-terminal sialic acids are found in glycoproteins from CHO and BHK (baby hamster kidney) cells. Indeed, these cell lines lack a functional copy of the gene coding for 2,6-ST (EC 2. 4. 99. 1). Sialic acid residues confer important properties to glycoproteins, due to their negative charge and to the terminal position. Specific receptors, such as the ASG receptor, have evolved in the liver and in macro-phages to bind and internalize glycoproteins devoid of terminal sialic acid molecules (Jefferis et al., 1998; Lund et al. 1990; Sarmay et al. 1992). Moreover, the binding of different members of the sialoadhesin family (Morgan et al. 1995) and CD22 ligand (Crocker et al., 1997) to terminal sialic acid residues is specific for either the $\alpha 2,3$ or $\alpha 2,6$ linkage.

Our group as well as others has demonstrated that the sialylation defect of CHO cells can be corrected by transfecting the 2,6-ST cDNA. Glycoproteins produced by such CHO cells display both $\alpha 2,6$- and $\alpha 2,3$-linked terminal sialic acid residues, similarly to human glycoproteins. Here, we have established a CHO cell line stably expressing $\alpha 2,6$-ST, providing a universal host for further transfections of human genes. Several relevant parameters of the universal host cell line were studied, demonstrating that the $\alpha 2,6$-ST transgene was stably integrated into the CHO cell genome, that transgene expression was stable in the absence of selective pressure, that the recombinant sialyltransferase was correctly localized in the Golgi and, finally, that the bioreactor growth parameters of the universal host were comparable to those of the parental cell line. A second step consisted in the stable transfection into the universal host of cDNAs for human glycoproteins of therapeutic interest such as IFN-γ.

The availability of DNA sequences coding for glycosyltransferases (Lee et al., 1989) has opened the possibility to genetically alter the glycosylation capabilities of a cell line. In particular, the cDNA for the rat liver α2,6-ST has been transiently or stably transfected into BHK (Grabenhorst et al., 1999) or CHO cells (Lee et al. 1989; Monaco et al., 1996) and the sialylation of a target recombinant protein has been confirmed to be in both α2,3- and α2,6-linkages. The plasmids pSfiSV-2,6-ST, pSfiSVneo and pSfiSVdhfr and pCISfiT: a new vector for in vitro amplification (Monaco et al. 1994) carrying the human CMV promoter was built from plasmid pCI (Promega, Madison, WI). First, one *Sfi*I site was inserted into the unique *Not*I site and the resulting plasmid (pCISfi) was further modified by introduction into the filled-in *BamH*I site of the oligonucleotides containing the 41 bp stretch corresponding to nucleotides +189 to +229 of the human gastrin gene, reported to act as a transcription terminator (Bragonzi et al. 2000). The resulting vector was named pCISfiT, and contained the SfiI site necessary for multimerization at the end of the multiple cloning site, plus a transcription terminator sequence downstream the poly-adenylation signal, to improve transcription efficiency from multimerized transcription units. Plasmid pCISfiT-IFN-γ contained the 1-kb open reading frame coding for human IFN-γ, excised from plasmid pSVL-IFN-γ (obtained from Glaxo Smith-Kline) by cleavage with *BamH*I, filling-in with the Klenow fragment of T4-DNA polymerase and cleavage with *Xho*I. The fragment was inserted between the *Xho*I and *Sma*I sites of the vector pCISfiT. DNA concatenamers for stable transfections were prepared according to the in vitro amplification method (Monaco et al., 1994) by mixing the SfiI-flanked expression units and the selectable marker unit at the following ratios: α2,6-ST transfection into CHO Dux B11 cells: pSfiSV-2,6-ST / pSfiSVneo = 10:1; and IFN-γ into the universal host: pCISfiT-IFN-γ / pSfiSVdhfr = 20:1. Ligation products were confirmed to extend between 50 and at least 200 kb by pulsed field electrophoresis extracted sequentially with phenol/chloroform, according to standard procedures and used for lipofection mediated transfections.

For Southern blot analysis CHO genomic DNA was extracted from cells by a salting-out procedure, and completely digested with *Sfi*I. 10 μg of digested DNA were separated on a 0. 8% DNA gel, and adequate amounts of the *Sfi*I-cleaved plasmids pSfiSV-2,6-ST and pCIS fiT-IFN-γ were loaded along with the samples, to allow estimation of band intensities. DNA was then blotted onto a HybondN[+] nylon membrane (Amersham Pharmacia Biotech, Upp-sala, Sweden) and hybridized to the ^{32}P-labeled, 607-bp BglII fragment of the rat α2,6-ST cDNA, according to standard techniques. The filter was then stripped and re-hybridized to the ^{32}P-labeled, 411-bp *Ssp*I fragment of the human IFN-γ cDNA. Adjacent cassettes of the α2,6-ST cDNA integrated in tandem into the genome were revealed as a single 3.5 kb band. Approximately ten copies of the 2,6-ST expression unit were detected in both UH and clone 54 (IFN-γ producing) cell lines and were also maintained in clone 54 cells grown in 2 μM methotrexate (MTX). A similar analysis was performed on UH cells grown without G418 for 2 months, and a comparable number of integrated copies was obtained indicating the stability of 2,6-ST expression in the absence of selection agent. Conversely, less than 10 copies of the IFN-γ cassette were integrated into the genome of clone 54, and as expected a dramatic increase in IFN-γ copy number was observed in clone 54 cells grown in 2 μM MTX amplification agent.

For transfection, 5×10^5 CHO-DxB11cells were plated onto 60 mm dishes and 10 μg of DNA concatenamers, were mixed with 2 mM ethanolic solution of DOGS (Transfectam®). Two days after transfection, cells were exposed to selective medium: either complete medium supplemented with 900 μg/mL G418 for neomycin resistance, or MEM alpha medium without nucleosides and deoxynucleosides for dhfr selection. Resistant colonies were obtained and individually grown in 96-wells plates.

Amplification in MTX was performed by exposing individual clones to increasing concentrations ranging from 10 nM to 2 μM. Cells were exposed to each MTX concentration for at least 3 weeks; resistant cells were pooled and expanded. Control UH cells were maintained in the above medium, but containing 500 μg/mL G418 for neomycin selection. Clone 54 cells, deriving from the stable transfection of the universal host with IFN-γ cDNA, and CHO clone 43 cells, stably producing IFN-γ (obtained from Glaxo Smith-Kline) were grown in MEM alpha medium without nucleosides and deoxynucleosides, and 2 μM MTX, the medium for clone 54 only also contained 500 μg/mL G418.

Flow cytometry using fluorescently labeled probes is a powerful method of searching for low-abundant clones expressing a particular phenotype from a mixed population of cells. We used FITC-labeled lectin from *Sambucus nigra* (SNA) that recognizes the Gal-α2,6-sialic acid epitope to isolate cells expressing the recombinant 2,6-ST enzyme transfected into CHO cells. The technique can be used in two modes, either to correct the sialylation defect in cells already making a recombinant protein of interest (in this case IFN-γ) or to correct the defect in the parental cell line which can subsequently be used as a "universal host" for expressing recombinant glycoproteins with both α2,3- and α2,6-sialylation (Bragonzi et al., 2000) . It could also be used to pick clones producing high amounts of the recombinant protein of interest (using a fluorescent antibody) and simultaneously check for authentic product quality (sialylated species using a different fluorophore covalently linked to SNA lectin). In this experiment, the presence of α2,6-sialic acid linkages was assessed in isolated clones by flow cytometry analysis of cells incubated with FITC-labeled SNA lectin, as described by [12]. 2,6-ST enzyme activity was again assayed in cell lysates from each clone by a lectin-based microtiter plate assay (Mattox et al., 1992).

CHO Dux B11 cells were stably transfected with DNA concatenamers constituted by an average of 10 expression units of rat 2,6-ST per neomycin resistance expression unit, using the in vitro amplification method. Twenty-seven G418-resistant colonies were screened by exploiting the specific binding of α2,6-sialic acid residues on plasma membrane glyco-proteins to SNA lectin, using flow cytometry analysis. Fourteen (52%) colonies displayed a positive binding of between 75 and 100% of each cell population. α2,6-ST enzyme activity was assayed on these colonies by the specific solid phase assay (Mattox et al. 1992) and ranged between 12 and 68 IU of α2,6-ST/mg protein. Figure 1 shows the correlation between flow cytometry positivity and α2,6-ST

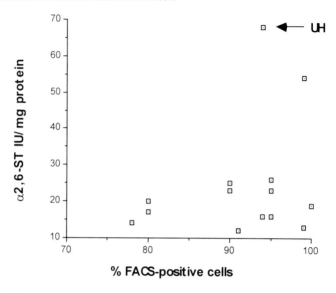

Figure 1. α2,6-ST expression in individual CHO Dux B11 clones. Correlation between positive FITC-SNA-lectin binding by flow cytometry analysis and α2,6-ST enzyme activity in clones of CHO Dux B11 cells stably transfected with α2,6-ST cDNA. The arrow indicates the clone chosen as universal host (UH).

activity, and the clone displaying the highest indices of both parameters was chosen for further characterization and named the universal host (UH). The stability of expression of α2,6-ST in the universal host was tested by growing the cells in the presence or in the absence of G418 selective pressure for 2 months, and binding to SNA lectin was again analyzed by flow cytometry. Iden-tical flow cytometry profiles were obtained in both cases indicating a stable expression of the 2,6-ST enzyme in the universal host.

In order to demonstrate that the α2,6-ST introduced into UH cells can act on the glycans of recombinant glycoproteins, the cDNA for human IFN-γ (placed under the hCMV pro-moter) was transfected into the universal host, using dhfr as the second selectable marker. Conditioned media from 200 individual clones were analyzed by an IFN-γ-specific ELISA, and displayed up to 1 pg IFN-γ/cell per 24 hours. To reach higher production levels, five of these clones were exposed to gradually increasing MTX concentrations, up to 2 μM. A steady increase in IFN-γ levels was observed, reaching a maximum of 37 pg/cell (18 mg/L) per 24 hours in clone 54. To test for expression stability, clone 54 cells were grown in the presence or in the absence of MTX for 2 months and assayed for IFN-γ production and for α2,6-ST activity. IFN-γ levels were completely abolished following removal of MTX whilst full 2,6-ST activity was maintained as determined by flow cytometry analysis, demonstra-ting that 2,6-ST expression was not affected by MTX treatment. The fact that no variation in copy number was observed after two months in culture without selective pressure indicates that the 2,6-ST concatenamer was stably integrated into the UH genome, and that no significant recombination events took place during the observed time span. The copy number of the ST transgene did not change even after transfection with IFN-γ and MTX treatment indicating that the two transgenes had integrated independently.

Naturally occurring sialyltransferases have been localized to the Golgi, by immunofluorescence (Lipsky, 1985) and immunoelectron microscopy (Marks et al., 1999). We utilized a rabbit antiserum to rat 2,6-ST (a kind of Prof. E. Berger, Zurich) and tetramethylrhodamine B isothiocyanate (TRITC)-conjugated polyclonal secondary antibody against rabbit Ig from DAKO to localize this enzyme. A trans-Golgi marker: NBD-C_6-ceramide from Molecular Probes was used to confirm the organelle location of 2,6-ST. Cells were grown for 2 days on glass coverslips, fixed in 2% paraformaldehyde, washed in PBS, saline and fatty acid-free bovine serum albumin. For α2,6-ST detection, cells were permeabilized with 0. 1% Triton X-100 for 10 minutes and incubated with the rat α2,6-ST for 1 hour followed by the secondary antibody for 30 minutes. Negative control experiments were also performed where the primary antibody was omitted. For Golgi staining, the permeabilization step was omitted and cells were incubated with 5 μM NBD-C_6-ceramide for 60 hours. Coverslips were mounted with a phenylene-diamine-mounting medium and analyzed at the fluorescence photomicroscope. The 2,6-ST antiserum strongly labeled vesicles in limited areas of the cytoplasm near the nucleus, consistent with a localization in the trans-Golgi apparatus, as demonstrated by the identical distribution of NBD-C_6-ceramide. The intracellular distribution of the recombinant 2,6-ST in the UH line was similar to the one observed for the natural enzyme in the human HepG2 cell line, known to express α2,6-ST.

The UH cell line was grown in a stirred tank bioreactor in order to verify that the integration of the exogenous gene coding for the α2,6-ST enzyme did not compromise the growth and metabolism of the cell line in suspension culture. The culture conditions have been described elsewhere (Hayter et al. 1993), except that medium was supplemented here with adenosine, cytidine, guanosine, uridine, 2'-deoxyadenosine, 2'-deoxycytidine, 2'-deoxy-guanosine and 2'-deoxyuridine to achieve a final concentration of 1 mg/L. The cell growth phase took place over a period of 50 hours, and cell viability stayed higher than 95% during this period, with a maximal viable cell concentration of 1. 2x10^6 cells/mL. The maximal cell specific growth rate (0. 04 h^{-1}) was high as compared with other CHO cell lines, but the uptake of nutrients (glucose and glutamine) and the production of toxic substances (lactate and ammonium) were similar to those observed with the parental cell line.

For IFN-γ purification, an IFN-γ affinity column was generated by coupling 20B8 monoclonal anti-IFN-γ antibody (Lonza Biologics) to CNBr activated Sepharose. The conformation of sialic acids in IFN-γ purified from either UH cells (clone 54) or CHO Dux B11 cells (clone 43) was determined in duplicate by reverse phase HPLC utilizing 2,3-sialic acid specific sialidase from Newcastle Disease Virus and the non-discriminating sialidase from *Arthrobacter ureafaciens* (Reuter & Schauer, 1994) The total amount of sialic acid was determined spectrophotometrically following acid-catalyzed hydrolysis of duplicate samples according to the method of Manzi and Varki (1994). The sialic acid conformation on IFN-γ purified from clone 54 (2,6-ST⁺) was 40.4% sialic acid in α2,6-linkage and 59.6% in α2,3-linkage. Conversely, IFN-γ from clone 43 (control, non-transfected) exclusively displayed α2,3-linkages, as expected. The sialic acid content was similar in both IFN-γ preparations (1.16 and 1.12 moles sialic acid per mole of protein respectively). The two IFN-γ preparations were therefore comparable in terms of extent of sialylation, and differ only for the presence of α-2,6-linked sialic acids in the protein produced in 2,6-ST+ (clone 54) cells. This indicates that the stable expression of 2,6-ST did not hamper the sialylation capability of the cells, and that the added sialyltransferase competes with the endogenous one for the pool of donor CMP-sialic acid under these culture conditions.

In a pharmacokinetic study, female Sprague-Dawley rats (160-270 g) were anaesthetized with Hypnorn and IFN-γ was administered by intravenous infusion (0.3 mg/kg). The tail was clipped and blood (100 μL) was then collected via the tail vein at 3, 5, 7, 10, 15, 20, 25, 30 and 35 minutes post-administration. Blood samples were centrifuged, plasma IFN-γ levels determined by ELISA, and data processed using the S-Plus statistical package from MathSoft Inc (Figure 2). The difference of log-AUC (Area Under Concentration curve) between the two groups was statistically significant (p=0.013) and indicates that the recombinant IFN-γ from UH cells persists longer in the blood compartment when compared to the same protein made in control cells. The clearance values obtained for the two groups were highly statistically significant, since the test-statistic, which was asymptotically normal, had a value of 3.39 (p <0.001).

To test whether the IFN-γ produced in the UH line as well as in normal CHO cells was biologically active, up-regulation of the endogenous expression of major histocompatibility complex (MHC) class II (Waldberger et al., 2000) in human colon carcinoma LoVo cells was assessed by flow cytometry analysis. 100 ng/ml each of IFN-γ purified from either clone 54 or clone 43 elicited an up-regulation in MHC class II of 3.00 ± 0.61 fold and 3.52 ± 0.52 fold, respectively. In comparison, 100 ng/ml of

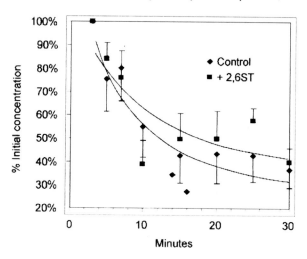

Figure 2. Clearance of IFN-γ glycoforms in rats. IFN-γ from the universal host (■) and from parental CHO (◆) cell lines were injected into rats and serum concentration was determined at the defined times.

purified, standard human IFN-γ at 10 IU/ng produced a 4.50 ± 0.49 fold increase in MHC class II expression in the same cell line (means ± SE for two measurements, run in quadruplicate). The observed differences between the two samples and between each sample and the standard were not statistically significant, as assessed by ANOVA analysis, indicating the biological activity of all three IFN-γ preparations was similar.

Several groups, including ours, have successfully transfected mammalian cells with the genes for different glycosyltransferases. Here, we have established and fully characterized a CHO universal host line with improved glycosylation capabilities, thus providing a relevant tool for the production of recombinant proteins for human therapy. Such a cell line is able to produce glycoproteins carrying both α2,6- and α2,3-linked terminal sialic acids, as found in human glycoproteins but not in products from normal CHO cells. The in vitro amplification transfection method, relying on the transfer of multiple concatenated copies of the transgene Monaco et al. 1994) was employed. In keeping with our previous results (Monaco et al. 1996), a high proportion (52%) of positive clones was obtained. The average number of expression cassettes per DNA concatenamer was increased in the present work from 5 to 10, in an attempt to enhance the expression of α2,6-ST, and clones with high α2,6-ST activity were obtained. The best of these clones, named "universal host" (UH), was studied to assess its suitability as a cellular host for the production of recombinant proteins.

In theory, an alternative strategy to the two-step process requiring preparation of the universal host line first, followed by transfection of the target glycoprotein, could consist in the co-expression of both the α2,6-ST and IFN-γ in one step, by transfection of DNA concatenamers including both expression cassettes and the selectable marker gene. However, preliminary transfections of CHO Dux B11 cells with IFN-γ DNA concatenamers including the neomycin resistance gene as selectable marker had failed to yield stable clones displaying satisfactory production levels. The use of the *dhfr* selectable marker gene coupled to MTX treatment was therefore judged mandatory. The two-step strategy discussed above was eventually preferred, to allow more flexibility in individually modulating the expression levels of the two genes. Indeed, continued selective pressure was not needed to maintain satisfactory α2,6-ST activity in the UH cell line. Conversely, high-level production of IFN-γ in MTX-amplified cells proved unstable, as often is the case in the CHO/dhfr expression system (Kaufman et al. 1985).

Recently, TNK-tPA, a variant of wild-type tissue plasminogen activator, was produced in CHO cells over-expressing α-2,3-ST (Weikert et al., 1999). The purified glycoprotein displayed increased sialylation levels, as compared to the control protein produced in wild-type CHO cells; this imparted improved pharmacokinetic properties to the protein, following iv. injection into rabbits. The prolonged permanence of TNK-tPA in circulation was ascribed to decreased recognition of the highly sialylated product by the asialoglycoprotein receptor. The differences in pharmacokinetics we have observed for IFN-γ are due to subtle changes in the stereochemistry of the glycan structure. Although further work is required to investigate other possible alterations in biological activity due to the presence of α2,6-sialic acid residues, this is already a positive indication of the effects of the linkage alteration we have introduced in the glycosylation properties of CHO cells.

In conclusion, we have established a CHO universal host cell line with improved sialylation capabilities. The UH line can be further engineered to produce therapeutically relevant glycoproteins, whose sialylation more closely resembles the pattern found in human products. Correctly sialylated and biologically active IFN-γ was produced by the UH line and displayed prolonged pharmacokinetics.

6. Sialylation of Human IgG-Fc Carbohydrate by Transfecting Rat α2,6-Sialyltransferase

The IgG molecule is comprised of four structural and functional regions: two identical antigen binding Fab regions linked through a flexible hinge to an Fc region that has binding/activation sites for effector ligands, e.g. C1q and Fcγ receptors (Shields et al., 2000; Sondermann et al., 2000). The Fc region is comprised of part of the hinge with inter-heavy chain disulfide bridges, two unpaired glycosylated C_H2 domains, and two non-covalently paired C_H3 domains. Glycosylation of the C_H2 domain at Asn-297 is essential to activation of effector functions (Jefferis, 1998) and together with structure at or proximal to the N-terminus of the C_H2 domain generates recognition sites for leukocyte Fcγ receptors and complement (Lund et al., 1990; Sarmay et al., 1992). Secreted IgG is a heterogeneous mixture of glycoforms exhibiting variable addition of the sugar residues fucose, galactose, sialic acid, and bisecting N-acetylglucosamine. The efficiency of effector function activation can vary between the glycoforms, e.g. galactosylation enhances recognition of IgG by complement or by human Fcγ receptors.

In a previous study, a recombinant IgG3 immunoglobulin against the hapten nitro-iodophenacetyl (NIP) with Phe-243 replaced by Ala (FA243) was expressed in a CHO-K1 parental cell line. The resulting IgG-Fc-linked carbohydrate was significantly α2,3-sialylated (53% of glycans), as indicated by normal and reverse phase HPLC analyses (Lund et al., 2000) in contrast to wild-type immunoglobulins which are poorly (<5%) sialylated. We transfected the rat 2,6-ST gene into this FA243 cell line, in order to investigate the biological properties of this new IgG3. (Jassal et al., 2001).

The vector used, pCISfiT containing an insert encoding for the rat 2,6-ST gene was a gift from Dr. Monaco. A *Kpnl/Xhol* digested fragment of pCISfiT containing the gene (1.7 Kb) was ligated into the single *Kpnl/Xhol* digested site of the pBK-CMV expression vector (Stratagene) and transformed into *E. coli* (TG1) competent cells by calcium shock. Plasmid DNA (pBK-CMV-α2,6-ST) was isolated from a number of individual kanamycin-resistant colonies, and shown by restriction with *Kpnl/Xhol* to contain the expression vector and the 1.7 Kb insert encoding the α2,6-ST gene. The CHO-K1 cell line FA243 was transfected with pBK-CMV-2,6-ST by spheroplast fusion. This cell line had previously been transfected with the genes encoding for FA243 IgG3 and selected on the basis of its resistance to mycophenolic acid (Lund et al., 2000). A 24 well plate was seeded with the transfected cells and left for 48 hours prior to selection with the antibiotic G418. The adherent cell lines were incubated in RPMI 1640 medium containing G418 (0.6 mM). After two weeks in selection medium transfected colonies were visible in all the wells, when limiting dilution was used to clone the transfectants. FA243 CHO-K1 cells that had not been transfected with the pBK-CMV-α2,6-ST vector construct failed to form colonies. Once the cloned cells reached confluence, they were checked for maintained IgG3 production by ELISA.

Clones having maintained levels of IgG production were assayed for 2,6-ST activity using the SNA-ELISA method reported previously (Mattox et al., 1992). The final FA243-ST clone was selected on the basis of optimal $\alpha 2,6$-ST activity (60.8 ± 6.6 mU/mg (n = 4)) relative to the FA243 cell line (14.1 ± 2.9 mU/mg (n = 4)), and amplified in roller bottle culture in order to produce IgG3 antibody. Cells were grown to a final cell density of 2.5 x 10^8 cells/ml in 2 litre roller bottles. The supernatant was collected and IgG3 antibodies were purified from cell supernatant on a column of NIP-Sepharose 4B and eluting with 0.5 mM NIP. SDS PAGE was used to check the purity of the IgG3 proteins.

To release oligosaccharides from IgG3 antibodies, 3 nmoles of purified antibody was incubated with PNGase F and the reducing ends of the released N-linked oligosaccharides were reductively aminated with the fluorophore 2-aminoacridone using sodium cyanoborohydride. Labeled glycans were analysed by normal phase HPLC using a GlycoSep N column (Oxford Glycosciences), detecting fluorescence using excitation and emission wavelengths of 428 and 525 nm. The sialic acid content and the nature of its linkage to the oligosaccharide were evaluated as described previously in this review.

All samples displayed a heterogeneous range of complex biantennary oligosaccharide structures with a single N-glycosylation site on each heavy chain at Asn-297 within the C_H2 domain of IgG, but the oligosaccharide profiles varied between the IgG3 antibodies. The glycans released from the wild-type IgG3 showed low levels of galactosylation (G0 + G0F), relative to those observed for the FA243 and FA243-ST IgG3 antibodies. Similarly, there was a much higher degree of sialylation (mono-sialyl + di-sialyl) for glycans from FA243 IgG3 and FA243-ST IgG3 than for the wild-type, which had minimal levels of sialylation. Thus, galactosylation and sialylation are inhibited for the nascent oligosaccharide chains of the wild-type IgG relative to its FA243 replacement. IgG-Fc linked glycans were sialylated (60% of glycans) such that the ratio of $\alpha 2,6$-:$\alpha 2,3$-linked sialic acids was almost equal (Table 1).

Table 1. Sialic acid composition and linkage of oligosaccharides derived from anti-NIP IgG3 wild-type and mutant antibodies.

Sialic Acid Linkage	Wild-type	Std. Dev.	FA243	Std. Dev.	FA243-ST+	Std. Dev.
% $\alpha 2,3$	2.4	0.1	56.7	0.4	35.4	3.5
% $\alpha 2,6$	0.0	0.0	1.9	0.6	30.1	3.0
Total	2.4	0.1	58.6	0.4	65.5	0.4

The nmoles sialic acid per nmole of IgG were determined in each case from 3 nmoles IgG, so that a theoretical yield of 100% corresponds to 6 nmoles sialic acid. Data are presented for three replicate determinations. The $\alpha 2.3$ and total sialic acids were determined following release with sialidase from Newcastle disease virus and *A. ureafaciens* respectively, and the $\alpha 2.6$ values determined by subtraction of $\alpha 2.3$ from the total value.

The FA243 replacement in IgG3 permits significant α2,6-sialylation, comparable to the physiological level found for human polyclonal IgG that has 27% ± 7% of oligosaccharide moieties that are α2,6 sialylated. One explanation for these findings is that sialylation is inhibited by virtue of extensive non-covalent interactions of the C_H2 domain with the nascent oligosaccharide chain (Deisenhofer et al., 1981) interactions which are reduced in the case of the replacement FA243. We suggest that for wild-type IgG3 expressed in CHO-K1 cells the extensive non-covalent interactions strongly inhibit galactosylation and sialylation, which is overcome to an extent in the case of the FA243 replacement, but still sufficient to exert the predominant limiting influence on the extent of sialylation. Thus sialylation is controlled primarily by the protein structure local to the carbohydrate, and that the two sialyltransferases compete to sialylate the nascent oligosaccharide.

Glycoproteins other than IgG have only minimal non-covalent interactions between the carbohydrate and the protein. An increased level of total sialylation (95%) has been reported (Hoffmann et al. 1994) for recombinant human trace β-protein expressed by baby hamster kidney cells relative to the native protein (60%) isolated from human cerebrospinal fluid. Thus, it is reasonable to conclude that different factors can limit α2,3-ST and α2,6-ST activity associated with these different glycoprotein substrates.

It has previously been shown that outer arm galactosylation enhances recognition of IgG by FcγRI, C1q and complement (Tsuchiya et al., 1989). We hypothesised that remodelling the glycans on the IgG3 would modulate its biological activities. The ability of the wild-type, NA297, FA243 and FA243-ST IgG3 antibodies to trigger lysis mediated through human complement was determined. Sheep red blood cells were NIP derivatized and tested with different antibody preparations followed by either Guinea pig or human serum to quantify complement-mediated cell lysis. In agreement with previous findings (Jefferis et al., 1996) the FA243 IgG3 required a 1.8-fold increased antibody concentration to give 50% lysis compared to the wild-type IgG3, however, FA243-ST+ IgG3 demonstrated a capacity to trigger complement mediated cell lysis similar to that of wild-type IgG3, indicating that the presence of the α2,6 sialylation linkage compensates for the effect of the replacement FA243. A similar trend was obtained using guinea pig complement such that FA243-ST required only a 1.4-fold increased IgG concentration to give 50% lysis relative to the wild-type IgG3, compared to 1.9-fold for the FA243 and 7-fold for NA297 IgG3.

Superoxide anion was measured as lucigenin enhanced chemiluminescence, using U937 cells treated with IFN-γ, and IgG3-sensitized NIP-derivatized human red blood cells in the presence of 0.25 mM lucigenin. This assays for ability to trigger a respiratory burst mounted through FcγRI expressed on γ-interferon stimulated U937 cells. FA243 IgG3 exhibited a reduced ability to trigger superoxide relative to the wild-type IgG3 (~ 70% of the maximal response). However, the FA243-ST IgG3 triggered a maximal response that was 1.3-fold greater than that observed for the wild-type protein. Thus, α2,6 sialylation contributes to recognition of IgG by human FcγRI.

Binding of antigen-complexed IgG3 to K562 cells via FcγRII receptors was determined by rosette formation (Lund et al., 1995) using human red cells were derivatized with NIP and sensitized with anti-NIP IgG3. The aglycosylated IgG3 antibody NA297 showed a greatly reduced capacity to rosette, confirming the requirement for carbohydrate to support recognition of IgG by human FcγRII. FA243 IgG3 exhibited a two-fold reduced capacity to rosette, relative to the wild-type IgG3. However, FA243-ST IgG3 showed a capacity to rosette that was similar to the wild-type IgG3, indicating that α2,6 sialylation contributes to recognition of IgG by human FcγRII.

In summary, these results suggest that the nature of the sialylation linkage can influence recognition by C1q/complement, FcγRI and FcγRII. A plausible explanation is that the α2,6-sialic acid linkage can affect the structural dynamics of the oligosaccharide moiety, in particular the primary and secondary GlcNAc residues that have a predominant influence on recognition of IgG by its effector ligands (Mimura et al, 2000). The oligosaccharide moieties may additionally affect the conformation of the lower hinge binding site for effector ligands on the IgG protein, stabilising conformations that are more tightly bound by the effector ligands.

7. Conclusions

In this review we have detailed the multiple sources of glycan heterogeneity that can be found in both natural and recombinant glycoproteins. Advances in carbohydrate analysis have facilitated detailed, site specific determinations of each glycoform and have paved the way to elucidating means by which glycans can be controlled and manipulated. We have reviewed several chemical approaches used to improve protein glycosylation, and have detailed two of our own case studies (using recombinant IFN-γ and IgG3) where genetic manipulation of sialyltransferase in the host CHO cell line resulted in significant changes to the properties of each molecule. A similar genetic strategy has also been used to incorporate bisecting GlcNAc residues into recombinant antibodies, which proved useful for enhancing antibody-dependent cytotoxicity (Umana et al., 1999) using the enzyme Glycosyltransferase III.

In future these chemical and genetic methods will also be assessed against the recently-introduced technique of post-fermentation manipulation of recombinant glycoproteins, e.g. using cheap, microbial derived sialyltransferases and sugar-nucleotide precursors invented by JC Paulsen and commercialised by Neose Technologies. Although these alternatives pose their own challenges in scale ability and enzyme elimination in the final product, time will tell if these prove more cost-effective methods manipulating recombinant glycoproteins.

8. References

Bianucci AM, Chiellini F. (2000) A 3D model for the human hepatic asialoglycoprotein receptor (ASGP-R).J Biomol Struct Dyn. 3: 435-51.

Borys MC, Linzer DI, Papoutsakis ET. (1993) Culture pH affects expression rates and glycosylation of recombinant mouse placental lactogen proteins by Chinese hamster ovary (CHO) cells. Biotechnology 6: 720-4.

Bragonzi A, Distefano G, Buckberry LD, Acerbis G, Foglieni C, Lamotte D, Campi G, Marc A, Soria MR, Jenkins N, Monaco L (2000) A new Chinese hamster ovary cell line expressing alpha-2,6-sialyltransferase used as universal host for the production of human-like sialylated recombinant glycoproteins. Biochim Biophys Acta 1474(3): 273-82.

Bush CA, Martin-Pastor M, Imberty A. (1999) Structure and conformation of complex carbohydrates of glycoproteins, glycolipids, and bacterial polysaccharides. Annu Rev Biophys Biomol Struct. 28: 269-93.

Crocker, P. R., Hartnell, A. , Munday, J. and Nath, D. (1997) The potential role of sialoadhesin as a macrophage recognition molecule in health and disease. Glycoconj. J. 14: 601-9.

Deisenhofer, J. (1981) Crystallographic refinement and atomic models of a human Fc fragment and its complex with fragment B of protein A from *Staphylococcus aureus* at 2.9- and 2.8Å resolution. Biochemistry 20: 2361-70.

Fagioli C, Sitia R. (2001) Glycoprotein quality control in the endoplasmic reticulum. Mannose trimming by endoplasmic reticulum mannosidase I times the proteasomal degradation of unassembled immunoglobulin subunits. J Biol Chem. 276: 12885-92.

Grabenhorst E, Schlenke P, Pohl S, Nimtz M, Conradt HS. (1999) Genetic engineering of recombinant glycoproteins and the glycosylation pathway in mammalian host cells. Glycoconjugate J. 16: 81-97.

Green, N.H., Hooker, A.D., James, D.C., Baines, A.J., Strange, P.G., Jenkins, N., and Bull, A.T. (1997). Control of interferon-gamma glycosylation by the addition of defined lipid supplements to batch cultures of recombinant chinese hamster ovary cells. In Animal Cell Technology, Basic & Applied Aspects, Vol 8. A. Funatsu, ed. (The Netherlands: Kluwer Academic Publishers), pp. 339-45.

Gu X, Wang DI (1998) Improvement of interferon-gamma sialylation in Chinese hamster ovary cell culture by feeding of N-acetylmannosamine. Biotechnol Bioeng 58(6): 642-8.

Hayter, P.M., Curling, E.M., Gould, M.L., Baines, A.J., Jenkins, N., Salmon, I., Strange, P.G., and Bull, A.T. (1993). The effect of dilution rate on CHO cell physiology and recombinant interferon-γ production in glucose-limited chemostat cultures. Biotechnol. Bioeng 42: 1077-85.

Hayter, P.M., Curling, E.M.A., Baines, A.J., Jenkins, N., Salmon, I., Strange,P.G. and Bull, A.T. (1991) Chinese hamster ovary cell growth and interferon production kinetics in stirred batch culture. Applied Micro Biotechnol 34: 559-64.

Hoffmann A., Nimtz, M., Wurster, U., and Conradt, H.S. (1994) Carbohydrate structures of β-trace protein from human cerebrospinal fluid- evidence for brain-type N-glycosylation. J. Neurochem. 63: 2185-96.

Hooker AD, Green NH, Baines AJ, Bull AT, Jenkins N, Strange PG, James DC (1999) Constraints on the transport and glycosylation of recombinant IFN-gamma in Chinese hamster ovary and insect cells. Biotechnol Bioeng 63: 559-72.

Jackson, P (1996) The analysis of fluorophore-labeled carbohydrates by polyacrylamide gel electrophoresis. Mol Biotechnol. 5: 101-23.

James, D. C., Goldman. M. H., Hoare, M., Jenkins, N., Oliver, R. W. A., Green, B. N. and Freedman, R. B. 1996. Post-translational processing of recombinant human interferon-gamma in animal expression systems. Protein Sci. 5: 331-40.

James, D.C., Freedman, R.B., Hoare, M., Ogonah, O.W., Rooney, B.C., Larionov, O.A., Dobrovolsky,V.N., Lagutin,O.V. and Jenkins,N. (1995). N-glycosylation of recombinant human interferon-γ produced in different animal expression systems. Bio/technology 13: 592-6.

Jassal R, Jenkins N, Charlwood J, Camilleri P, Jefferis R, Lund J. (2001) Sialylation of human IgG-Fc carbohydrate by transfected rat alpha2,6-sialyltransferase. Biochem Biophys Res Commun. 286: 243-9.

Jefferis, R., Lund, J., and Goodall, M. (1996) Modulation of FcγR and human complement activation by IgG3-core oligosaccharide interactions. Immunol. Lett. 54: 101-4.

Jefferis, R., Lund, J., and Pound, J.D. (1998) IgG-Fc mediated effector functions: molecular definition of interaction sites for effector ligands and the role of glycosylation. Immunol. Rev. 163: 59-76.

Jenkins, N. (1996). Role of physiology in the determination of protein heterogeneity. Curr. Opinion Biotechnol. 7: 205-8.

Jenkins, N., Parekh, R. B. and James, D. C. 1996. Getting the glycosylation right - implications for the biotechnology industry. Nature Biotechnology 14: 975-81

Kaufman RJ, Wasley LC, Spiliotes AJ, Gossels SD, Latt SA, Larsen GR, Kay RM. (1985) Coamplification and coexpression of human tissue-type plasminogen activator and murine dihydrofolate reductase sequences in Chinese hamster ovary cells. Mol Cell Biol. 5: 1750-9

Kemp, P. A., Jenkins, N., Clark, A. J. and Freedman, R. B. 1996. The glycosylation of human recombinant alpha-1-antitrypsin expressed in transgenic mice. Biochem. Soc. Trans. 24: 339

Kornfeld R, Kornfeld S. (1985) Assembly of asparagine-linked oligosaccharides. Annu Rev Biochem. 54: 631-64.

Lamotte, D., Eon-Duval, A., Acerbis, G., Distefano, G., Monaco, L., Soria, M., Jenkins, N., Engasser, J.M., and Marc, A. (1997). Controlling the glycosylation of recombinant proteins expressed in animal cells by genetic and physiological engineering. In Animal Cell Technology. M.J.T. Carrondo, ed. (Netherlands: Kluwer Academic Publishers), pp. 761-5.

Lee KB, Loganathan D, Merchant ZM, Linhardt RJ. (1990) Carbohydrate analysis of glycoproteins. A review. Appl Biochem Biotechnol. 23: 53-80.

Lee, E.U., Roth, J., and Paulsen, J.C. (1989) Alteration of terminal glycosylation sequences on N-linked oligosaccharides of chinese hamster ovary cells by expression of β-galactoside α2,6-sialyltransferase. J. Biol. Chem. 264: 13848-55.

Lipsky NG, Pagano RE. (1985) A vital stain for the Golgi apparatus. Science. 228: 745-7.

Lund, J., Takahashi, N., Popplewell, A., Goodall, M., Pound, J.D., Tyler, R., King D.J., and Jefferis, R. (2000) Expression and characterization of truncated forms of humanized L243 IgG1: Architectural features can influence synthesis of its oligosaccharide chains and affect superoxide production triggered through human Fcγ receptor I. Eur. J. Biochem. 267: 7246-56.

Lund, J., Takahashi, N., Pound, J.D., Goodall, M., and Jefferis, R. (1996) Multiple interactions of IgG with its core oligosaccharide can modulate recognition by complement and human Fcγ Receptor I and influence the synthesis of its oligosaccharide chains. J. Immunol. 157: 4963-9.

Lund, J., Takahashi, N., Pound, J.D., Nakagawa, H., Goodall, M., and Jefferis, R. (1995) Oligosaccharide-protein interactions in IgG can modulate recognition by Fcγ receptors. FASEB J. 9: 115-9.

Lund, J., Tanaka, T., Takahashi, N., Sarmay, G., Arata, Y., and Jefferis, R. (1990) A protein structural change in aglycosylated IgG3 correlates with loss of huFcγRI and huFcγRII binding and/or activation. Molec. Immunol. 27: 1145-53.

Manzi, A. E. and Varki, A. (1994) in Glycobiology, a practical approach (Fukuda, M. and Kobata, A. , ed.) pp. 34-7, IRL Press, Oxford, UK

Marks DL, Wu K, Paul P, Kamisaka Y, Watanabe R, Pagano RE. (1999) Oligomerization and topology of the Golgi membrane protein glucosylceramide synthase. J Biol Chem. 1999 Jan 1;274(1): 451-6.

Mattox, S., Walrath, K., Ceiler, D., Smith, D.F., and Cummings, R.D. (1992) A solid-phase assay for the activity of CMPNeuAc-Gal β1,4GlcNAc-R α2,6-sialyltransferase. Anal. Biochem. 206: 430-6.

Mimura, Y., Church, S., Ghirlando, R., Ashton, P.R., Dong, S., Goodall, M., Lund, J., and Jefferis, R. (2000) The influence of glycosylation on the thermal stability and effector function expression of human IgG1-Fc: properties of a series of human glycoforms. Molec. Immunol. 37: 697-706.

Monaco L, Tagliabue R, Soria MR, Uhlen M. (1994) An in vitro amplification approach for the expression of recombinant proteins in mammalian cells. Biotechnol Appl Biochem. 20: 157-71.

Monaco, L., Marc, A., Eon-Duval, A., Acerbis, G., Distefano, G., Lamotte, D., Engasser, J.-M., Soria, M., and Jenkins, N. (1996) Genetic engineering of α2,6-sialyltransferase in recombinant CHO cells and its effects on the sialylation of recombinant interferon-γ. Cytotechnology 22: 197-203.

Morgan, A., Jones, N. D. Nesbitt, A.M., Chaplin, L., Bodmer, M.W., and Emtage, J.S. (1995) The N-terminal end of the C$_H$2 domain of chimeric human IgG1 anti-HLA-DR is necessary for C1q, FcγRI and FcγRII binding. Immunology 86: 319-24.

Murray K, Dickson AJ. (1997) Dichloroacetate inhibits glutamine oxidation by decreasing pyruvate availability for transamination. Metabolism. 46: 268-472.

Ogonah, O.W., Freedman, R.B., Jenkins, N., Patel, K. and Rooney, B.C. 1996. Isolation and characterization of an insect-cell line able to perform complex N-linked glycosylation on recombinant proteins. Bio/technology 14: 197-202.

Regoeczi E, Kay JM, Chindemi PA, Zaimi O, Suyama KL.(1991) Transferrin glycosylation in hypoxia. Biochem Cell Biol. 69: 239-44.

Reuter, G., and Schauer, R. (1994) Determination of sialic acids. Meth. Enzymol. 230: 168-99.

Sarmay, G., Lund, J., Rozsnyay, Z., Gergely, J., Jefferis, R. (1992) Mapping and comparison of the interaction sites on the Fc region of IgG responsible for triggering antibody dependent cellular cytotoxicity (ADCC) through different types of human Fc receptor. Molec. Immunol. 29: 633-9.

Shields, R.L., Namenuk, A.K., Hong, K., Meng, Y.G., Rae, J., Briggs, J., Xie, D., Lai, J., Stadlen, A., Li, B., Fox, J.A., and Presta, L.G. (2000) High resolution mapping of the binding site on human IgG1 for FcγRI, FcγRII, FcγRIII and FcRn and design of IgG1 variants with improved binding to FcγR. J. Biol. Chem. 276: 6591-604.

Sondermann, P., Huber, R., Oosthuizen, V., and Jacob, U. (2000) A structural basis for immune complex recognition: the 3.2Å crystal structure of the human IgG1 Fc-fragment-FcγRIII complex. Nature 406 : 267-73.

Sowa Y, Sakai T. (2000) Butyrate as a model for "gene-regulating chemoprevention and chemotherapy.". Biofactors. 12: 283-7.

Sutton CW, O'Neill JA, Cottrell JS. (1994) Site-specific characterization of glycoprotein carbohydrates by exoglycosidase digestion and laser desorption mass spectrometry. Anal Biochem. 218: 34-46.

Tsuchiya, N., Endo, T., Matsuta, K., Yoshinoya, S., Aikawa, T., Kosuge, E., Takeuchi, F., Miyamoto, T., and Kobata, A. (1989) Effects of galactose depletion from oligosaccharide chains on immunological activities of human IgG. J. Rheumatol. 16: 285-90.

Umana P, Jean-Mairet J, Moudry R, Amstutz H, Bailey JE. (1999) Engineered glycoforms of an antineuroblastoma IgG1 with optimized antibody-dependent cellular cytotoxic activity. Nat Biotechnol. 17: 176-80.

Waldburger JM, Masternak K, Muhlethaler-Mottet A, Villard J, Peretti M, Landmann S, Reith W. (2000) Lessons from the bare lymphocyte syndrome: molecular mechanisms regulating MHC class II expression. Immunol Rev. 2000, 178: 148-65.

Weikert S, Papac D, Briggs J, Cowfer D, Tom S, Gawlitzek M, Lofgren J, Mehta S, Chisholm V, Modi N, Eppler S, Carroll K, Chamow S, Peers D, Berman P, Krummen L. (1999) Engineering Chinese hamster ovary cells to maximize sialic acid content of recombinant glycoproteins. Nat Biotechnol.;17, 1116-11121.

Yang M, Butler M. (2000) Effect of ammonia on the glycosylation of human recombinant erythropoietin in culture. Biotechnol Prog. 16: 751-9

6. TARGETING OF GENETICALLY ENGINEERED GLYCOSYLTRANSFERASES TO IN VIVO FUNCTIONAL GOLGI SUBCOMPARTMENTS OF MAMMALIAN CELLS

E. GRABENHORST, M. NIMTZ, and H.S. CONRADT

Protein Glycosylation, GBF - Gesellschaft für Biotechnologische Forschung mbH, Mascheroder Weg 1, D-38124 Braunschweig, Germany

1. Introduction

The importance of the posttranslational modification of polypeptides with N- or O-linked oligosaccharides is well documented by their implication in numerous biological phenomena (Varki, 1993). It has already been recognized early at the beginning of the area of recombinant DNA technology based expression of human therapeutics (Berman and Lasky, 1985) that only mammalian host cells meet the criteria for an appropriate biotechnological development of recombinant human glycoprotein therapeutics since in principle, they provide most of the posttranslational modifications that are necessary for their *in vivo* use in humans, *e.g.* the proper masking of Gal with NeuAc residues for prolonged circulation half-life or the absence of antigenic carbohydrate structural motifs which are produced in plants. The interest in the development of therapeutic agents by many biopharmaceutical companies has led to the attractive new research area of the biotechnology of mammalian cells as factories for medicinal glycoproteins. It must be envisaged that a major application field will also evolve from the interest in functional genomics where the biological phenomena associated with posttranslational modifications will contribute to our understanding of regulation.

Protein-linked oligosaccharides control the intracellular trafficking and tissue targeting of polypeptides, their half-life *in vivo* and their dynamic interaction with other proteins on the cell surface or in body fluids. Carbohydrate structures of glycoproteins are typically polypeptide-specific, and it has been shown that each individual glycosylation site of a glycoprotein may contain its own characteristic pattern of oligosaccharide chains. Apart from the three dimensional structure and the transport kinetics of a glycoprotein that govern its decoration with carbohydrate chains by a given cell, also the tissue or cell type that synthesizes a glycoprotein plays an important role in the known phenomenon of microheterogeneity of protein glycans (c.f. Figure 1) examples of structural polymorphism]. This microheterogeneity results from the regulated expression of a characteristic repertoire of glycosidases and glycosyltransferases which differs in the various cells / tissues of an organism and also between different species (see Figure 2). For example, human transferrin secreted from liver cells into the blood stream contains

M. Al-Rubeai.(ed.), Cell Engineering. 149-170.
© 2002 *Kluwer Academic Publishers. Printed in the Netherlands.*

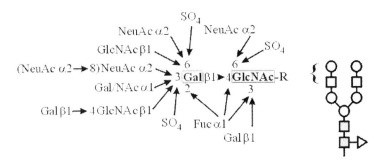

Figure 1: Examples of possible terminal modifications of the Galβ1-4GlcNAc-R motif as part of a complex-type biantennary N-glycan (scheme on the left). Taken into account that different modifications can be attached to each of the antenna, a tremendous oligosaccharides heterogeneity can be found on a single glycoprotein. Each of the terminal modifications is carried out at least by <u>one</u> special glycosyltransferase which potentially could compete with other cellular enzymes for the same acceptor substrate. It is evident, that highly complex regulation mechanism of these biosynthetic enzymes must be operating in the Golgi of cells.

serum-type oligosaccharides with mostly nonfucosylated diantennary chains masked with terminal α2,6-linked NeuAc. The same protein synthesized by brain tissue and isolated from human cerebrospinal fluid shows the presence of preponderantly *asialo* as well as *asialo-agalacto* diantennary forms with proximal fucose and with bisecting GlcNAc. The asialo-transferrin from cerebrospinal fluid exhibits the typical *brain-type* glycosylation characteristics of polypeptides synthesized by brain tissues (Hoffmann, et al., 1994; Hoffmann, et al., 1995; Pohl, et al., 1997).

In the field of the biotechnological production of recombinant glycoprotein therapeutics or recombinant retrovirus vectors as well as in areas aiming at the *ex vivo* expansion of human primary cells for medicinal treatment, one has to consider potential different interactions of differently glycosylated cell / virus surface glycoconjugates or of soluble glycotherapeutics with cellular receptors and a subsequent altered modulation of intracellular signalling cascades. During the past 15 years many publications have described the oligosaccharide structures of recombinant glycoproteins which have been expressed from different mammalian and non-mammalian expression systems, and there is an abundance of data concerning the glycosylation of a great number of glycoproteins from the most frequently used host cell lines. Presently, a great deal of efforts is going into investigations how to improve recombinant host cell lines, and here especially mammalian cells, for the manufacturing of therapeutic glycoproteins and therapeutic retroviral vectors with improved or novel *in vivo* properties. It seems promising to explore the advantages of new generations of therapeutics with respect to higher *in vivo* stability and, in addition, of new products with carbohydrate-based tissue-targetable addressing signals and beneficial biological properties. For this, the currently most frequently used host cell lines must be genetically engineered with newly introduced glycosyltransferase genes. The transferases must be stably directed into the appropriate subcellular compartment for their efficient function within the protein glycosylation pathway of the selected host cell line. To achieve this, we need to understand in more detail the regulatory mechanisms underlying the individual steps in protein synthesis and posttranslational modification reactions of polypeptides localizing in the transport compartments of animal cells.

2. Considerations for the Expression of Recombinant Human Glycoproteins from Mammalian Host Cell Lines

Recombinant mammalian cells are now used for the production of many human glyco-protein therapeutics and are cultured in large bioreactor systems by the pharmaceutical industry. In most of these biotechnological processes Chinese hamster ovary cells (CHO) and baby hamster kidney cells (BHK-21) are being used as hosts, and most of our present knowledge about the culture conditions that can affect the fine structural characteristics of recombinant glycoproteins produced in bioprocesses has been obtained from studies with these two hosts cells (Conradt, et al., 1990; Goochee and Monica, 1990; Goochee, et al., 1991; Jenkins, et al., 1996). In several cases (see Table 1), the recombinantly manufactured glycoprotein therapeutics have terminal carbohydrate motifs that are different from those found on their natural counterparts. Furthermore, some carbohydrate structural motifs might be present in a product that are not ideal for the intended therapeutic use in humans. Newly discovered biological / medicinal aspects of certain carbohydrate motifs on glycoproteins will lead to investigate alternative expression cells or genetically improved host cell lines as also discussed elsewhere in this volume (see chapter by Krummen et al.).

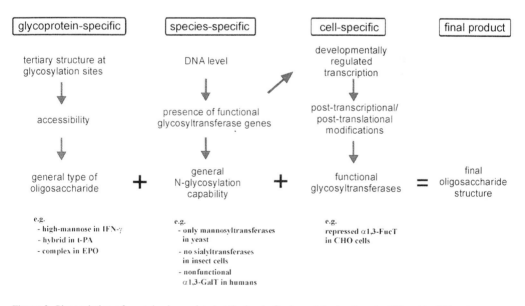

Figure 2. Glycosylation of proteins is regulated at the level of polypeptide structure, is different in different organisms and may vary within the different cell types of the same organism.

Different secretory *N*-glycoproteins synthesized by a cell presumably must pass the same members of the oligosaccharide processing machinery *en route* through the Golgi but they may acquire a different *polypeptide-specific* modification. For example, human erythropoietin is decorated with preponderantly tetraantennary oligosaccharides with one, two or three N-acetyllactosamine repeats (Nimtz, et al., 1993), human antithrombin III is modified by a mixture of di- and triantennary glycans and some tetraantennary chains (Björk, et al., 1992) whereas human β-TP contains only highly sialylated diantennary N-acetyllactosamine-type structures (Grabenhorst, et al., 1995; Grabenhorst, et al., 1998) when expressed from BHK-21 or CHO cells (see Figure 3). This indicates that for a given host cell the final oligosaccharide structures present on a glycoprotein are largely determined by the polypeptide itself, and thus, the glycosylation of a product reflects the accessibility of the catalytic domains of the Golgi membrane-anchored glycosyltransfer-ases to the glycosylation domain of the protein (Grabenhorst, Schlenke, et al., 1998). However, since each individual cell line selected for expression of a glycoprotein contributes with its own specific glycosylation capacity to the final oligosaccharide structure of the polypeptide, the heterogeneity of the same glycoprotein obtained from two different host cell lines may differ in several aspects and the product may have different biological and hence pharmaceutical properties.

We have found that a constitutively secreted glycoprotein expressed at a level of 0.1 μg/ml from BHK-21 or CHO cells has the same carbohydrate structure as has the protein that is expressed from a clone at a 200-fold higher level. As a general rule, it appears that the glycosylation machinery of the expression host cell is not a critical factor for

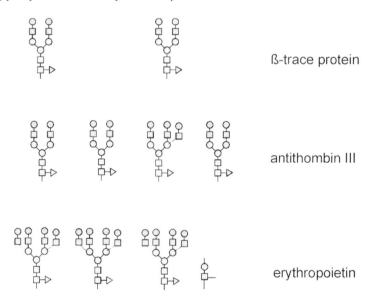

ß-trace protein

antithombin III

erythropoietin

Figure 3. Schematic presentation of the different type of oligosaccharides of three different glycoproteins expressed from the same BHK-21 or CHO host cell line: the branching (di-, tri-, or tetraantenary) of oligosaccharides is *polypeptide specific.*

efficient posttranslational modification of a polypeptide with carbohydrates. However, it can not be excluded that for proteins exhibiting abnormal intracellular trafficking behaviour unexpected glycosylation patterns could be encountered. It should be noted that in rare cases the selection procedures used for the isolation of single high expression cell clones may lead to a variant cell clone with unusual glycosylation characteristics, as has been recognized in our laboratory in a few cases with cells lines where e.g. complete loss of complex-type glycosylation properties was observed.

It has been recognized from the work published over the past 15 years, that CHO and BHK-21 cells basically show a very similar glycosylation characteristics for recombinantly expressed glycoproteins (Grabenhorst, Schlenke, et al., 1998). Table 1 summarizes the main carbohydrate structural features of recombinant glycoproteins expressed from murine and from hamster cell lines. The type of branching and the N-acetyllactosamine content of N-linked glycans on the same recombinant glycoprotein expressed from CHO, BHK-21 or from the murine cell lines will be very similar and this can also be expected for the oligosaccharide patterns at individual glycosylation sites. However, there will be a significantly higher microheterogeneity of outer antennae carbohydrate motifs in the recombinant products formed by mouse host cells (e.g. α2,3- as well as α2,6-linked

Table 1. Major carbohydrate motifs synthesized by different host cell lines with various recombinant glycoproteins.

carbohydrate structure	host cell line			
	CHO	**BHK-21**	**C127**	**Ltk⁻**
proximal α1,6-linked fucose	+	+	+	+
terminal motifs				
Fuc(α1-2)Gal-R[1]	+	+	?	?
α2,6-NeuAc	-	-	+	+
α2,3-NeuAc	+	+	+	+
NeuAc(α2-8)NeuAc(α2-3)R	+	+	-	+
NeuGly[1]	+	+/-	+	+
Gal(β1-4)GlcNAc repeats	+	+	+	+
Gal(β1-3)GlcNAc-R	-	+	-	-
sulfated glycans[2]	+	+	+	+
Gal(α1-3)Gal	+	-	+	+
GalNAc(β1-4)GlcNAc	-	+[3]	-	-
antennarity				
triantennary	+	+	+	+
tetraantennary	+	+	+	+
branched repeat structure	?	-	+	+
mannose 6-phosphate[4]	+	+	?	?
bisecting GlcNAc	-	-	+	+

[1]detectable in trace amounts; may increase up to 25% of all structures
[2]sulfate linked to Gal or GlcNAc
[3]detected in large amounts in the BHK-21A variant cell line [14-15]
[4]in some cell clones after selection (Man$_{5-6}$GlcNAc$_2$)

NeuAc, different NeuGly content, presence of larger amounts of Gal(α1-3)Gal-R structures, Gal(β1-3)GlcNAc-R, the amount and type of sulfated structures as well as branched antennae in murine C127 cells and Ltk^- cells). Therefore, the hamster cell lines provide a more favourable expression host cell system when aiming at a lower glycosylation heterogeneity of the recombinant product.

In most of the publications dealing with carbohydrate structural analysis of recombinant glycoproteins, the results have been obtained with highly purified glycoprotein preparations manufactured for use in patients. Therefore, these products represent only a minor fraction of the total product synthesized by the pertinent cell line and are highly enriched in selected glycoforms with the preferred high therapeutic efficacy. *E.g.*, in the case of recombinant human erythropoietin from CHO and BHK cells, only the highly sialylated isoforms of the recombinant product is manufactured for medical use representing only about 20-25% of the total erythropoietin originally synthesized by the host cell line. However, for a detailed knowledge of the glycosylation characteristics of an expression host cell, a quantitative purification the secretion product, *e.g.* by immunoaffinity isolation applying specific antibodies, is indispensible.

3. Genetic Engineering of Host Cell Lines with Novel Glycosylation Properties

As is clear from Table 1, several biological important glycosylation motifs found on human proteins cannot be synthesized by CHO or BHK cells. Amongst them are the masking of Gal(β1-4)GlcNAc-R antennae with α2,6-linked N-acetylneuraminic acid or the modification of Gal(β1-4)GlcNAc-R with α1,3/4-linked fucose (resulting in Lewis[x,a] or sialyl Lewis[x,a] epitopes). This can be accomplished by introducing the appropriate foreign glycosyltransferase genes into the parental cell. It can be envisaged that any introduction of a new (preferably human) glycosyltransferase into the biosynthetic modification pathway carries the potential risk of profound overall alterations of the final carbohydrate structures of a glycoprotein by uncontrolled competition with endogenous enzymes for the same acceptor substrate. Furthermore, several distinct carbohydrate structural motifs can be synthesized by more than one glycosyltransferase (see Table 2 for overview of sialyltransferases and fucosyltransferases), and it is advantageous to identify the appropriate enzyme for the intended modification. Another complication arises from our only incomplete knowledge about the *in vivo* specificity of many glycosyltransferases since their substrate activities have been determined by *in vitro* assays using small nonphysiological acceptor substrates.

4. Assessment of the *in vivo* Specificity of Glycosyltransferases by Coexpression Together with a Reporter Glycoprotein

A final description and comparison of the *in vivo* specificity of the individual glycosyltransferases can only be achieved by structural analysis of the product(s) that have passed along the biosynthetic and transport pathway of a cell. As schematically

Table 2. Description and properties of human sialyltransferases and fucosyltransferases.

Enzyme		cDNA		gen. organization		protein		in vivo specificity (acceptor underlined)
		GI acc. no.	transcripts [kb]	exons in ORF	gene locus	N-terminus	size [aa]	
Sialyltransferases (ST)								
ST3Gal I	α2,3-ST	3169563		>1	8q24.3	MVTLRKRTL..	340	Galβ1-3GalNAc-O-protein (glycolipids?)
ST3Gal II	α2,3-ST	13648950		>1	16	MKCSLRVWF..	350	Galβ1-3GalNAc-O-protein (glycolipids?)
ST3Gal III	α2,3-ST	388014		>1	1p34-33	MGLLVFVRN..	375	Galβ1-3GlcNAc-R / Galβ1-4GlcNAc-R
ST3Gal IV	α2,3-ST	414890		9-10	11q23-24	MVSKSRWKL..	329	Galβ1-3GalNAc-R / Galβ1-4GlcNAc-R
ST3Gal V	α2,3-ST	4262429			2q11.2	MRRPSLLLK..	362	Galβ1-4Glc-Cer
ST3Gal VI	α2,3-ST	4827246	3.0/1.8		3	MRGYLVAIF..	331	Galβ1-4GlcNAc-R
ST6Gal I	α2,6-ST	13648462			3q27-28	MIHTNLKKK..	406	Galβ1-4GlcNAc-R / GalNAcβ1-4GlcNAc-R
ST6GalNAcT 1-6	α2,6-ST	6 enzymes, not further specified here						(Siaα2-3)Galβ1-3GalNAc-O-protein
ST8Sia I	α2,8-ST	13651660		>1	12p12.1-11.2	MAVLAWKFP..	341	Siaα2-3Galβ1-4Glc-Cer
ST8Sia II	α2,8-ST	5174676		>1	15q26	MQLQFRSWM..	375	Siaα2-3R / Siaα2-8Siaα2-3R (certain glycoproteins)
ST8Sia III	α2,8-ST	6798484		>1	18q21	MRNCKMARV..	380	Siaα2-3R / Siaα2-8Siaα2-3R (certain glycoproteins)
ST8Sia IV	α2,8-ST	11416842		>1	5q21	MRSIRKRWT..	359	Siaα2-3R / Siaα2-8Siaα2-3R (certain glycoproteins)
ST8Sia V	α2,8-ST	9558746		>1	18	MRYADPSPN..	376	Sia in gangliosides
Fucosyltransferases (FucT)								
FucT-I	α1,2-FucT	4503804		1	19q13.3	MWLRSHRQL..	365	Galβ1-4GlcNAc-R / Galβ1-3GlcNAc-R
FucT-II	α1,2-FucT	4503806	3.4	1	19q13.3	MSSLLPTAV..	346	Galβ1-3GlcNAc-R / Galβ1-4GlcNAc-R
FucT-III	α1,3/4-FucT	4503808	2.7/2.4	1	19p13.3	MDPLGAAKP..	361	[(Siaα2-3)?]Galβ1-3GlcNAc-R / Siaα2-3Galβ1-4GlcNAc-R
FucT-IV	α1,3-FucT	182070	6.0/3.0/2.3	1 (?)	11q21	MRRLWGAAR..	530	Galβ1-4GlcNAc-R
FucT-V	α1,3/4-FucT	4503810		1	19p13.3	MDPLGPAKP..	374	Siaα2-3Galβ1-4GlcNAc-R (/ Siaα2-3Galβ1-3GlcNAc-R ?)
FucT-VI	α1,3-FucT	4503814	3.5/2.5	1	19p13.3	MDPLGPAKP..	359	Siaα2-3Galβ1-4GlcNAc-R / Galβ1-4GlcNAc-R
FucT-VII	α1,3-FucT	4758405	3.1/2.3/2.0	1	9q34.3	MNNAGHGPT..	342	Siaα2-3Galβ1-4GlcNAc-R
FucT-VIII	α1,6-FucT	4758407		>1	14q24.3	MRPWTGSWR..	575	R-GlcNAc-N-protein (= proximal Fuc)
FucT-IX	α1,3-FucT	5729831	12/3.0/2.5	1	6q16	MTSTSKGIL..	359	Galβ1-4GlcNAc-R (in vitro)

shown in Figure 4, we suggest the recombinant expression of the full length form of human glycosyltransferases along with a suitable reporter glycoprotein(s) at a constant expression level in a heterologous mammalian host cell line that is devoid of the pertinent enzyme activity. This approach is considered to serve as a valuable model providing the information on the *in vivo* specificities of different members of a glycosyltransferase family (Grabenhorst, et al., 1995; Grabenhorst, et al., 1998; Schlenke, et al., 1998; Costa, et al., 1997) and should allow for the identification of the favourable enzyme to be used in glycosylation engineering of a host cell line for the subsequent production of a glycoprotein with defined altered glycosylation characteristics. Basic information can be expected from such studies concerning the intracellular organization of the protein glycosylation machinery and the temporal and spatial distribution of the transferases in the *in vivo* biosynthetic compartments. Prerequisites for such an *in vivo* assay system are: (1) reproducible transfection methods applying high expression vectors and efficient selection as well as isolation of stable clones, (2) providing a defined level of an acceptor substrate along the biosynthetic pathway in form of a reporter glycoprotein, (3) achieve a reproducible expression level of the newly introduced glycosyltransferases, (4) establish a suitable rapid and quantitative isolation method for the cellular product, (5) establish an efficient and detailed quantitative oligosaccharide micro-analytical technique. Using a glycoprotein with defined structural characteristics, the above approach of analyzing the oligosaccharide chains of a reporter polypeptide from stable transfected cells should provide detailed information about the *in vivo* substrate specificity of the newly introduced member of glycosyltransferase family acting on the same precursor substrate. Transient expression experiments are of limited value, since cell damage and cell leakage resulting from the transfection procedures is considered to lead to artefacts. As an example, the results that were obtained with the analysis of the cellular *in vivo* specificity of human α1,3/4- fucosyltransferases expressed in BHK-21 cells will be described in the following chapter.

5. The *in vivo* Specificity of Human α1,3/4-Fucosyltransferases Expressed in Transfected BHK-21 Cells Coexpressing Reporter Glycoproteins

Within the family of human fucosyltransferases [see Table 2] the α1,3/4-fucosyltransferases III-VII are characterized by adding fucose to the GlcNAc residue in either sialylated or unsialylated Gal(β1-3)GlcNAc-R (type I structures) or Gal(β1-4)Glc NAc-R (type II structures) antennae of glycoproteins. The enzymes participate in the regulation of the synthesis of the Lewis[x] (Lex) and sialyl Lewis[x] (SLex) type ligands that are involved in the initiation of recruitment of leukocytes during inflammatory processes mediated by the selectin family of cell surface receptors (Varki, 1994; Lasky, 1995; McEver, et al., 1995; Kansas, 1996). Fucosylated glycoconjugates have also been reported to play a central role in other important biological phenomena, *e.g.* differentiation and tumorigenesis. Significantly increased levels in human serum glycoproteins with peripheral fucose have been well documented in inflammatory states in patients.

Several human tissues and cells express more than one fucosyltransferase at the same time and it has been difficult to obtain pure enzyme preparations from natural sources

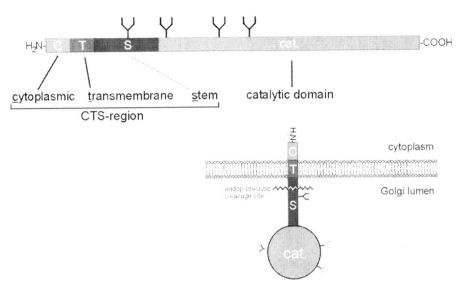

Figure 4. Schematic drawing illustrating the polypeptide domain structure of mammalian glycosyltransferases; as depicted the N-terminal cytoplasmic region is attached to the transmembrane region which spans the Golgi membrane and is believed to prevent the movement of the enzymes to the plasma membrane. The stem region links the catalytic domain facing the Golgi lumen; the stem region of several glycosyltransferases is susceptible to endoproteolytic attack and thus may presumably involved also in the intracellular turnover of some transferases.

for the unequivocal assessment of the specificity of individual transferases. The cloning of the α1,3/4-fucosyltransferases III-VII and their expression in recombinant form has provided a tool for isolating substantial amounts of the individual pure enzymes for studying their substrate specificity *in vitro* separately. However, from the *in vitro* data, the involvement of each individual fucosyltransferase in the generation of the different selectin ligands was not be fully assessed.

All mammalian glycosyltransferases share a common polypeptide domain structure as type II transmembrane Golgi proteins [see Figure 4]. According to current opinion, their *transmembrane region* is responsible for the retention of the enzymes in the proper Golgi compartment. Several transferases contain *N*-glycosylation sites in their stem region and/or their catalytic domain; however, no information is available if, or to what extent, the modification with *N*-glycans is involved in their *in vivo* activity or specificity.

A number of publications have appeared that describe the recombinant expression of human fucosyltransferases (de Vries, et al., 1995; de Vries, et al., 1997; Shinoda, et al., 1997; Britten, et al., 1998; Natsuka, et al., 1994) mostly as soluble forms lacking the cytoplasmic, the transmembrane and some part of the stem region. In several cases, recombinant chimeras containing N-terminally fused polypeptide fragments (*e.g.* of protein A) have been constructed to facilitate recombinant enzyme purification and it cannot be ruled out that the substrate specificity of these forms may differ from the natural transferases.

Coexpression of β-TP as a reporter glycoprotein from cells transfected with a human α1,3/4-fucosyltransferase therefore should yield oligosaccharides with the Lewis[X] or the sialyl Lewis[X] motifs or mixtures of the two motifs depending on the *in vivo* specificity of the transfected fucosyltransferase. Human β-TP is a 168 amino acid protein which contains two N-glycosylation sites that are occupied with almost exclusively diantennary complex-type chains (Hoffmann, et al., 1994; Grabenhorst, et al., 1995; Grabenhorst, et al., 1998). Similar to human transferrin described before, β-TP isolated from human cerebrospinal fluid exhibits "brain-type" glycosylation character-istics, *i.e.*, mainly truncated asialo chains, bisecting GlcNAc, complete proximal and some peripheral fucosylation besides small amounts of α2,3/6-sialylated N-glycans (Hoffmann, et al., 1994; Pohl, et al., 1997). Recombinant human β-TP expressed from wild-type BHK-21 cells is also modified with almost exclusively diantennary oligosaccharides at each of its two N-glycosylation sites, however, as shown in Figure 3, the oligosaccharide pattern here is very homo-genous, the structures contain either two or one α2,3-linked NeuAc and only small amounts of asialo chains are present (Grabenhorst, et al., 1995; Grabenhorst, et al., 1998). In total, 12 different diantennary N-linked oligosaccharides can be expected in β-TP from BHK-21 cells expressing human FucT-VI, four each of asialo, mono- and disialo chains with no, one or two peripheral fucose residues, respectively.

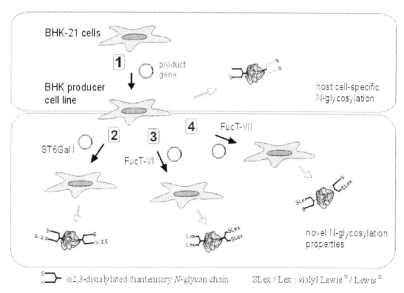

⊃— α2,3-disialylated diantennary N-glycan chain   SLex / Lex : sialyl Lewis[X] / Lewis[X]

Figure 5. Scheme illustrating the glycosylation engineering of cells and in-vivo specificity analysis of glycosyltransferases: Step **1** = generation of the reporter glycoprotein expressing host cell line (highly sialylated diantennary oligosaccharides with α2-3-linked NeuAc; which subsequently serves as the host for subsequent transfection with selected glycosyltransferase genes; **2** = coexpression of human α2-6 sialyltransferase results in detection of α2-6-linked NeuAc preponderantly at the Man-3 branch indicating the substrate specificity of the ST; **3** = coexpression of FucT-VI results in a mixture of oligosaccharides terminating with Lex and sLex; **4** = coexpression of human FucT-VII results in detection of exclusively the sLex motif attached to the reporter glycoprotein.

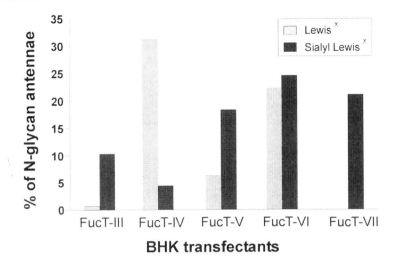

Figure 6. In-vivo specificity of human fucosyltransferases III – VII as determined by coexpression of the enzymes together with human β-trace protein from stable BHK-21 cells [13].

Using the approach shown in Figure 5, our laboratory has investigated in detail the *in vivo* biosynthetic activity of the human α1,3/4-fucosyltransferases III-VII in BHK-21 cells (Grabenhorst, et al., 1998). We found that each human α1,3/4-fucosyltransferase is characterized *in vivo* by the synthesis of an individual ratio of SLex : Lex with FucT-VII forming exclusively SLex and FucT-IV preponderantly (90%) Lex, whereas FucT-VI expression results in a 1.1 : 1 mixture of SLex and Lex motifs in the oligosaccharides of the coexpressed re-porter glycoprotein β-TP. The *in vitro* specificity data of the enzymes clearly support the exclusive SLex-forming specificity of the FucT-VII catalytic domain and the Lex-forming specificity of the FucT-IV catalytic domain [c.f. Figure 6]. Consequently, in order to get access to its Gal(β1-4)GlcNAc-R substrate, FucT-IV should be localized in a cellular subcompartment before α2,3-sialylation occurs, since, according to all data available, the human α2,3-sialyltransferase ST3Gal III does not transfer NeuAc to Lex (compare Figure. 7). Likewise, FucT-VII action strictly depends on the proper supply with α2,3-sialylated acceptors. FucT-VII should either colocalize with ST3Gal III in the same functional area, or, more preferable, should reside in a later subcompartment than ST3Gal III in order to get access to highly sialylated acceptor substrate concentrations. The human FucT-VI catalytic domain recognizes both, *in vivo* and *in vitro*, the sialylated or unsialylated acceptor motifs with a high efficiency (Grabenhorst, et al., 1998; Nimtz, et al., 1998), therefore, this enzyme should either colocalize with ST3Gal III resulting in competition for the common Gal(β1-4)GlcNAc-R substrate, or should have a broader subcompartmental distribution. We have also shown that a variant of FucT-VI constructed by replacement of its cytoplasmic, transmembrane and stem (CTS)-region with the signal peptide sequence of human interleukin-2 is efficiently secreted from cells but does *not show in vivo* function-al activity when expressed at a total activity level comparable to *wt*-FT6 cells (Grabenhorst, et al., 1998).

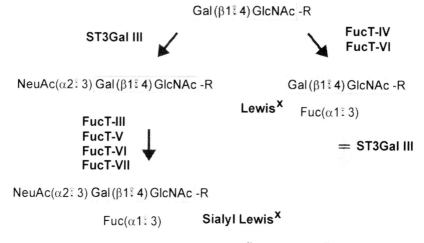

Figure 7. Scheme describing the biosynthesis of Lewis[X] and sialyl Lewis[X] determinants on glycoconjugates based on the *in vivo* specificity of human FucT-III-VII towards N-glycoproteins based on previously reported work [13]. In agreement with current concepts, α2,3-sialylation by ST3Gal III does not occur with α1,3-fucosylated N-acetyllactosamine antennae.

6. Targeting of Glycosyltransferases to *in vivo* Golgi Subcompartments by Genetic Engineering of their CTS-Region (Cytoplasmic, Transmembrane and Stem Region)

According to the current models, the members of the glycosyltransferase family of enzymes are thought to be arranged in a sequential manner within the individual Golgi subcompartments. However, the control mechanisms that underly this sequential distribution of glycosyltransferases into different Golgi subcompartments are not fully understood. Some key Golgi enzymes like α-mannosidase II and GnT-I have been localized in the medial- and *trans*-Golgi whereas several terminal glycosyltransferases (GalT-I, ST6Gal I, ST3Gal III, FucT-V, FucT-VI) have been localized in the *trans*-Golgi or trans Golgi network (TGN). There is some evidence provided in the published papers that most glycosyltransferases appear to have an overlapping distribution into more than only one morphological defined subcompartment. For example GnT-I has been localized to the medial <u>and</u> *trans*-Golgi whereas GalT-I and ST6Gal I localize to the *trans*-Golgi as well as to the TGN (Nilsson, et al., 1993; Rabouille, et al., 1995). The transmembrane region (compare Figure 4) as well as the flanking domains of type II Golgi resident glycosyltransferases have been identified to maintain Golgi retention (Colley, 1997). The *bilayer thickness model* for Golgi retention of glycosyltransferases (Bretscher and Munro, 1993; Munro, 1995) postulates that the length of the transmembrane region of transferases mediates Golgi retention. A second hypothesis (Nilsson, et al., 1993; Nilsson, Slusarewicz, and Warren, 1993) proposes a disulfide-linked *homo-/hetero-oligomerization* of the enzymes to function as a Golgi retention signal by preventing the large complexes to be delivered to secretory vesicles and ongoing transport to the plasma membrane. The two models do not provide sufficient information about the mechanisms that control the *in vivo* functional organization of the different members of the glycosyltransferase families and how their sequential arrangement within different subcompartments might be accomplished. Immunochemical localization techniques lack the sensitivity to resolve in detail the distribution of, *e.g.*, the many late acting Golgi glycosyltransferases in the functional network of the *trans*-Golgi / TGN (Gleeson, 1998). A further complication with immunodetection of enzymes in defined cellular subcompartments results from the migration of the newly synthesized membrane-bound glycosyltransferases from the endoplasmic reticulum through the compartments of the secretory pathway until they arrive at their final destination in individual functional Golgi stacks. It has been shown that glycosyltransferases them-selves undergo complex-type N-glycosylation including terminal sialylation (Grabenhorst, et al., 1998; Costa, et al., 1997; Bosshart and Berger, 1992). In addition, there are several reports describing different levels of intracellular proteolytically cleaved forms of certain glycosyltransferases that might change under different physiological conditions in different cells (Strous and Berger, 1982; Weinstein, et al., 1987; Paulson and Colley, 1989) (Johnson, Donald, and Watkins, 1993.

We have shown previously that a variant of FucT-VI constructed by replacement of its CTS-region with the signal peptide sequence of human interleukin-2 is efficiently secreted from cells but does not show *in vivo* functional activity when expressed at a total activity level comparable to *wt*-FT6 cells (Grabenhorst, et al., 1998). This result as well as reports by other groups suggesting the transmembrane and flanking regions of several glycosyl-transferases (ST6Gal I, β4GalT1, GnT-I and α1,2-fucosyltransferase) as playing an important role in their Golgi retention (see references 28-40 in Grabenhorst and Conradt, 1999) prompted us to investigate the properties of glycosyltransferase CTS-regions in the *in vivo* functional targeting of the human FucT-VI catalytic domain to different biosynthetically active Golgi subcompartments as described below.

If localizing to *early* Golgi compartments, the FucT-VI catalytic domain would be expected to encounter low levels of sialylated *N*-glycans, and in agreement with its previous characterized specificities would preferentially transfer fucose residues to Gal(β1-4)GlcNAc-R resulting in increased Lewis^x synthesis in the secreted product.

Likewise, its targeting to a *later* compartment would be expected to lead to preferential formation of sialyl Lewis^x motifs by the enzyme from the availibility of

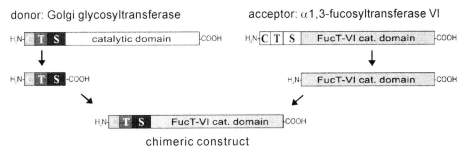

Figure 8. Construction of chimeric glycosyltransferases by fusion of CTS-regions from different donor glycosyltransferases and the catalytic domain of human fucosyltransferase VI.

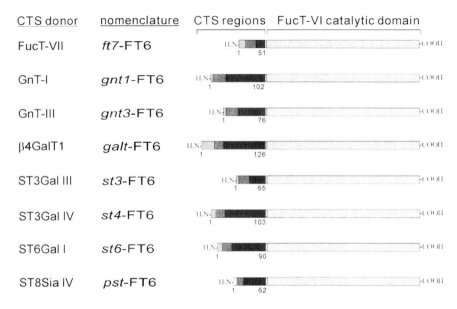

Figure 9. Genetic engineering of human FucT-VI chimeras by fusing the CTS-regions of different glycosyltransferases to the catalytic domain of FucT-VI. The numbering of the CTS amino acid sequence is from the N-terminus of the CTS donor transferase; the nomenclature for chimeric constructs is used in the text.

already α2,3-sialylated precursor substrates. Some examples of genetically engineered chimeras with the human FucT-VI catalytic domain and the CTS-regions of different human glycosyltransferases are depicted in Figure 9.

7. *In Vivo* Specificity of FucT-VI Chimeras Expressed in BHK-21 Cells Together with β-Trace Protein

The design of the FucT-VI CTS-variants is described in detail in (Grabenhorst and Conradt, 1999). Based on *an in vitro* assay using cellular extracts, almost identical enzyme activities were detected in all stable transfected cell lines. FucT-VI catalytic activity was detected in all cell supernatants but not in the medium of cells transfected with gnt3-FT6, ft7-FT6 and st4-FT6. For detemination of the *in vivo* specificity of the chimeras, the β-trace protein was isolated from each cell supernatant and was subjected to glycosylation analysis including MALDI/TOF-MS of oligosaccharides before or after reduction and permethylation as well as by HPAEC-PAD mapping before and after removal of NeuAc from glycans (Grabenhorst, et al., 1998; Grabenhorst and Conradt, 1999). The results of the analysis are depicted in Figure 10.

The increasing amount of Lewis[x] glycans formed by *gnt1*-FT6 is paralleled by a concomitant decrease in the overall sialylation of β-TP, since the BHK cell endogenous ST3 Gal III does not act on Lewis[x] structures (Grabenhorst, et al., 1998). Therefore, the substrate specificity of *wt*-FT6 towards both, neutral and α2,3-sialylated oligosaccharides,

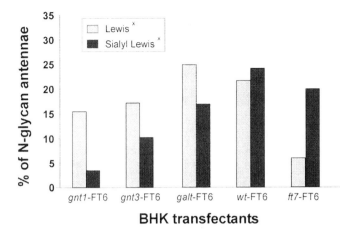

BHK transfectants

Figure 10. Ratio of Lewis[X] and sialyl Lewis[X] motifs in β-TP oligosaccharides expressed from cell stable lines transfected with various glycosyltransferase chimeras. The results obtained indicate that the ratio of Lex/SLex motifs increases when CTS regions from early acting donor glycosyltransferases (e.g. GnT-transferases) is fused to the FucT-VI (FT6) catalytic domain whereas an decrease of Lex/SLex is observed when the CTS-region from a late acting transferase is added. This then is in agreement with the view that backward targeting (towards ER compartment) as well as a forward targeting of the FT6 catalytic domain can be achieved.

appears to be an intrinsic property of its catalytic domain, since an enzyme form lacking the CTS-region also acts on both substrates *in vitro* (Grabenhorst, et al., 1998; Nimtz, et al., 1998). From our results obtained with the *gnt1*-FT6 variant, we conclude that the GnT-I CTS-region prevents the forward movement of the enzyme to the late Golgi subcompartment where *wt*-FT6 exerts its biosynthetic activity and, therefore, *gnt1*-FT6 has access to only limited amounts of the Gal(β1-4)GlcNAc-R substrate. This result is corroborated by our observation that, with a *gnt1*-FT6 cell line expressing 1/10 of the *wt*-FT6 activity, no peripheral fucosylation of β-TP oligosaccharides was observed. This *in vivo* activity for *gnt1*-FT6 then is in aggreement with the subcompartmental localization reported for GnT-I and β4GalT1. Two thirds of GnT-I have been immunolocalized in the medial-Golgi of HeLa cells and 1/3 of the enzyme co-immunolocalize in the *trans*-Golgi together with 50% of the cellular β4GalT1 in HeLa cells (Nilsson, et al., 1993; Rabouille, et al., 1995). In context with discussions of the kin recognition hypothesis, human GnT-I has been shown to oligomerize with α-mannosidase II by interactions mediated by their stalk regions (Nilsson, et al., 1996) which retain the two proteins in the same two Golgi compartments. Since our *gnt1*-FT6 variant comprises the human GnT-I stalk region (aa 1-102), our *in vivo* specificity data could in fact be explained by such an interaction of this chimera with cellular α-mannosidase II.

The *ft7*-FT6 variant, like wild-type FucT-VII described earlier (Grabenhorst, et al., 1998), was found to be completely resistant towards intracellular proteolysis as is the case for the *gnT3*-FT6 chimera, whereas large amounts of FucT-VI activity were measured in the medium of most of the other stably transfected cell lines including those where a sensitivity to intra-cellular endoproteolysis has already been described for the wild-type enzymes (ST6Gal I, β4GalT1 and FucT-III). From the detection of intracellular proteolysis of *st3*-FT6 and *pst*-FT6 we speculate that wild-type sialyltransferases are also susceptible to cleavage, although no data have been reported so far in the literature.

Apparently, there are at least three important signals contained in the CTS-regions of glycosyltransferases mediating [see also Figure 11]:

(i)   *their Golgi retention*

(ii)   *their targeting to specific in vivo functional areas*

(iii)   *the susceptibility of the enzymes towards intracellular endoproteolytic cleavage*

The later characteristics of some glycosyltransferases might constitute a intrinsic property of the Golgi enzymes for regulation of their intracellular switch-off, *e.g.* in cellular differentiation processes where new activities might enter and be involved in important biological phenomena.

Our results suggest that the *in vivo* functional distribution of glycosyltransferases can be mapped by the stable expression of CTS-variants and the analysis of their biosynthetic products. The results obtained for the *ft7*-FT6 variant indicate a forward targeting of this chimera into a functional Golgi compartment later than *wt*-FT6 where the *ft7*-FT6 catalytic domain has access to higher levels of already α2,3-sialylated precursor *N*-glycans. In the case of *wt*-FT6, the earlier α1,3-fucosylation of β-TP asialo oligosaccharides by the FucT-VI catalytic domain leads to decreased overall α2,3-sialylation of the reporter glycoprotein. In contrast, β-TP from *ft7*-FT6 cells used here exhibits a higher overall degree of sialylation, similar to the value detected in β-TP secreted from wild-type BHK-21 cells that do not coexpress fucosyltransferase(s) (Grabenhorst, et al., 1995; Grabenhorst, et al., 1998).

In the biosynthesis of fucosylated ligands for the selectin family of carbohydrate receptors, the different cellular sialyl- and fucosyltransferase activities play a crucial role since, *e.g.*, α1-3-fucosylation of Gal(β1-4)GlcNAc-R motifs prevents subsequent sialylation (and thus formation of sialyl Lewis^X) by ST3Gal III, and α2,6-sialylated oligosaccharides are not a substrate for α1-3/4-fucosyltransferases (Grabenhorst, et al., 1998). The expression of different sialyl- and fucosyltransferases in the same cell could lead to a competition for common acceptor substrates, therefore it was of interest to identify possible different targeting signal properties of sialyltransferase CTS-regions compared to those of fucosyltransferases. Our results reported previously (Grabenhorst and Conradt, 1999) for the stably expressed *st6*-FT6 would point to an earlier *in vivo* functional localization of this variant compared to *wt*-FT6. The high proportion of Lewis^x already detectable in the *st3*-FT6 product also points to a retention in an earlier compartment of this chimera when compared to *wt*-FT6. However, by comparison of the functional activity of FucT-VI CTS-variants expressed at different enzyme levels, it is clear that in principal increasing amounts of the enzyme lead to increased synthesis of Lewis^x oligosaccharides, most likely due to an increased concentration of active enzyme forms already during their transport within early subcompartments of the biosynthetic pathway. Our results perfectly agree with the suggested functional colocalization and *in vivo* competition of the two enzymes for their common substrate Gal(β1-4)GlcNAc-R (Grabenhorst, et al., 1995).

The *pst*-FT6 variant produced a SLex : Lex ratio intermediate between *st3*-FT6 / *st6*-FT6 and *wt*-FT6 (Grabenhorst and Conradt, 1999). It is important to note that the length of transmembrane region (13 aa) assigned to polysialyltransferase (ST8Sia IV, PST) is significantly shorter than the same region of all other transferases used in this study (17-21 aa). PST is involved in the biosynthesis of polysialylated (NeuAc(α2-8))$_n$NeuAc(α2-3) Gal(β1-4)GlcNAc-R struc-tures which are considered as important signals in tissue developmental processes (Nakayama, et al., 1998). One would expect that the enzyme functionally localizes together with the cellular ST3Gal III in the same or in a later functional Golgi area since PST adds NeuAc in α2,8-linkage to already α2,3-sialylated substrates. The *in vivo* data resulting from the expression of the *pst*-FT6 variant support such a hypothesis and clearly indicate a compartmental localization earlier than both FucT-VI and FucT-VII.

In summary, the overall α1,3-fucosylation of the reporter glycoprotein β-TP and as well as the determination of the SLex : Lex ratio of its oligosaccharides allow for a mapping of the functional localization of the chimeras either before or later than the *wt*-FT6. We therefore hypothesize that this sequential arrangement represents also the sequential distribution of the CTS-donor glycosyltransferases in the order from early to late Golgi subcompartments:

GnT-I < (ST6Gal I, ST3Gal III) < GnT-III < ST8Sia IV < β4GalT1 < FucT-VI < ST3Gal IV < FucT-VII

It is noteworthy that the secreted β-TP from all transfected cell lines contained exclusively oligosaccharides with intact N-acetyllactosamine antennae. No truncated oligosaccharides in any of the β-TP preparations isolated from the supernatant of the different CTS-variant cells was observed. It appears that the partition of the new *galt*- and *gnt1*-FT6 variants into the pre-existing gradient of the endogenous glycosyltransferases in the compartments does not interfere with the regular oligosaccharide modification pathway of the cells.

This finding is in agreement with the view that it would be difficult to saturate the mechanism underlying Golgi retention of transferases (Gleeson, 1998). We have obtained some evidence that the *in vivo* functional activity of FucT-VI can be modulated by the over-expression of the enzyme, since in a single cell clone expressing 3-times higher intracellular FucT-VI activity than was obtained routinely, 75% of the β-TP oligosaccharides were modified with peripheral Fuc, 2/3 of which contained two Lex motifs. This observation can easily be explained by the early modification of Gal(β1-4)GlcNAc-R chains encountered by the high level of fucosyltransferase molecules exhibiting functional activity already during their transport through the early Golgi compartments. Consequently, the early fucosylation of acceptor substrate leads to reduced amounts available for subsequent α2,3-sialylation by ST3Gal III and also to substrate depletion for a sub-sequent SLex formation by FucT-VI polypeptides that might have been gated into later subcompartmental areas in the high expression cell line. The detection of any such late residing enzyme subpopulation that might have resulted from a hypothetical saturation of the subcompartmental retention mechanism, of course, is prevented by the *in vivo* activity approach used in our investigation. However, our data obtained with chimeras expressing similar intracellular FucT-VI activities clearly indicate the presence of signals contained in the CTS-regions that can cause a forward (*ft7*) or a backward (*gnt1, st3, st6, gnt3*) targeting of enzymes operating in the late bio-synthetic glycosylation pathway and thus provide a regulatory means for the spatial separation of enzyme activities competing for the same substrates in the same compartments.

The model for Golgi retention of glycosyltransferases (Bretscher and Munro, 1993; Munro, 1995) suggests a lipid-mediated sorting mechanism to be responsible for preventing Golgi membrane proteins to be transported to the plasma membrane since the length of the transmembrane domain (TMD) of 17-22 aa of most Golgi proteins is about 5 aa shorter than those of the plasma membrane (Gleeson, 1998). The cholesterol concentration gradient formed throughout the secretory compartments would result in a lipid bilayer with increasing diameter across the Golgi where the Golgi-resident proteins partition into different lipid / glycolipid microdomains when compared to plasma membrane proteins (Munro, 1995). However, in view of our mapping results, and considering the different TMDs of PST (13 aa), ST6Gal I (17 aa) and GnT-I (21 aa), it is difficult to understand how this model can be applied to all Golgi enzymes. Also computer

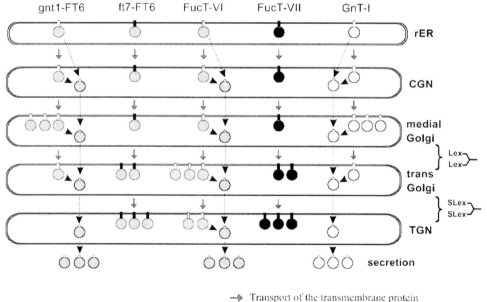

Figure 11. Model describing the regulation of the targeting and intra compartmental stability of glycosyltransferases. The model is based on the observation that proteolytically cleaved catalytic domains released from the Golgi membrane into the lumen of the subcompartments <u>does not</u> have any *in vivo* activity. *Right part:* trafficking of wild-type glycosyltransferases mediated by their CTS-region. *Left par* : altered trafficking of chimeric FucT-VI variants mediated by the CTS-regions of the CTS-donor transferases FucT-VII and GnT-I. The release of proteolytically cleaved forms from the membrane may occur in any Golgi compartment.

modeling of the different glycosyltransferase TMDs does not lead to satisfying results, therefore it is conceivable that the flanking domains also contribute to the partition of glycosyltransferases into lipid microdomains and the enzymes are in fact targeted to nonequivalent *in vivo* functional areas of the late glycosylation compartment by the signals contained in the CTS-region. The significance of the entire CTS-region of glycosyltransferases for their *in vivo* function is also emphasized by our observation that this polypeptide moiety must contain signals mediating resistance (*ft7, gnt3*) or susceptibility to intracellular cleavage (*gnt1, st3, st6, galt, pst, ft6*) by yet unidentified endoprotease(s). The physiological significance of this phenomenon may be considered as playing a role for the rapid elimination of glycosyltransferase activities when, *e.g.*, new enzyme genes become activated during differentiation or in developmental processes.

8. Application of Golgi Glycosyltransferase Engineering

Recently, this approach has been successfully used to reduce the expression of the Gal(α1,3)Gal(β1-4)GlcNAc epitope synthesized by NIH3T3 PA317 packaging cell line (Hansen, et al., 2001). The Gal(α1-3)Gal motif on surface proteins of murine cells is considered one of the major factors of retrovirus vectors derived from these hosts being the target of attack by the human complement system. Serum-resistant packaging cells were engineered by transfection with fucosyltransferase / chimeric sialyltransferase enzymes resulting in competition of the endogenous α1,3-galactosyltransferase. The stable cells were found to exhibit a significantly reduced biosynthesis of the Gal(α1αGal motif on cellular and viral glycoproteins (Hansen, 2000). Enzyme competition resulted in production of retrovirus vectors showing an up to 3.5 fold increase in serum stability when compared to retroviruses from the unmodified parental wild-type PA317 mouse packaging cell line. Thus, these results demonstrate the successful application of the technology of glycosyltransferase engineering for the improvement of retroviral vectors for *in vivo* gene therapy and furthermore suggests also its applicability in the modification of cells/tissues in the area of xenotransplantation. A further field of application lies in the production of pharmaceutical glycoproteins from transgenic animals. It has been shown that different transgenic animals (e.g. cows, pigs, goats) produce proteins with a different degree of glycostructure polymorphism as is the case for recombinant mammalian host cell lines; some of the structures may have important functional consequences or are antigenic in humans.

9. Perspectives

The engineering/design of new mamalian cell lines with altered oligosaccharide structures can be achieved by expression of (terminal) glycosyltransferases complementing the wild-type glycosylation repertoire of cells since the pertinent nucleotide-sugar donor substrates are provided by the cellular metabolic and transport machinery.

However in many cases the newly introduced enzyme might severely compete with endogenous cellular activities for the same acceptor substrate on a glycoprotein. In these cases, a successful engineering can only be achieved by directing the intended enzyme activity into a subcellular compartment where such a competition is minimized. Thus, engineering of the intracellular glycosyltransferase targetting properties can increase the yield of the final product and reduce its microheterogeneity. Based on the known *in vitro* specificity even branch-specific modifications can be achieved. Taken together, these aspects are of importance in view of the emerging field of functional genomics / functional proteomics where our understanding of the fine-tuning of biological activities of polypeptides by posttranslational modification (glycoproteins, viral therapeutic vectors and cells / tissues) is essential for the development of improved human therapeutics.

10. References

Berman PW. Lasky LA (1985) Engineering glycoproteins for use as pharmaceuticals. *Trends Biochem. Tech.* 3: 51-3.

Björk I. Ylinenjarvi K. Olson ST. Hermentin P. Conradt HS. Zettlmeissl G (1992) Decreased affinity of recombinant antithrombin for heparin due to increased glycosylation. *Biochem. J.* 286: 793-800.

Bosshart H. Berger EG (1992) Biosynthesis and intracellular transport of alpha-2,6-sialyltransferase in rat hepatoma cells. *Eur. J. Biochem.* 208: 341-9.

Bretscher MS. Munro S (1993) Cholesterol and the Golgi apparatus. *Science* 261: 1280-1.

Britten CJ. van den Eijnden DH. McDowell W. Kelly VA. Witham SJ. Edbrooke MR. Bird MI. de Vries T. Smithers N (1998) Acceptor specificity of the human leukocyte α3-fucosyltransferase: role of FucT-VII in the generation of selectin ligands. *Glycobiology* 8: 321-7.

Colley KJ (1997) Golgi localization of glycosyltransferases: more questions than answers. *Glycobiology* 7: 1-13.

Conradt HS. Hofer B. Hauser H (1990) Expression of human glycoproteins in recombinant mammalian cells: Towards genetic engineering of *N*- and *O*-glycoproteins. *Trends Glycosci. Glycotechnol.* 2: 168 -81.

Costa J. Grabenhorst E. Nimtz M. Conradt HS (1997) Stable expression of the Golgi form and secretory variants of human fucosyltransferase III from BHK-21 cells: Purification and characterization of an engineered truncated form from the culture medium. *J. Biol. Chem.* 272: 11613-21.

Gleeson PA (1998) Targeting of proteins to the Golgi apparatus. *Histochem. Cell Biol.* 109: 517-32.

Goochee CF. Gramer MJ. Andersen DC. Bahr JB. Rasmussen JR (1991) The oligosaccharides of glycoproteins: bioprocess factors affecting oligosaccharide structure and their effect on glycoprotein properties. *Bio Technology* 9: 1347-55.

Goochee CF. Monica T (1990) Environmental effects on protein glycosylation. *Bio/Technology* 8: 421-7.

Grabenhorst E. Conradt HS (1999) The cytoplasmic, transmembrane and stem regions of glycosyltransferases specify their in vivo functional sublocalization and stability in the Golgi. *J. Biol. Chem.* 274: 36107-16.

Grabenhorst E. Hoffmann A. Nimtz M. Zettlmeissl G. Conradt HS (1995) Construction of stable BHK-21 cells coexpressing human secretory glycoproteins and human Galβ1-4GlcNAc-R α2,6-sialyltransferase: α2,6-linked NeuAc is preferably attached to the Galβ1-4GlcNAcβ1-2Manα1-3-branch of biantennary oligosaccharides from secreted recombinant β-trace protein. *Eur. J. Biochem.* 232: 718-25.

Grabenhorst E. Nimtz M. Costa J. Conradt HS (1998) In vivo specificity of human α1,3/4 fucosyltransferases III-VII in the biosynthesis of Lewis X and sialyl Lewis X motifs on complex-type *N*-glycans: Coexpression studies from BHK-21 cells together with human β-trace protein. *J. Biol. Chem.* 273: 30985-94.

Grabenhorst E. Schlenke P. Pohl S. Nimtz M. Conradt HS (1998) Genetic engineering of recombinant glycoproteins and the glycosylation pathway in mammalian host cells. *Glycoconjugate J.* 16: 81-97.

Hansen W. Grabenhorst E. Nimtz M. Conradt HS. Wirth M (2001) Formation of the α1,3gal-epitope in retrovirus producing mouse cells can be reduced by directed enzyme competition of α1,3-galactosyltransferase activity and results in serum-stabilized mouse retrovirus (submitted)

Hansen W (2000) Glykodesign und Herstellung sicherer Verpackungszellinien zur Erzeugung rekombinanter, serumstabiler Retroviren für die Gentherapie. Doctoral Thesis. Technical University of Braunschweig.

Hoffmann A, Nimtz M, Getzlaff R., Conradt HS (1995) "Brain-type" *N*-glycosylation of asialotransferrin from human cerebrospinal fluid. *FEBS Lett.* 359: 164-8.

Hoffmann A, Nimtz M, Wurster U, Conradt HS (1994) Carbohydrate structures of β-trace protein from human cerebrospinal fluid: Evidence for "brain-type" *N*-glycosylation. *J. Neurochem.* 63: 2185-96.

Jenkins N, Parekh RB, James DC (1996) Getting the glycosylation right: Implications for the biotechnology industry. *Nature Biotechnology* 14: 975-81.

Johnson PH, Donald ASR, Watkins WM (1993) Reassessment of the acceptor specificity and general properties of the Lewis blood-group gene associated alpha-3/4-fucosyltransferase purified from human milk.*Glycoconj. J.* 10: 152-64.

Kansas GS (1996) Selectins and their ligands: current concepts and controversies. *Blood* 88: 3259-87.

Lasky LA (1995) Selectin-carbohydrate interactions and the initiation of the inflammatory response. *Ann. Rev. Biochem.* 64: 113-39.

Munro S (1995) An investigation of the role of transmembrane domains in Golgi protein retention. *EMBO J.* 14: 4695-704.

McEver RP, Moore KL, Cummings RD (1995) Leukocyte trafficking mediated by selectin-carbohydrate interactions. *J. Biol. Chem.* 270: 11025-8.

Nakayama J, Angata K, Ong E, Katsuyama T, Fukuda M (1998) Polysialic acid, a unique glycan that is developmentally regulated by two polysialyltransferases, PST and STX, in the central nervous system: from biosynthesis to function. *Pathol. Int.* 48: 665-77.

Natsuka S, Gersten KM, Zenita K, Kannagi R, Lowe JB (1994) Molecular cloning of a cDNA encoding a novel human leukocyte α1.3-fucosyltransferase capable of synthesizing the sialyl Lewis x determinant. *J. Biol. Chem.* 269: 16789-94.

NilssonT, Pypaert M, Hoe MH, Slusarewicz P, Berger EG, Warren G (1993) Overlapping distribution of two glycosyltransferases in the Golgi apparatus of HeLa cells. *J. Cell Biol.* 120: 5-13.

Nilsson T, Rabouille C, Hui N, Watson R, Warren G (1996) The role of the membrane-spanning domain and stalk region of N-acetylglucosaminyltransferase I in retention, kin recognition and structural maintenance of the Golgi apparatus in HeLa cells. *J. Cell Sci.* 109: 1975-89.

Nilsson T, Slusarewicz P, Warren G (1993) Kin recognition. A model for the retention of Golgi enzymes. *FEBS Lett.* 330: 1-4.

Nimtz M, Grabenhorst E, Gambert U, Costa J, Wray V, Morr M, Thiem J, Conradt HS (1998) In vitro α1-3 or α1-4 fucosylation of type I and type II oligosaccharides with secreted forms of recombinant human fucosyltransferases III and VI. *Glycoconjugate J.* 15: 873-83.

Nimtz M, Martin W, Wray V, Klöppel K-D, Augustin J, Conradt HS (1993) Structures of sialylated oligosaccharides of human erythropoietin expressed in recombinant BHK-21 cells. *Eur. J. Biochem.* 213: 39-56.

Paulson JC, Colley K (1989) Glycosyltransferases. Structure, localization, and control of cell type-specific glycosylation. *J. Biol. Chem.* 264: 17615-8.

Pohl S, Hoffmann A, Rüdiger A, Nimtz M, Jaeken J, Conradt HS (1997) Hypoglycosylation of a brain glycoprotein (β-trace protein) in CDG syndromes due to phosphomannomutase deficiency and *N*-acetylglucosaminyltransferase II deficiency. *Glycobiology* 7: 1077-84.

Rabouille C, Hui N, Hunte F, Kieckbusch R, Berger EG, Warren G, Nilsson T (1995) Kin recognition between medial Golgi enzymes in HeLa cells. *J. Cell Sci.* 108: 1617-27.

Schlenke P, Grabenhorst E, Nimtz M, Conradt HS (1998) Construction and characterization of stably transfected BHK-21 cells with human-type sialylation characteristics: transfection with the human α2,6-sialyltransferase gene leads to the biosynthesis of recombinant glycoproteins bearing NeuAc(α2-6)Gal(β1-4)GlcNAc-R and NeuAc(α2-6)GalNAc(β1-4)GlcNAc-R *N*-glycans motifs. *Cytotechnology* 30: 17-25.

Shinoda K, Morishita Y, Sasaki K, Matsuda Y, Takahashi I, Nishi T (1997) Enzymatic characterization of human α1.3-fucosyltransferase Fuc-TVII synthesized in a B cell lymphoma cell line. *J. Biol. Chem.* 272: 31992-7.

Strous GJAM, Berger EG (1982) Biosynthesis, intracellular transport, and release of the Golgi enzyme galactosyltransferase (lactose synthetase A protein) in HeLa cells. *J. Biol. Chem.* 257: 7623-8.

Varki A (1993) Biological roles of oligosaccharides: all of the theories are correct. *Glycobiology* 3: 97-130.

Varki A (1994) Selectin ligands. *Proc. Natl. Acad. Sci. U.S.A.* 91: 7390-7.

de Vries T, Srnka CA, Palcic MM, Swiedler SJ, van den Eijnden DH, Macher BA (1995) Acceptor specificity of different length constructs of human recombinant α1,3/4-fucosyltransferases. Replacement of the stem region and the transmembrane domain of fucosyltransferase V by protein A results in an enzyme with GDP-fucose hydrolyzing activity. *J. Biol. Chem.* 270: 8712-22.

de Vries T, Palcic MP, Schoenmakers PS, van den Eijnden DH, Joziasse DH (1997) Acceptor specificity of GDP-Fuc:Galβ1-4GlcNAc-R α3-fucosyltransferase VI (FucT-VI) expressed in insect cells as soluble, secreted enzyme. *Glycobiology* 7: 921-7.

Weinstein J, Lee EU, McEntee K, Lai P-H, Paulson JC (1987) Primary structure of beta-galactoside alpha 2,6-sialyltransferase. Conversion of membrane-bound enzyme to soluble forms by cleavage of the NH2-terminal signal anchor. *J. Biol. Chem.* 262: 17735-43.

7. A METABOLIC SUBSTRATE-BASED APPROACH TO ENGINEERING NEW CHEMICAL REACTIVITY INTO CELLULAR SIALOGLYCOCONJUGATES

K.J. YAREMA

Whitaker Biomedical Engineering Institute, GWC School of Engineering, Johns Hopkins University, Baltimore MD 21218, U.S.A.

1. Introduction

Current examples of cellular engineering, the "purposeful modification of intermediary metabolism" (Cameron and Tong, 1993) or the "improvement of cellular activities" (Bailey, 1991), share two common features. First is the reliance on recombinant DNA technology to modulate the proteome of the cell. Advances in protein design and engineering, genomics, and proteomics (Hatzimanikatis et al., 1999; Kao, 1999; Schilling et al., 1999), coupled with sophisticated mathematical algorithms and computer modeling (Gombert and Nielsen, 2000; Schuster et al., 2000), promise to provide the cellular engineer of the future with exquisite recombinant DNA-based tools for manipulation of the metabolic machinery and architecture of a cell. The sheer complexity of biological macromolecules, however, has encouraged the exploration of alternate technologies. One such alternative is an emerging substrate-based approach where small-molecule metabolites, instead of the enzymes that process these compounds, are manipulated to effect changes in cellular properties. More specifically, substrate-based cellular engineering exploits the naturally occurring biosynthetic pathways of the cell to incorporate unnatural metabolic intermediates into cellular components (Figure 1).

The emerging field of substrate-based cellular engineering differs from established approaches not only in methodology, but also in applications and aims. In addition to a reliance on recombinant DNA techniques, most current examples of cellular engineering share the common goal of enhancing the ability of a cell to produce metabolic products, ranging from small-molecule metabolites to recombinant proteins, deemed useful to humankind. This objective, the production of a useful biomolecule, typically reduces secondary effects on the cellular metabolism to a matter of minor concern. By contrast, the primary goal of substrate-based cellular engineering, so far, has been to alter the composition of the cellular architecture in a precise manner for a defined purpose. An attractive feature of substrate-based cellular engineering is the ability to mediate exact metabolic changes amidst an overwhelmingly complex cellular environment. By selecting the properties of the unnatural metabolic intermediate carefully, it is possible to target a particular biosynthetic pathway to the exclusion of all other thereby specifically engineering the designated cellular component.

M. Al-Rubeai.(ed.). Cell Engineering. 171-196.

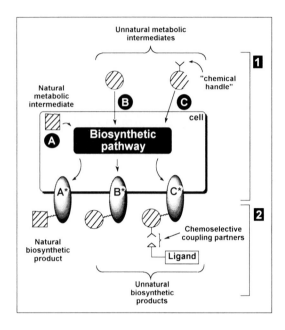

Figure 1. A substrate-based approach to cellular engineering. Substrate-based cellular engineering exploits the naturally occurring biosynthetic pathways of the cell to incorporate unnatural metabolic intermediates into a cellular component. In the example shown, the endogenous metabolic intermediate **A** (square cross-hatched symbol) is metabolically converted into a natural biosynthetic product, in this case the cell surface marker **A***. In certain situations it is possible to intercept the pathway with an unnatural intermediate (**B**, circle). **B** competes with the endogenous substrate **A** for conversion to the cell surface marker, in this case the unnatural biosynthetic product **B***. By selecting a substrate with the appropriate modification (compound **C**), the corresponding metabolic product can be engineered to contain a "chemical handle" (cell surface product **C***) capable of further elaboration with the appropriate chemical coupling partner. Consequently, the substrate-based methodology presented in this paper can be considered a two step approach (1) biosynthetic incorporation of an unnatural metabolite into the architecture of a cell followed by (2) subsequent chemical ligation with an externally-delivered reagent.

Biosynthetic enzymes inherently possess a high degree of substrate specificity. A benefit of this stringency is that metabolic pathways generally utilize "correct" substrates and thereby ensure the fidelity of biological processes. The reluctance of most pathways to accept unnatural analogs of metabolic intermediates, however, severely hinders the development of substrate-based cellular engineering. The challenge of this approach, therefore, is to identify pathways with sufficiently relaxed substrate specificity to allow the successful incorporation of modified metabolic intermediates into the targeted cellular components (Figure 1).

2. The Sialic Acid Biosynthetic Pathway: A Model for Substrate-Based Cellular Engineering

Considering the challenge of identifying a pathway permissive for unnatural substrates, the discovery that the sialic acid biosynthetic pathway offered relaxed stringency (Schultz and Mora, 1975; Schwartz et al., 1983; reviewed in Goon and Bertozzi, 2001) provided an enticing opportunity to experimentally establish the metabolic substrate-based approach to cellular engineering. As described in this report, the sialic acid pathway has served as a model to illustrate both the theoretical concepts underlying a substrate-based approach as well as the practical applications of this new technology.

The potential biological impact afforded by engineering cell surface components was demonstrated when substrate-mediated modifications to cell surface sialoglycoconjugates were shown to directly alter cellular behavior (Keppler et al., 2001; Goon and Bertozzi, 2001; Yarema, 2001). Subsequently, the development of physiologically compatible chemoselective ligation reactions (discussed below) now allows further chemistry to be performed on the engineered cellular component by the external delivery of a complementary reaction partner (Figure 1). This second step in the metabolic substrate-based cellular engineering process has greatly increased the versatility of this emerging technology, allowing modulation of processes well beyond those strictly mediated by sialic acids (Mahal et al., 1997; Yarema and Bertozzi, 1998; Yarema et al., 1998; Bertozzi and Kiessling, 2001).

2.1. ManNAc ANALOGS ARE INCORPORATED INTO CELL SURFACE SIALOSIDES

Exogenously supplied N-acetylmannosamine (ManNAc) analogs can intercept the sialic acid pathway and, by the sequential action of several enzymes, be converted to their corresponding sialic acid counterparts and displayed on the cell surface (Figure 2) as the terminal residue of an oligosaccharide (Figure 3). To date, ManNAc analogs designed for cell surface display most commonly have been modified at the N-acyl position. Various substitutions (**R**) for the naturally occurring N-acyl constituents (Figure 3B) are shown in Figure 3. Initially, the engineering of cell surface sialoglycoconjugates was accomplished by using ManNAc analogs with chemically inert N-acyl side chains (eg., ManProp, ManBut, and ManPent; Figure 3C). When converted to their cell surface sialic acid counterparts (eg. SiaPent, Figure 4A) these unnatural metabolic precursors alter a variety of important cellular processes (Keppler et al., 2001).

2.2. APPLICATIONS OF CELL SURFACE SIALOSIDE ENGINEERING

When displayed on the cell surface, sialic acids play defining roles in critical processes that include immunorecognition, cell adhesion and growth, and the binding of certain pathogens (Kelm and Schauer, 1997, Varki, 1997). In addition, a number of human disease states, including cancer, are characterized by aberrant sialic acid expression (Sillanaukee et al., 1999). In some cases, the role of sialic acid in mediating recognition and binding events is limited to the contribution of its negatively charged carboxylic

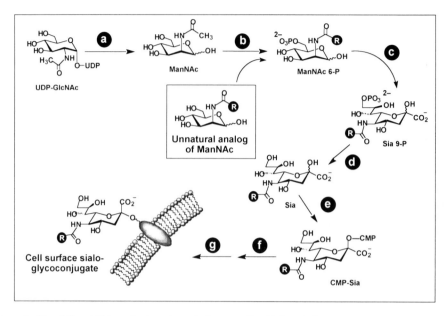

Figure 2. The sialic acid biosynthetic pathway in human cells. Sialic acid is a generic term for a family of unique 9-carbon monosaccharides. Metabolic conversion of UDP-GlcNAc into a cell surface sialoglyco-conjugate proceeds by the sequential action of **a,b**: UDP-GlcNAc 2-epimerase/ManNAc 6-kinase, **c**: sialic acid 9-phosphate synthase, **d**: sialic acid 9-phosphatase, **e**: CMP-sialic acid synthetase, **f**: CMP-sialic acid Golgi transporter, **g**: sialyltransferase (several in human). The most abundant sialic acid (Sia) in humans is N-acetylneuraminic acid (NeuAc, R = CH₃). The natural metabolic intermediates along this pathway are uridine diphosphate-N-acetylglucosamine (UDP-GlcNAc), N-acetylmannosamine (ManNAc), N-acetylmannosamine 6-phosphate (ManNAc 6-P); N-acetylneuraminic acid 9-phosphate (Sia 9-P); N-acetylneuraminic acid (Sia); CMP-N-acetylneuraminic acid (CMP-Sia); and glycoconjugate-bound N-acetylneuraminic acid. Interception of the pathway with unnatural analogs of ManNAc results in the cell surface presentation of the corresponding unnatural sialic acid (refer to Figure 3 and Keppler et al., 2001 for examples of the unnatural "**R**" moiety).

acid to the overall characteristics of a larger carbohydrate epitope (Wong et al., 1997; Yarema and Bertozzi, 1998). In these situations alteration of the N-acyl side chain by a substrate-based engineering approach would not be expected to evoke a biological response. In other situations, however, biological responses depend on the **exact** structure of the sialic acid residue. In these cases, even an extremely subtle modification such as the extension of the N-acetyl group of ManNAc with chemically inert alkyl side chains (e.g., ManProp, ManBut, or ManPent, Figure 3C) results in dramatic changes in cellular behavior when the corresponding sialoglycoconjugates are displayed on the cell surface. It should be noted that even natural sialic acids, such as N-glycolylneuraminic acid, when displayed in inappropriate biological contexts by exploiting metabolic substrate-based methods, can also significantly modify cellular behavior (Collins et al., 2000).

Specific biological responses elicited by engineering unnatural chemical structures into cell surface sialoglycoconjugates range from altered viral binding, modified cell adhesion and morphology, and stimulation of the immune system. In one study, human B-

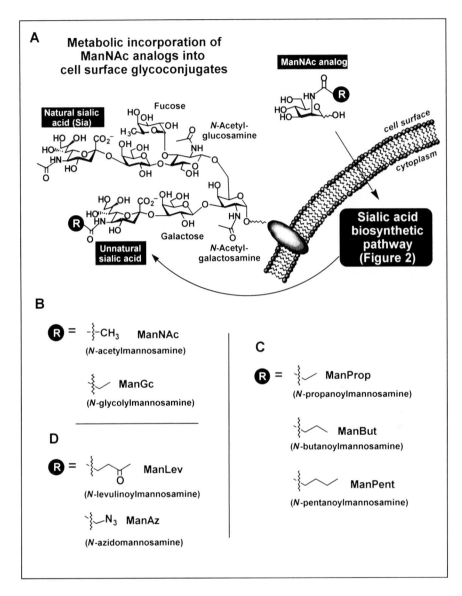

Figure 3. The sialic acid biosynthetic pathway: a model for substrate-based cellular engineering. **A.** Exogenously supplied N-acetylmannosamine (ManNAc) analogs can intercept the sialic acid pathway and, by the sequential action of several enzymes (see Figure 2), be converted to their corresponding sialic acid counterparts and displayed on the cell surface as the terminal residue of an oligosaccharide. ManNAc **analogs** utilized in sialoglycoconjugate engineering have been most commonly modified at the N-acyl position. Shown in panel **B** are "R" groups for expression of the naturally occurring sialic acids NeuAc (N-acetylneuraminic acid) and NeuGc (N-glycolylneuraminic acid). Unnatural N-acyl substitutions are shown in panel **C** for non-reactive analogs and in panel **D** for analogs containing a "chemical handle" (as further discussed in the text and Figure 4).

lymphoma and green monkey kidney epithelium cells displaying the unnatural sialic acids SiaProp, SiaBut, or SiaPent experienced altered susceptibility to infection by polyoma virus, a pathogen that uses sialic acid as a ligand for binding prior to infection (Keppler et al., 1995). This finding raises the intriguing possibility that ManNAc analogs could find a role as antiviral agents. In another study, human diploid fibroblast cells expressing similarly altered sialosides lost their sensitivity to contact inhibition of growth (Wieser, 1996). It is interesting to note that even ManProp, the analog with the most subtle structural change, is capable of eliciting profound biological effects when incorporated into its sialic acid counterpart, SiaProp. In addition to the effects mentioned above, SiaProp selectively stimulates the proliferation of particular types of cells found in the rat nervous system (Schmidt et al., 1998; Schmidt et al., 2000). Furthermore, when presented in context of the tumor-associated antigen, polysialic acid (the $\alpha2,8$-linked homopolymer of sialic acid), SiaProp is capable of stimulating the immune system of the host resulting in decreased metastatic potential of the tumor (Liu et al., 2000).

Many applications of substrate-based sialoglycoconjugate engineering are incredibly specific, depending on (1) the exact structural modification imposed on the sialic acid residue and (2) the particular cell type under study. To give an example of the first point, polyoma viral binding was increased by certain ManNAc analogs and decreased by others (Keppler et al., 1995). The second point, the influence of cell type on the exact response, was demonstrated in an experiment where cells isolated from rat brain were incubated with ManProp. Although all cell types converted this analog to SiaProp, only certain cell types (astrocytes and microglia) experienced enhanced growth rates (Wieser et al., 1996; Schmidt et al., 1998). Additional study is required to understand the underlying molecular mechanisms responsible for the exquisite ability of cells to respond differently to such minor alterations in the overall cell surface characteristics. Once they are better understood, these substrate-based methods promise to be an incredibly precise tool to modulate sialic acid-specific biological events (Keppler et al., 2001).

2.3. CHEMOSELECTIVE LIGATION TECHNOLOGY EXPANDS THE SCOPE OF SIALOSIDE ENGINEERING

The potential applications of sialoglycoconjugate engineering have been dramatically expanded by the recent development of metabolic precursors that allow the engineered sialic acid to be endowed with a "chemical handle" such as the ketone functionality of SiaLev (Figure 4B). The goal of cell surface display of this type of unnatural sialoglycoconjugate is to exploit the "chemical handle" to decorate the cell surface with ligands that can affect a wide range of cellular properties, not just those that are sialic acid specific (Mahal et al., 1997; Yarema, 2001). Considering that sialic acids are terminal residues on cell surface glycoconjugates (Figure 3), they are poised on the outer periphery of the cell and are ideally situated for further elaboration by the attachment of an externally delivered reagent.

The ability to install and exploit "chemical handles" on the cell surface is made possible by the recent development of chemoselective ligation reactions. The objective of these transformations is to achieve the tremendous selectivity of non-covalent

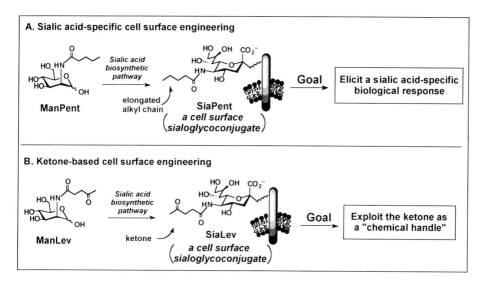

Figure 4. Goals of sialic acid-mediated cell surface engineering. Unnatural sialoglycoconjugates introduced onto the cell surface can directly elicit sialic acid-specific biological responses (panel A). Sugars with chemically inert N-acyl side chains such as ManProp, ManBut, and ManPent (Figure 3) fit into this category, and when converted to their sialic acid analog (eg. SiaPent, shown), can alter viral binding, modify cell adhesion and morphology, and stimulate the immune system (see discussion in main text). When the engineered sialoglycoconjugate contains a "chemical handle" (such as the ketone functionality of SiaLev, panel B), this new chemical property can be exploited to install ligands on the cell surface designed to modulate a wide range of cellular properties, not just those that are sialic acid specific.

recognition events, such as antibody-antigen binding, while overcoming the limitations that are inherent in using large macromolecules as experimental tools (Rideout et al., 1990; Lemieux and Bertozzi, 1998; Saxon and Bertozzi, 2000). A chemoselective ligation reaction requires two participating functional groups that have reactivities finally tuned to avoid unwanted side reactions that have the potential to occur in a chemically complex environment such as a human cell. In the context of substrate-based cellular engineering, the goal is to biosynthetically incorporate one of the chemical coupling partners into a cellular component by including it in the design of an unnatural metabolic intermediate. This coupling partner can then be further elaborated by the attachment of a complementary, externally delivered reagent.

2.4. CHEMOSELECTIVE LIGATION REACTIONS APPLIED TO KETONE-CONTAINING SIALOGLYCOCONJUGATES

The development of chemoselective ligation partners for use in a physiological environment requires addressing several specific considerations. Ideally the reactive partners are abiotic, form stable adducts under physiological conditions, and recognize each other while ignoring their cellular surroundings (Lemieux and Bertozzi, 1998; Saxon and Bertozzi, 2000). In addition, one of the partners should be sterically undemanding, allowing incorporation by the cellular biosynthetic machinery that, as disccussed above, has limited tolerance for altered metabolic precursors. A relative simple modification can be made to the ManNAc analogs with extended N-alkyl side chains to endow their downstream sialoglycoconjugates metabolic products with properties required of a chemoselective coupling partner. Specifically, an inert N-alkyl substituent can be converted into a potential chemoselective coupling partner by addition of a carbonyl group to the penultimate carbon of the N-alkyl chain of ManPent (Figure 3D). This alteration, while sterically undemanding, provides the substrate with a new chemical functionality, the ketone (Figure 4B). While not abiotic, ketones are relatively inert toward other compounds found within a cell and are generally absent from the cell surface. Importantly, they are electrophilic, thereby presenting a new chemical reactivity on the largely nucleophilic cell surface. Therefore, successful biosynthetic conversion of ManLev to its cell surface sialic acid counterpart SiaLev presents a chemoselective coupling partner, the ketone, on the cell surface (Mahal et al., 1997; Yarema et al., 1998).

2.5. CHEMOSELECTIVE REACTION OF AMINOOXY AND HYDRAZIDE CONJUGATES WITH THE KETONE FUNCTIONALITY OF SiaLev

Once presented on the cell surface, ketone-containing sialoglycoconjugates can participate in chemoselective ligation reactions with complementary aminooxy or hydrazide functional groups (Figure 5). Both of these nucleophiles are abiotic and unreactive towards the natural constituents of the cell surface allowing selective ligation to metabolically incorporated ketone groups (Mahal et al., 1997; Yarema et al., 1998). The ligands attached to these externally delivered nucleophilic coupling partners are not sterically constrained by the requirement of enzymatic processing and can be, in theory, any peptide, carbohydrate or small-molecule compatible with physiological conditions. The ketone "chemical handle" of SiaLev, upon decoration with any biologically compatible ligand, expands sialoglycoconjugate engineering beyond sialic acid-specific applications to now encompass a wide range of cellular properties.

2.6. APPLICATIONS OF KETONE-BASED CELL SURFACE ENGINEERING

In the relatively short time since ketone-based cell surface engineering was first described (Mahal et al., 1997), several applications have already been demonstrated. One application is the chemical construction of new glycosylation patterns on cells (Yarema et al., 1998). In theory, this technique allows any complex carbohydrate that can be synthesized by chemical methods to be appended to the cell surface, thereby potentially changing the molecular identity and biological activity of the cell. In another application, a ManLev-based approach has shown promise in targeting either diagnostic (Lemieux et al., 1999) or therapeutic (Mahal et al., 1997; Mahal et al., 1999) agents selectively to tumor cells based on the aberrantly high sialic acid metabolism of many these cells (Sell 1990; Takana et al., 1994; Lemieux et al., 1999). ManLev-mediated installation of SiaLev has allowed this unnatural sialic acid to function as a novel viral (Lee et al., 1999). Finally, a ManLev-based forward genetics scheme has allowed rarely

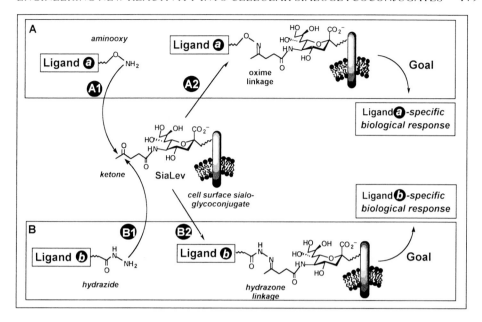

Figure 5. Chemoselective reaction of aminooxy and hydrazide conjugates with the ketone functionality of SiaLev. **A.** The aminooxy functionality can react selectively with the ketone moiety of SiaLev (step **A1**) to form a covalent oxime linkage (step **A2**). Hydrazide conjugates can undergo similar chemoselective coupling reactions with the ketone to produce a stable hydrazone conjugate (panel **B**). Depending on the exact composition of the ligand (eg., **a** or **b**), these conjugates can evoke a wide range of biological responses.

receptor and has facilitated viral-mediated gene delivery to otherwise refractory cells occurring cells with "masked" metabolic defects to be selected from a large wild-type population thereby offering novel insights into the regulation of the cellular glycosylation machinery (Yarema and Bertozzi, 2001; Yarema et al., 2001).

Considering that utilization of a substrate-based cellular engineering approach for any practical application first requires a thorough characterization of expression parameters for the resulting unnatural biomarker, a particularly useful chemoselective coupling partner for cell surface ketones is biotin hydrazide. This reagent, upon further reaction with fluorescently-labeled avidin, allows rapid kinetic and quantitative evaluation of SiaLev presentation on the cell surface by flow cytometry (Yarema et al., 1998). Most potential applications of a substrate-base cellular engineering approach are dependent on the number of SiaLev residues displayed on the cell surface, how soon they appear after ManLev incubation begins, how long lived are they, as well as how accessible they are to the externally delivered agent. The clinical success of delivering therapeutic or diagnostic agents to sialoglycoconjugate-resident ketones, for example, is just one application that critically depends on a thorough evaluation of each of these parameters (Mahal et al., 1997; Mahal et al., 1999; and Lemieux et al., 1999). To provide guidelines for addressing these considerations this report next discusses variables that determine the quantitative and temporal aspects of SiaLev expression.

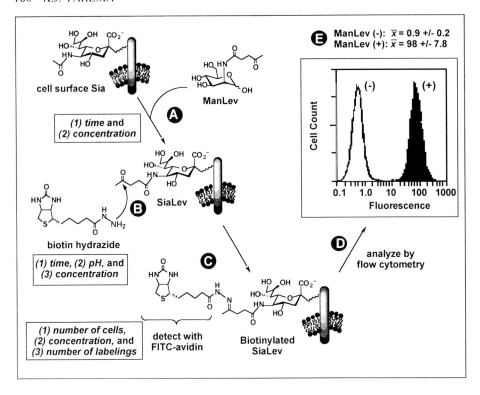

Figure 6. Overview of variables that contribute to the expression and detection of cell surface SiaLev. Factors that influence the expression of cell surface SiaLev include the (1) length of incubation with ManLev as well as the (2) concentration of this precursor (step **A**). Factors that affect quantitation of SiaLev include the (1) concentration of biotin hydrazide during the ketone-labeling step (step **B**) as well as (2) the pH and (3) duration of the reaction. FITC-avidin staining of the newly biotinylated cells (step **C**) is influenced by (1) total number of cells, (2) the concentration of the fluorophore, and the (3) number of times the staining procedure is repeated. Analysis by flow cytometry (step **D**) allows quantitation of the number of ketones that were successfully labeled on individual cells. Flow cytometry results (**E**) are given in arbitrary fluorescence units that, upon appropriate control experiments, can be converted into the number of fluorophores on each cell.

"Standard" experimental conditions for the entire process of (1) ManLev incubation, (2) biotin hydrazide labeling, (3) FITC-avidin staining and (4) flow cytometry quantitation of SiaLev expression (Figure 6), have already been described in detail (Jacobs et al., 2000). This reference (Jacobs et al., 2000) provides step by step instructions sufficient to allow this work to be easily replicated in any appropriately equipped laboratory and will not be repeated here. Rather, the current report will focus on how variations from "standard" conditions affect the expression and subsequent exploitation of engineered cell surface sialoglycoconjugates. The goal of thorough characterization is be able to "dial in" any desired number of cell surface epitopes and subsequently react a defined portion of the newly installed molecular handles with an incoming ligand with precision and reproducibility. The following description is narrowly focused on metabolic conversion of ManLev to SiaLev, and subsequent detection with biotin hydrazide and FITC-avidin. The experimental variables addressed herein, however, can be easily generalized to encompass virtually any metabolic substrate and detection scheme.

3.   Experimental Considerations for Sialoglycoconjugate Engineering Applications

3.1. FACTORS THAT AFFECT SiaLev EXPRESSION AND DETECTION

The experimental focus of this report is on ManLev and subsequent ketone-based chemoselective ligation reactions. All reagents are commercially available; in particular ManLev can be purchased from Molecular Probes (Eugene, OR) while other reagents are generally available from multiple suppliers (in particular Sigma, St. Louis MO offers many of the biological and chemical reagents required and cell lines are available from the American Type Culture Collection (ATCC), Manassas, VA). Many of the other ManNAc analogs described in this report are not currently commercially available and must be synthesized as described in detail elsewhere (Angelino et al., 1995; Yarema et al., 1998).

3.2. FACTORS THAT DETERMINE THE CONVERSION OF ManLev TO CELL SURFACE SiaLev

To achieve expression of SiaLev, the appropriate volume of ManLev (typically dissolved in PBS as a 500 mM stock solution) is simply added to the normal growth media of cells. The presence of antibiotics and the concentration of serum in the media generally have no affect on the utilization of ManLev by the cells. Various human cell lines, three of which are described in detail in this report, HL-60, HeLa, and Jurkat, were incubated with various concentrations of ManLev for intervals up to seven days and the time dependency of the resulting cell surface ketone presentation is shown in Figure 7A. The number of ketones reaches equilibrium after approximately five days (Figure 7B) and can be maintained indefinitely at high levels without noticeable adverse effects on the cells (Yarema et al., 1998). The dose-dependency for ketone expression is shown in detail for each cell line Figure 7C; the three cell lines are compared in panel D.

The parameters for SiaLev expression (Figure 7) establish several important points. First, the number of cell surface ketones can be "dialed in" with considerable precision by adjusting the concentration of ManLev in the culture media and the length of time that cells are incubated with this analog. Second, it should be emphasized that the level of ketones expressed on the cell surface is cell line dependent. This point can be clearly seen when data from all three human cell lines are plotted on the same graph (Figure 7B and D). HeLa cells display approximately four-fold more ketones than either the HL-60 or Jurkat cells do. The molecular basis for this disparity has not been firmly established but possibly is due to the fact that HeLa cells are proficient in the expression of both N-linked and O-linked glycans whereas Jurkat cells are deficient in O-linked, and HL-60 cells are deficient in N-linked glycoconjugates. Because SiaLev can be displayed in

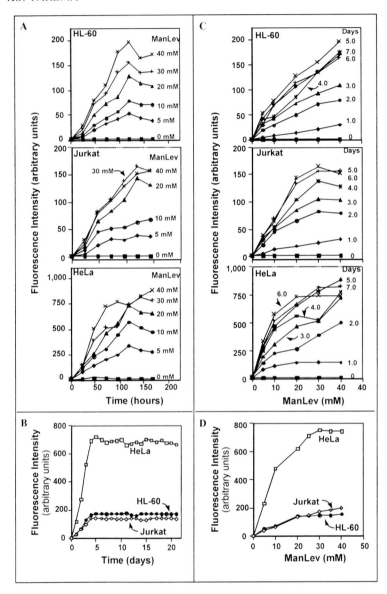

Figure 7. Parameters for cell surface SiaLev expression. **A.** The expression of SiaLev is shown for HL-60, Jurkat, and HeLa cells incubated for time intervals up to 168 h (seven days) with various concentrations of ManLev. Each data point represents the average of three experiments following the ketone detection scheme presented in Figure 6 (1.0 mM BH labeling for 2.0 h at pH 6.5, FITC-avidin staining at 5.6 g/mL). **B.** Each cell line was incubated with 25 mM ManLev and an aliquot of each culture was assayed on a daily basis for SiaLev expression. **C.** As a variation of the data shown in panel **A**, the expression of SiaLev is shown for HL-60, Jurkat, and HeLa cells with various ManLev concentrations for the indicated number of days. **D.** The three human cell lines were incubated for three days with the indicated concentration of ManLev, and analyzed for ketone expression as described above.

both linkages, HeLa cells would be expected to have more expression (Yarema et al, 1998). In addition, HeLa cells are physically larger than either of the other two cell line allowing additional expression when measured on a per cell basis.

The fluorescence data shown in Figure 7 are presented in "arbitrary units" obtained from flow cytometry analysis. Control experiments established that, for the conditions used in collection of the data presented in the current report, an arbitrary unit of fluorescence represents 1.3×10^6 ketone groups that were successfully biotinylated and subsequently detected by FITC-avidin (Yarema et al., 1998). Based on this conversion factor, HL-60 and Jurkat cells display approximately 2.0×10^6, and HeLa cells about 10^7, SiaLev residues under saturating conditions (>30 mM ManLev for > five days). However, these numbers are based on a particular labeling and detection protocol and, as discussed below, represent only a portion of the total number of SiaLev residues present on the cell surface.

3.3. FACTORS THAT AFFECT BIOTIN HYDRAZIDE LABELING OF CELL SURFACT KETONES.

The SiaLev expression data, presented in Figure 7 and discussed above, are based on "standard" biotin hydrazide labeling conditions. Specifically, these conditions entail incubation of ManLev-treated cells with 1.0 mM biotin hydrazide for 2.0 h in PBS adjusted to a pH value of 6.5. Changing any of these variables can modify the signal obtained upon staining with fluorescently labeled avidin and analysis by flow cyto-metry. In particular, detection of SiaLev is highly dependent on the concentration of biotin hydrazide (Figure 8A). The signal increases linearly up to 6.0 mM of this rea-gent, suggesting that there are at least six fold more ketones present on the cell surface than depicted in Figure 7 (where the data was obtained by using 1.0 mM biotin hydra-zide). Solubility issues prevent higher concentrations of biotin hydrazide from being used in the labeling reaction, precluding establishment of an upper limit for cell surface ketone expression by this method.

A cursory mathematical analysis of the biotin hydrazide labeling reaction suggests that there should be sufficient reagent present to react with all the ketones displayed on the cell surface. There are approximately 5.0×10^8 glycoconjugate-bound sialic acid res-idues per Jurkat cell (Yarema et al., 2001). Assuming that every cell surface sialic acid residue is replaced with SiaLev upon incubation with ManLev, there would by a maxi-mum of 5.0×10^{14} cell surface ketone groups present in a typical biotin hydrazide labeling reaction (based on 1.0×10^6 cells). Standard reaction conditions (500   L volume with 1.0 mM biotin hydrazide) include 3.0×10^{17} molecules of biotin hydrazide, a 600-fold molar excess. Clearly, the bimolecular coupling reaction between the hydra-zide and ketone is relatively inefficient under "standard" conditions. Longer reaction times to allow the ligation to reach completion might be expected to give an enhanced signal. However, the reaction reaches apparent equilibrium after 1.5 to 2.0 h (Figure 8B). Considering the long-lived nature of the stable covalent hydrazone linkage (Figure 4B), this equilibrium most likely reflects membrane turnover and recycling events to a greater degree than the reverse reaction to form the ketone and hydrazide components.

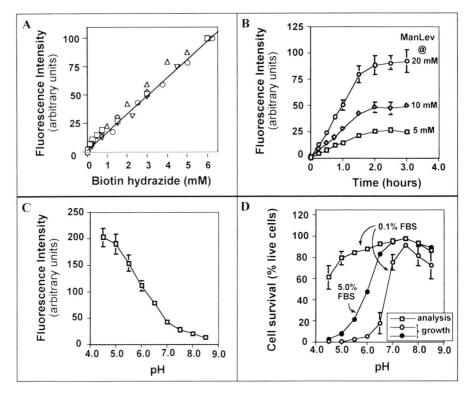

Figure 8. Factors that affect biotin hydrazide labeling of cell surface ketones. **A.** Jurkat cells were incubated with 2.5 mM ManLev for three days and cell surface SiaLev was determined by the ketone-detection strategy (Figure 6) utilizing various concentrations of biotin hydrazide. **B.** Jurkat cells were incubated with 2.5, 5.0 or 10 mM ManLev for three days and incubated with 1.0 mM biotin hydrazide for various time intervals up to three hours. **C.** Jurkat cells were incubated with 5.0 mM ManLev for three days and labeled with 2.0 mM biotin hydrazide over a range of pH values from 4.5 to 8.5 (adjusted by adding HCl to the PBS buffer prior to addition of the cells). In panels **A** through **C** cell surface levels of biotin were determined by FITC-avidin staining followed by flow cytometry analysis. Cell survival after biotin hydrazide labeling at various pH values is given in panel **D**. Data represented by square symbols indicate the number of events that "gate" as live cells by flow cytometry immediately after the labeling process; circles indicate longer term cell survival analyzed by trypan blue exclusion 12 h after labeling. Solid lines indicate data obtained with 0.1% FBS in the biotin hydrazide labeling reaction and the dotted line indicates survival with 5.0% FBS. The dashed vertical line in panels **C** and **D** indicates a pH of 6.5.

Consequently longer reaction times do not achieve the goal of complete biotinylation of cell surface SiaLev.

The chemoselective ligation ketone-hydrazide reaction is also pH dependent. Lowering the pH increases the reaction rate and results in a considerably higher cell surface signal (Figure 8C). Unfortunately optimal chemical reactivity occurs at conditions too acidic (pH <5.0) to be compatible with living cells (Figure 8D). Consequently a pH of 6.5 (dashed line on Figure 8C and D) represents a compromise between cell mortality (minimized at pH 7.5) and chemical reactivity (maximized below pH 5.0) and is routinely used in standard biotin hydrazide labeling reactions. It should be emphasized that even though not all cell surface ketones are readily conjugated with the incoming chemoselective coupling partner, the number that do react ($> 10^6$ per cell) is sufficient for many applications (see discussion of application above).

3.4. FACTORS THAT AFFECT FITC-AVIDIN STAINING OF BIOTINYLATED CELLS

Once cells are biotinylated, the next step in the detection of SiaLev is staining with FITC-avidin. Factors that affect this process are the FITC-avidin concentration in the staining reaction and the number of cells present (Figure 9, A and B respectively). While neither variable has an overwhelming effect on the ultimate signal, conditions

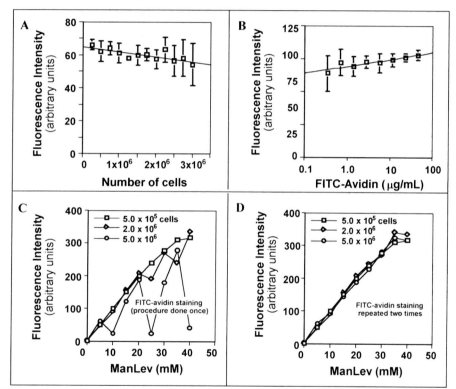

Figure 9. Factors that affect FITC-avidin staining of biotinylated cells. Jurkat cells were incubated with 5.0 mM ManLev for three days and labeled with 2.0 mM biotin hydrazide. They were then either stained with various concentrations of FITC-avidin (at a constant cell concentration of 1.5×10^6 cells/mL; panel **A**) or at various cell concentrations (with FITC-avidin maintained constant at 5.6 μg/mL; panel **B**). In both panels (**A** and **B**) the cells were stained twice with FITC-avidin; the effects of repeating the avidin staining step are shown in panel **C** (one staining) and panel **D** (procedure repeated twice; i.e. three times in total).

should be maintained constant from experiment to experiment to ensure a high level of reproducibility. Irreproducibility is particularly problematic at lower FITC-avidin concentrations and higher cell numbers (Figure 9A and B). Further analysis of this variability shows that increasing the number of cells per sample from 0.5×10^6 to 5.0×10^6 cells results in increasingly erratic outcomes (Figure 9C). One source of this error is biotin hydrazide that becomes physically trapped in the cell pellet during the washing steps. Considering that five to six orders of magnitude more biotin hydrazide is used to label the cells than FITC-avidin is used to subsequently stain them, it is crucial to efficiently remove all the biotin reagent from the cells. Otherwise, unremoved free biotin hydrazide can tritrate the incoming FITC-avidin away from cell surface resulting in the observed erratic results (Figure 9C). This problem is particularly acute when processing large numbers of cells due to a large pellet size but can be avoided by repeating the FITC-avidin step one or more times (Figure 9D).

3.5. ACETYLATED ManLev IS EFFICIENTLY INCORPORATED INTO THE SIALIC ACID PATHWAY.

The millimolar concentrations of ManLev needed for high levels of SiaLev expression (Figure 7) require unacceptably large amounts of this substrate when scaling up experiments beyond a volume of a few milliliters. This limitation prevents easy extension of substrate-based cellular engineering approaches from in vitro cell culture to in vivo animal models. Therefore strategies to increase the delivery efficiency of exogenously supplied metabolites to their target pathways are critical for the widespread application of substrate-based approaches. Applied to carbohydrate engineering, this goal of increased uptake can be achieved by the addition of hydrophobic acetyl groups to the metabolic precursor that increases its membrane permeability and cellular uptake (Sarkar et al., 1995, Lemieux and Bertozzi, 1999; Collins et al., 2000). In the ManLev example, the fully acetylated counterpart ($Ac_4ManLev$) has enhanced ability to transit the membrane (Figure 10A). Once inside a cell, the acetyl groups are removed (most likely by the enzymatic action of non-specific esterases) producing non-acetylated ManLev. This form of ManLev, identical to the exogenously delivered compound used in the previously described experiments, can access the sialic acid pathway and be converted to cell surface SiaLev.

Cell surface expression of SiaLev, beginning with $Ac_4ManLev$ rather than ManLev, requires an additional enzymatic step to produce the non-acetylated form of this sugar. This step, however, is not kinetically limiting. Instead, cell surface SiaLev expression is actually achieved four to eight hours faster with $Ac_4ManLev$ compared to the free monosaccharide (Figure 10B) demonstrating the limiting nature of transport of the free sugar into the cell. On a molar basis, acetylation renders $Ac_4ManLev$ 200 to 500-fold more potent than ManLev as a precursor for SiaLev presentation (Figure 10C). Enhanced access to the sialic acid pathway is seen for a wide range of ManNAc analogs (data not shown) suggesting that the use of acetylated sugars will be a valuable tool in future substrate-based cellular engineering endeavors. Acetylated sugars, however, do have the drawback of toxicity at high concentrations. Unlike ManLev, a compound that is not toxic even at very high concentrations (> 40 mM, Yarema et al., 1998), the

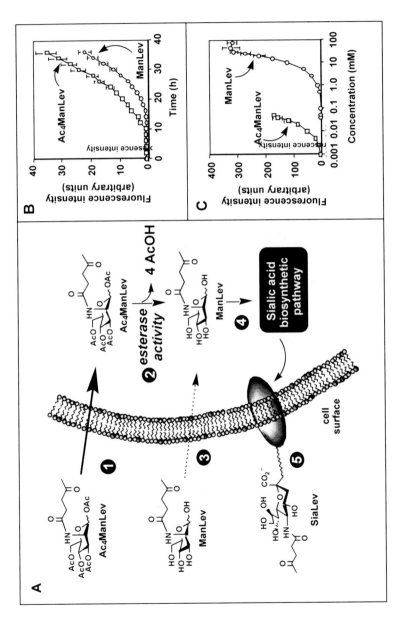

Figure 10. Acetylated ManLev is efficiently incorporated into the sialic acid pathway. **A.** Acetylated ManLev (Ac₄ManLev) has increased membrane permeability (**1**) compared to ManLev (**2**) due to its increased hydrophobicity. Once inside a cell, esterases remove the acetate groups (**3**) producing non-acetylated ManLev (**4**) that can be converted into SiaLev by the sialic acid biosynthetic pathway (**5**). **B.** Jurkat cells were incubated with 10 mM ManLev or 25 μM Ac₄ManLev and cell surface expression of SiaLev was determined at the indicated time points. **C.** Jurkat cells were incubated with various concentrations of ManLev or Ac₄ManLev for three days and the cell surface expression of SiaLev was determined as described (Figure 6).

acetylated derivative causes inhibition of cell growth and eventual toxicity for prolonged incubation above 25 µM in most human cells. This feature prevents saturation of the cell surface with SiaLev similar to that achieved with ManLev (Figure 10C). Nevertheless it should be emphasized that low $Ac_4ManLev$ concentrations (< 20 µM) that cause no adverse cellular effects are sufficient to provide the cell surface with an adequate number of SiaLev residues (>10^7/cell) for most subsequent ketone-based ligation applications.

3.6. REVERSIBILITY OF SiaLev EXPRESSION

One advantage of a substrate-based approach to cellular engineering, compared to the more established genetic approaches, is that cellular changes can be easily reversed simply by removing the unnatural analog thereby offering enhanced control of cellular manipulations. SiaLev expression, for example, rapidly diminishes upon removing ManLev from the culture medium. Under cell culture conditions the subsequent removal of ketones from the cell surface closely follows a dilution pattern expected from cell doubling (Yarema et al., 1998). SiaLev can also be removed from the cell surface by addition of the natural precursor ManNAc even in the continued presence of ManLev (Figure 11). Consequently, by providing the correct combination of the natural or unnatural metabolic substrate, the downstream expression of the unnatural biosynthetic product can be controlled with remarkable precision (Yarema et al., 1998).

The ability to competitively remove SiaLev from the cell surface with ManNAc is of limited practical significance in cell culture conditions where substrate-containing media can be easily replaced with substrate-free media. In this situation, subsequent cell growth ensures that cell surface expression is rapidly lost. In an animal model, however, cells are typically long-lived and may not normally recycle their all of their membrane components at a rapid rate. Furthermore, while free sugars are rapidly

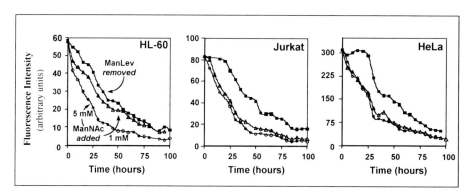

Figure 11. Turnover of SiaLev from the cell surface. HL-60, Jurkat, and HeLa cells were incubated with 5.0 mM ManLev for seven days to achieve equilibrium expression of SiaLev (Figure 7B). Removal of ManLev from the culture medium (at time = 0 h) results in a rapid decrease in SiaLev expression following approximately a one day lag period in Jurkat and HeLa cells; and almost immediately in HL-60 cells. SiaLev removal from the cell surface can be expedited by the addition of ManNAc (1.0 or 5.0 mM at time = 0 h) in the continued presence of 5.0 mM ManLev in the culture media.

cleared in vivo, acetylated derivatives may partition into lipid components and contribute to long-lasting cellular expression. Both of these factors increase the value of ManNAc as a potent in vivo competitor for rapid removal of unnatural sialosides from the cell surface of even slowly growing cells in continued presence of the unnatural precursor (Yarema et al., 1998).

4. Prospects and Applications for Substrate-Based Cellular Engineering

This article has focused on providing experimental details for the single example of ManLev-based cellular engineering. Characterization of the parameters of SiaLev can be broadly adapted to develop conditions for a host of substrate-based applications that are currently expanding in multiple directions. In addition to extension from cell culture to whole animal work as mentioned above, variations of this technique now being pursued include (1) new detection strategies, (2) new chemoselective coupling partners, (3) exploitation of additional biosynthetic pathways, and (4) the development of host cells for production of substrate-engineered bioproducts (Figure 12).

4.1. NEW STRATEGIES FOR THE DETECTION AND EXPLOITATION OF ENGINEERED CELL SURFACE EPITOPES

As described above, engineered sialoglycoconjugate epitopes have already been exploited to deliver biologically compatible compounds to the cell surface via the ketone-based chemoselective ligation reaction. This strategy shares similarities with immune-based approaches where antibody conjugates are exploited to deliver compounds, including diagnostic and therapeutic agents, to the cell surface. One application of these immune methods is delivery of toxins to tumor associated antigens as a means to selectively target cancer cells. Unfortunately antibody targeting approaches are hindered by structural diversity of epitopes, antibody cross-reactivity in vivo, and low density of antigenic determinants ($< 10^5$ per cell) (Lauffenburger and Linderman, 1993; Gupta and Weissleder, 1996; Lemieux et al., 1999). Considering that sialic acid is often over-expressed on cancer cells, one advantage to the targeting of SiaLev, compared to naturally occurring tumor associated antigens, is the increased number of the engineered epitopes (eg., 5×10^6 SiaLev residues) present on the cell surface compared to the naturally occurring antigens (eg., 50,000, Lemieux and Bertozzi, 1999).

In practice, the numerical advantage of the engineered sialic acid eptiopes may be offset by the relatively low chemical reactivity of a hydrazide or hydroxylamine probe at physiological pH (see discussion above) compared to the rapid binding of a biological probe such as an antibody. The ability to raise antibodies against unnatural sialic acid epitopes raises the intriguing possibility that they could be used as sensitive probes for the cellular presentation of these markers (Figure 12A) (Liu et al 2000, Lemieux et al 2001). Therefore, regarding an engineered sialoglycoconjugate, such as SiaLev, as an antigen rather than a "chemical handle" combines the quantitative advantages that these unnatural epitopes offer with the kinetic advantages of antibody-mediated recognition and binding events.

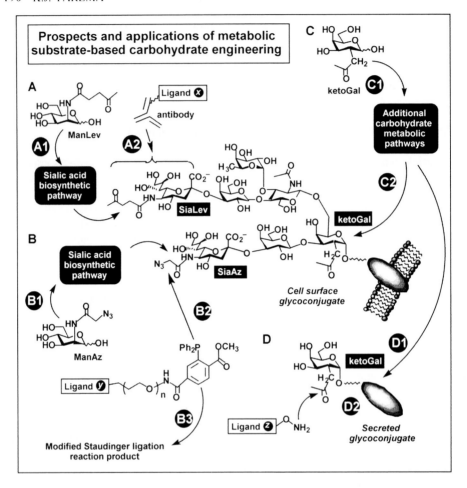

Figure 12. Prospective applications of substrate-based cellular engineering.   A. New detection strategies include antibody mediated recognition and binding of the engineered SiaLev epitope. **B.** New chemoselective ligation reactions exploit the sialic acid pathway to install the azide-containing sialoglycoconjugate "SiaAz" into cellular components. The azide can then react with a ligand containing a phosphine via a modified Staudinger reaction (see Saxon and Bertozzi, 2000 for details). **C.** Additional biosynthetic pathways can be accessed by compounds such as "ketoGal" (Hang and Bertozzi, 2001). **D.** Substrate-based methods can also be exploited for production of glycoproteins, in the example shown ketoGal is used to produce a glycopeptide containing a single unnatural sugar reside proximal to the peptide backbone that can be further elaborated with synthetically produced oligosaccharide coupling partners (Figure 13).

4.2. NEW CHEMOSELECTIVE LIGATION STRATEGIES

Ketone ligation reactions are useful for cell surface engineering applications but have limited intracellular use due to competition with endogenous ketone-containing metabolites. Development of chemoselective ligation reactions in which both coupling partners are abiotic and orthogonal toward all chemical functionalities found within a cell is required to extend substrate-based approaches to the intracellular millieu. Recently a second chemoselective ligation reaction based on the Staudinger reaction has been developed that meets these criteria (Saxon and Bertozzi, 2000). The Staudinger reaction occurs between a phosphine and an azide to produce an aza-ylide. These coupling partners are both completely abiotic and the azide, like the ketone, is a sterically constrained functional group tolerated by the sialic acid biosynthetic machinery. The azide can be appended to the N-acyl position of ManNAc to provide a vehicle for delivery into cellular glycoconjugates in the form of the corresponding azido-sialic acid (Figure 12B). Once incorporated into a cellular component, the azide is capable of facile chemical reaction with a phosphine to form an aza-ylide. In the classical Staudinger reaction this intermediate hydrolyzes spontaneously in an aqueous environment to yield a primary amine and phosphine oxide. Therefore, to render this reaction compatible with physiological conditions, the phosphine coupling partner was designed to include an appropriately situated electrophilic trap that enables the aza-ylide intermediate to rear-range to a stable amide bond (Saxon and Bertozzi, 2000). This modification of the Stau-dinger reaction extends substrate-based approaches from the cell surface to the intra-cellular environment.

4.3. EXTENSION OF SUBSTRATE-BASED APPROACHES TO ADDITIONAL BIOSYNTHETIC PATHWAYS

Despite the remarkable utility of sialoglycoconjugate engineering, analogous exploitation of additional metabolic pathways will even further enhance the ability of the cellular engineer to precisely manipulate the complex carbohydrates of a cell. One promising step in this direction is the expression of "ketoGal", a ketone-containing analog of N-acetylgalactosamine (GalNAc), on the surface of mammalian cells (Figure 12C) (Hang and Bertozzi, 2001). Once incorporated into a cellular glycoconjugate, the ketone moiety of ketoGal can be used as a coupling partner in chemoselective ligation reactions similar to those described for SiaLev. Unlike sialic acid residues that usually occur at the outer termini of complex carbohydrates, GalNAc is situated at an internal position, proximal to the peptide chain in O-linked glycoconjugates, as well as in the repeating units of the polymeric oligosaccharide chondroitin sulfate. The ongoing development of metabolic analogs for presentation in a variety of structural contexts continues to broaden the scope of substrate-based engineering approaches.

This report has focused on methods to incorporate unnatural substituents into the carbohydrate component of cells by exploiting the substrate permissivity of the naturally occurring enzymes of the cell. Protein engineering techniques now available include directed evolution methods that expedite the development of enzymes with extended substrate specificity (Fong et al., 2000; Zhang et al., 1997). This ability to custom design enzymes to process unnatural substrates that are currently unpalatable for the native enzymes of a cell will allow metabolic substrate-based cellular engineering endeavors to be extended to additional biosynthetic pathways. The combination of chemical and genetic approaches will provide a powerful tool to direct the construction of cellular components with well-defined oligosaccharide landscapes of novel composition (Yarema and Bertozzi, 1998; Bertozzi and Kiessling, 2001). Finally, cellular engineering efforts to combine chemistry and genetics are not limited to the field of glycobiology. In particular, protein engineers are currently exploring use of unnatural substrates in organisms harboring expanded genetic repertoires in their quest to incorporate amino acids analogs into the proteins of a cell (Liu and Schultz, 1999; Kiick and Tirrell, 2000; Sharma et al., 2000; Wang et al., 2001; Döring et al., 2001).

4.4. INDUSTRIAL PRODUCTION OF SUBSTRATE-ENGINEERED BIOMOLECULES

A major focus of current cellular engineering efforts is the development of cells that function as bioreactors for industrial production of primary or secondary metabolites as well as recombinant proteins. A limitation of microbial protein expression systems is their inability to install many of the post-translational modifications necessary for correct functioning of the protein (Fussenegger et al., 1999). In some cases, these problems are alleviated by use of eukaryotic systems including yeast (Wiseman, 1991), insect (Goosen et al., 1993) or even animal (most often rodent) (Cartwright 1994) cells as production hosts. In other situations, most notably the production of glycoproteins, the installation of the "correct" oligosaccharide is highly species-specific and finding the appropriate cell to function as a production host can be a formidable challenge. Extension of metabolic substrate-based techniques, from the engineering of cellular components as described in this report, to the installation of unnatural constituents in secreted products may represent a significant advance in the ability of the biotechnology industry to produce biologically relevant glycoproteins (Figure 12D).

GalNAc analogs (Figure 12C) are ideally suited for engineering the carbohydrate portion of a glycoprotein. First, rodent cells that are typically better suited to protein overproduction than human cells are, utilize GalNAc analogs much more efficiently than ManNAc analogs (Yarema et al., 1998; Hang and Bertozzi, 2001). Importantly, glycosylation deficient variations of rodent cell lines (such as Chinese hamster ovary) are available that display truncated carbohydrates on their glycoproteins. Such cell lines may facilitate the production of glycoproteins with their glycosylation sites decorated with a single, unnatural sugar residue (Figure 13). If these secreted proteins, analogous to substrate-engineered cell surface glycoconjugates, feature a chemical handle as part of the unnatural monosaccharide (Figure 12D), they can be further elaborated by the chemoselective ligation methods previously discussed.

The approach where a monosaccharide appended to a peptide backbone is further elaborated by attachment of an additional carbohydrate has already been demonstrated in a wholly synthetic model system. Specifically, an antibacterial insect glycopeptide exhibited biological activity even though the carbohydrate structure contained an un- natural glycosidic connection (Figure 13). (Rodriguez et al., 1997; Winans et al., 1999). Prospects for the future include development of methods to extend the size of the pep- tide (currently limited by synthetic concerns) to full size proteins (available by recom- binant DNA expression in host cells compatible with substrate-based approaches). Sub- sequently, installation of the increasingly complex oligosaccharides now available by sophisticated synthetic schemes onto protein scaffolds engineered by metabolic substrate-based methods will allow production of large glycoproteins not attainable by present techniques.

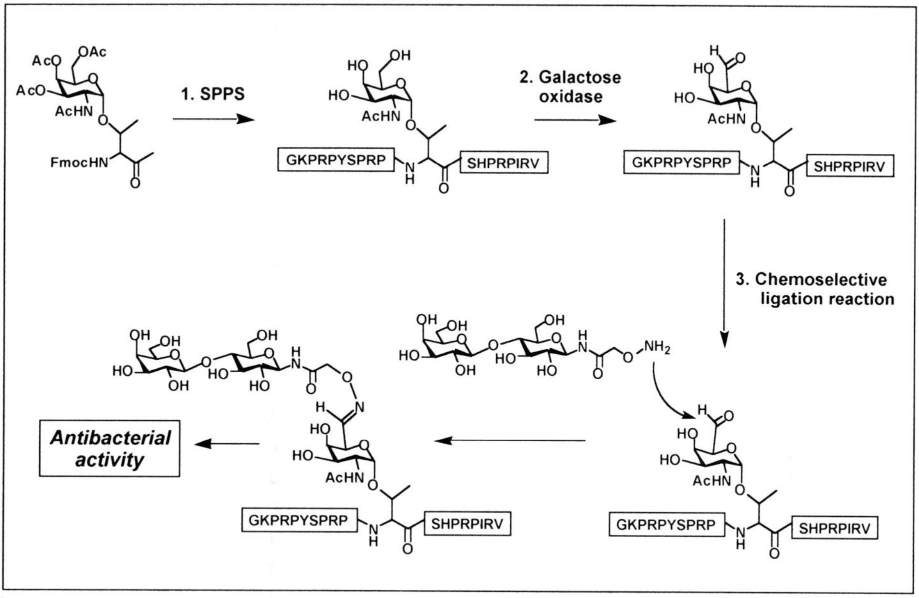

Figure 13. Emerging strategies for production of complex glycoproteins. The complete chemical synthesis of the glycopeptide drosocin by solid phase peptide synthesis (SPPS, step 1), was followed by enzymatic oxidation to produce the C-6 aldehyde (step 2). The aldehyde, similar to the ketone, can chemoselectively couple with an aminooxy producing a stable oxime linkage (refer to Figure 5). Subsequent reaction with an incoming lactose hydroxylamine disaccharide (step 3) produces a biologically active form of glycosylated drosocin (adapted from Rodriguez et al., 1997). A combination of recombinant DNA protein expression and metabolic substrate-based methods promise to expedite the first two steps of this process. Such a streamlined strategy will facilitate the production of full-size, singly glycosylated proteins that contain an unnatural sugar residue poised for reaction with any synthetic oligosaccharide containing the requisite chemical functionality for chemoselective ligation.

The ability to produce homogenous glycoproteins for characterization and functional studies is an important concern considering the heterogenous mixture of glycoforms typically obtained from in vitro expression systems is a significant factor hindering the development of glycoprotein-based pharmaceutical compounds (Cumming, 1991). Metabolic substrate-based methods, therefore, are a valuable component of emerging semi-synthetic protein engineering technology now being developed to facilitate the production of high quality, homogeneous glycoproteins (Macmillan and Bertozzi, 2000).

5.   References

Angelino N.J., Bernacki R.J., Sharma M., Dodson-Simmons O., and Korytnyk W. (1995) Versatile intermediates in the selective modification of the amino function of 2-amino-2-deoxy-D-mannopyranose and the 3-position of 2-acetamido-2-deoxy-D-mannose: potential membrane modifiers in neoplastic control. Carbohydr Res. 276: 99-115.
Bailey J.E. (1991) Toward a science of metabolic engineering. Science 252: 1668-75.

Bertozzi C.R., and Kiessling L.L. (2001) Chemical glycobiology. Science 291: 2357-64.

Cameron D.C., and Tong I.T. (1993) Cellular and metabolic engineering. An overview. Appl. Biochem. Biotechnol. 38: 105-40.

Cartwright T. (1991) Animal cells as bioreactors. Cambridge University Press, Cambridge, England.

Collins B.E., Fralich T.J., Itonori S., Ichikawa Y., and Schnaar, R.L. (2000) Conversion of cellular sialic acid expression from N-acetyl- to N-glycolylneuramininic acid using a synthetic precursor, N-glycolylmannosamine pentaacetate: inhibition of myelin-associated glycoprotein binding to neural cells. Glycobiology 10: 11-20.

Cumming D.A. (1991) Glycosylation of recombinant protein therapeutics: control and functional implications. Glycobiology 1: 115-30.

Döring V., Mootz H.D., Nangle L.A., Hendrickson T.L., de Crécy-Lagard V., Schimmel P., and Marlière P. (2001) Enlarging the amino acid set of Escherichia coli by infiltration of the valine coding pathway Science 292: 501-4.

Fong S., Machajewski T.D., Mak C.C. and Wong.C.-H. (2000) Directed evolution of D-2-keto3-deoxy-6-phosphogluconate aldolase to new variants for the efficient synthesis of D- and L-sugars. Chem. Biol. 7: 873-83.

Fussenegger M., Bailey J.E., Hauser H., and Mueller P.P. (1999) Genetic optimization of recombinant glycoprotein production by mammalian cells. Trends Biotechnol. 17: 35-42.

Gombert A.K., and Nielsen J. (2000) Mathematical modelling of metabolism. Curr. Opin. Biotechnol. 11: 180-6.

Goon S., and Bertozzi C.R. (2001) Metabolic substrate engineering as a tool for glycobiology. In: Glycobiology: Principles, Synthesis, and Applications. Wang P.G., and Bertozzi C.R. (Eds.) Marcel Dekker, Inc., New York, NY, 641-74.

Goosen M.F.A. (1993) Insect cell culture engineering: an overview. In: Insect cell culture engineering. Goosen M.F.A., Daugulis A.J., and Faulkner P. (Eds.) Marcel Dekker, Inc., New York, NY, 1-16.

Gupta H. and Weissleder R. (1996) Targeted contrast agents in MR imaging. Magn. Reson.Imaging. Clin. N. Am. 4: 171-84.

Hang H.C. and Bertozzi C.R. (2001) Ketone isosteres of 2-N-acetamindosugars as substrates for metabolic cell surface engineering. J. Am. Chem. Soc. 123: 1242-3.

Hatzimanikatis V., Choe L.H., and Lee K.H. (1999) Proteomics: theoretical and experimental considerations. Biotechnol. Prog. 15: 312-8.

Jacobs C.L., Yarema K.J., Mahal L.K., Nauman D.A., Charters N.W., and Bertozzi C.R. (2000) Metabolic labeling of glycoproteins with chemical tags through unnatural sialic acid biosynthesis. Methods Enzymol. 327: 260-75.

Kao, C.M. (1999) Functional genomic technologies: creating new paradigms for fundamental and applied biology. Biotechnol. Prog. 15: 304-11.

Kayser H., Zeitler R., Kannicht C., Grunow D., Nuck R., and Reutter W. (1992) Biosynthesis ofa nonphysiological sialic acid in different rat organs using N-propanoyl-D-hexosamines as precursors. J. Biol. Chem. 267: 16934-8.

Kelm S., and R. Schauer R. (1997) Sialic acids in molecular and cellular interactions. Int. Rev.Cytol. 175: 137-240.

Keppler O.T., Horstkorte R., Pawlita M., Schmidt C., and Reutter, W. (2001) Biochemical engineering of the N-acyl side chain of sialic acid: biological implications. Glycobiology 11: 11R-18R.

Keppler O.T., Stehling P., Herrmann M., Kayser H., Grunow D., Reutter W., and Pawlita M. (1995) Biosynthetic modulation of sialic acid-dependent virus-receptor interactions of two primate polyoma viruses. J. Biol. Chem. 270: 1308-14.

Kiicka K.L. and Tirrell D.A. (2000) Protein engineering by in vivo incorporation of non-natural amino acids: control of incorporation of methionine analogues by Methionyl-tRNA Synthetase Tetrahedron 56: 9487-93.

Lauffenburger D.A., and Linderman, J.J. (1993) Receptors: Models for binding, trafficking, and signaling Oxford University Press, Oxford, England.

Lee, J.H., Baker T.J., Mahal L.K., Zabner J., Bertozzi C.R., Wiemer D.F., and Welsh M.J. (1999) Engineering novel cell surface receptors for virus-mediated gene transfer. J. Biol. Chem. 31: 21878-84.

Lemieux, G.A., and Bertozzi C.R. (1998) Chemoselective ligation reactions with proteins,oligosaccharides and cells. Trends Biotechnol. 12: 506-13.

Lemieux G.A., and Bertozzi C.R. (2001). Modulating cell surface immunoreactivity by metabolic induction of unnatural carbohydrate antigens. Chem Biol. 8: 265-75.

Lemieux, G.A., Yarema K.J., Jacobs C.L. and Bertozzi C.R. (1999) Exploiting differences in sialoside expression for selective targeting of MRI contrast reagents. J. Am. Chem. Soc. 121: 4278-80.

Liu D.R., and Schultz P.G. (1999) Progress toward the evolution of an organism with an expanded genetic code. Proc. Natl. Acad. Sci. U.S.A. 96, 4780-5.

Liu T., Guo Z., Yang Q., Sad S., and Jennings H.J. (2000) Biochemical engineering of surface   2-8 polysialic acid for immunotargeting tumor cells. J. Biol. Chem. 275: 32832-6.

Mahal L.K., Yarema K.J., and Bertozzi C.R. (1997) Engineering chemical reactivity on cell surfaces through oligosaccharide biosynthesis. Science 276: 1125-8.

Mahal L.K., Yarema K.J., Lemieux G.A., and Bertozzi C.R. (1999) Chemical approaches toglycobiology: engineering cell surface sialic acids for tumor targeting. In Sialobiology and Other Novel Forms of Glycosylation. Inoue Y., Lee Y.C., and Troy II F.A. (Eds.) Gakushin Publishing Company, Osaka, Japan.

Marcaurelle L.A. and Bertozzi C.R. (2001) Chemoselective elaboration of O-linked glycopeptide   mimetics by alkylation of 3-thioGalNAc. J. Am. Chem. Soc. 123: 1587-95.

Macmillan D., and Bertozzi C.R. (2000) New directions in glycoprotein engineering. Tetrahedron, 56: 9515-25.

Rodriguez E.C., Winans K.A., King D.S., and Bertozzi C.R. (1997) A strategy for the chemoselective synthesis of O-linked glycopeptides with native sugar-peptide linkages, J. Am. Chem. Soc. 119: 9905-6.

Rideout, D., Calogeropoulou T., Jaworski J., and McCarthy M. (1990) Synergism through direct covalent bonding between agents: a strategy for rational design of chemotherapeuticcombinations. Biopolymers 29: 247-62.

Sarkar A.K., Fritz T.A., Taylor W.H., and Esko J.D. (1995) Disaccharide uptake and priming in animal cells: inhibition of sialyl Lewis X by acetylated Gal beta 1-->4GlcNAc beta-O-naphthalenemethanol. Proc. Natl. Acad.Sci. U.S.A. 92: 3323-7.

Saxon E. and Bertozzi C.R. (2000) Cell surface engineering by a modified Staudinger reaction. Science 287: 2007-10.

Schilling C.H., Schuster S., Palsson B.O., and Heinrich R. (1999) Metabolic pathway analysis: basic concepts and scientific applications in the post-genomics era. Biotechnol. Prog. 15: 296-303.

Schmidt, C., Ohlemeyer C., Kettenmann H., Reutter W. and Horstkorte R. (2000) Incorporation of N-propanoylneuraminic acid leads to calcium oscillations in oligodendrocytes upon the application of GABA. FEBS Lett. 478: 276-80.

Schmidt C., Stehling P., Schnitzer J., Reutter W., and Horstkorte R. (1998) Biochemical engineering of neural cell surfaces by the synthetic N-propanoyl-substituted neuraminic acid precursor. J. Biol. Chem. 273: 19146-52.

Schultz A.M., and Mora P.T. (1975) Inhibition of the metabolism of amino sugars with 2-deoxy-2-(2-fluoroacetamido)- -D-glucopyranose. Carbohydrate Res. 40: 119-27.

Schuster S., Fell D.A., and Dandekar T. (2000) A general definition of metabolic pathways useful for systematic organization and analysis of complex metabolic networks. Nat. Biotechnol. 18: 326-33.

Schwartz E.L., Hadfield A.F., Brown A.E., and Sartorelli A.C. (1983) Modification of sialic acid metabolism of murine erythroleukemia cells by analogs of N-acetylmannosamine.Biochim. Biophys. Acta 762: 489-97.

Sell S. (1990) Cancer-associated carbohydrates identified by monoclonal antibodies. Hum. Pathol. 21: 1003-19.

Sharma N., Furter R, Kast P, and Tirrell D.A. (2000) Efficient introduction of aryl bromide functionality into proteins in vitro. FEBS Lett. 467: 37-40.

Sillanaukee P., Ponnio M., and Jaaskelainen I.P. (1999) Occurrence of sialic acids in healthy humans and different disorders. Eur. J. Clin. Invest. 29: 413-25.

Takano R., Muchmore E., and Dennis J.W. (1994) Sialylation and malignant potential in tumour cell glycosylation mutants. Glycobiology 4: 665-74.

Varki A. (1997) Sialic acids as ligands in recognition phenomena. FASEB J. 11: 248-55.

Wang L., Brock A., Herberich B., and Schultz P.G. (2001) Expanding the genetic code of Escherichia coli Science 292: 498-500.

Wieser J.R., Heisner A., Stehling P., Oesch F., and Reutter W. (1996) In vivo modulated N-acyl side chain of N-acetylneuraminic acid modulates the cell contact-dependent inhibition of growth. FEBS Lett. 395: 170-3.

Winans K.A., King D.S., Rao V.R., and Bertozzi C.R. (1990) A chemically synthesized version of the insect antibacterial glycopeptide, diptericin, disrupts bacterial membrane integrity. Biochemistry 38: 11700-10.

196   K.J. YAREMA

Wiseman, A. (1991) Genetically-engineered proteins and enzymes from yeast: production control. Ellis Horwood Limited, Chichester, West Sussex, England.

Wong C.-H., Moris-Varas R., Hung S.-C., Marron T.G., Lin C-C., Gong K.W., Weitz-Schmidt G. (1997) Small molecules as structural and functional mimics of sialyl Lewis x tetrasaccharide in selectin inhibition: A remarkable enhancement of inhibition by additional negative charge and/or hydrophobic group. J. Am. Chem Soc. 119: 8152-8.

Yarema, K.J. (2001) New directions in carbohydrate engineering: A metabolic substrate-based approach. Biotechniques, in press.

Yarema K.J., and Bertozzi C.R. (1998) Chemical approaches to glycobiology and emerging carbohydrate-based therapeutic agents. Curr. Opin. Chem. Biol., 2: 49-61.

Yarema K.J., and Bertozzi C.R. (2001) Characterizing glycosylation pathways. GenomeBiology 2: r4.1-r4.10.

Yarema K.J., Goon S., and Bertozzi C.R. (2001) Metabolic selection of glycosylation defects in human cells. Nat. Biotechnol. in press.

Yarema K.J., Mahal L.K., Bruehl R.E, Rodriguez E.C., and Bertozzi C.R. (1998) Metabolic delivery of ketone groups to sialic acid residues. Applications to cell surface glycoform engineering. J. Biol. Chem. 273: 31168-79.

Zhang J.H., Dawes G., and Stemmer W.P. (1997) Directed evolution of a fucosidase from a galactosidase by DNA shuffling and screening. Proc. Natl. Acad. Sci. U.S.A. 94: 4504-9.

8. ADDRESSING INSECT CELL GLYCOSYLATION DEFICIENCIES THROUGH METABOLIC ENGINEERING

S.M. LAWRENCE and M.J. BETENBAUGH*
Department of Chemical Engineering, The Johns Hopkins University, 400 N. Charles St., Baltimore, MD 21218 U.S.A.
*Corresponding Author: (410)516-5461, beten@jhu.edu

1. Introduction

1.1. RECOMBINANT PROTEINS AND BIOTECHNOLOGY

As the biotechnology industry matures, recombinant proteins are becoming increasingly valuable pharmaceutical products. These recombinant proteins, such as insulin or erythropoietin, are complex, high molecular weight heteropolymers of amino acids. Consequently, fermentation and cell culture remain the best methods for their large-scale production unlike more traditional, lower molecular weight pharmaceuticals that are often generated by chemical synthesis.

Living cells produce proteins through the process of translation which results in the primary amino acid structure of a protein. While many hosts are available for recombinant protein production (Jenkins *et al.*, 1996), each of these cell systems has advantages and disadvantages (Prokop *et al.*, 1991) and range in evolutionary sophistication from bacteria to human cells from specific tissues. Table 1 summarizes the benefits and drawbacks relevant to biotechnology of the most commonly used recombinant hosts. In general, the more evolutionarily advanced the cell line used for a particular protein expression system, the more sophisticated the product as far as proper folding, post-translational modification, and the likelihood of acceptance by the immune system when taken therapeutically. However, more complex cell lines typically have lower protein yields and tend to have higher production costs.

1.2. THE BACULOVIRUS EXPRESSION SYSTEM

According to the criteria of Table 1, insect cell culture, taken in this review to mean insect cell lines used with the baculovirus expression system, is an intermediate system. The cost of insect cell culture is higher than that for yeast and bacterial expression systems. The yields of the baculovirus system, however, are often much higher than mammalian cells and approach the yields of bacterial culture (O'Reilly *et al.*, 1992).

In the baculovirus expression system, foreign genes are expressed by genetic engineering of the foreign gene into the genome of the baculovirus. Foreign genes replace native baculovirus genes that are non-essential to viral propagation in cell culture

M. Al-Rubeai.(ed.), Cell Engineering, 197-214.
© 2002 *Kluwer Academic Publishers. Printed in the Netherlands.*

Table 1. Comparison of common recombinant protein expression systems.

Expression System	Cost	Expression Levels	Post-translational Modifications	N-Glycosylation	Secretion	Other Notes
Bacteria	$	High	No	No	Into Periplasm	Correct folding problematic
Yeast	$$	Low/ Moderate	Most	High Mannose	Yes	
Insect	$$$	Moderate/ High	Yes	Typically high mannose or paucimannosidic	Yes	No FDA approved process for therapeutic proteins
Mammalian	$$$$	Low/ Moderate	Yes	Complex	Yes	Many lines require serum or cannot be grown in suspension

(O'Reilly *et al.*, 1992). Proteins are produced in very high yields during the late stages of infection when expressed under the strong *polyhedrin* (*polh*) or *p10* promoters, although other promoters are available that vary in extent and chronology of expression (O'Reilly *et al.*, 1992). Foreign genes can be cloned into a baculovirus using a number of commercially available systems, and the resulting recombinant baculoviruses are used to infect cell lines derived from insects including the fall army worm (*Spodoptera frugiperda*; Sf9 or Sf21), the cabbage looper (*Trichoplusia ni*; High-Five), *Mamestra brassicae* (Mb-0503), or *Estigmene acrea* (Ea4). During the infection, the virus takes over the cellular machinery and produces the recombinant protein under control of the *polh* or other promoters. While baculovirus infection is lytic and the host cells die of the infection within days, new budded viruses are produced which can be used to infect more cells (O'Reilly *et al.*, 1992).

1.3. N-GLYCOSYLATION

The baculovirus system produces recombinant proteins with many post-translational modifications including proteolytic cleavage, glycosylation, acylation, amidation, phosphorylation, prenylation, and carboxymethylation (O'Reilly *et al.*, 1992). However, the post-translational modifications such as glycosylation are often not as "complex" as those produced in mammalian cell lines. The degree of complexity or processing of the final glycosylation moiety can affect many protein properties including solubility, activity, and antigenicity (Goochee *et al.*, 1991; Jenkins *et al.*, 1996), and these properties are all of considerable concern to the biotechnology industry.

During glycosylation, carbohydrates are added to and removed from specific amino acids of a protein. In *O*-glycosylation, carbohydrates are added to serine (Ser) or threonine (Thr) residues. In *N*-glycosylation, carbohydrates are added to asparagine (Asn) residues of the consensus sequence Asn-X-Ser/Thr. *N*-Glycosylation starts with the transfer of a specific polysaccharide structure from a dolichol-phosphate donor in the endoplasmic reticulum (ER). The polysaccharide is then modified by the removal of sugars with glycosidase enzymes and by the addition of sugars with glycosyltransferase enzymes. In mammalian cells, this process has been extensively studied (Kornfeld and Kornfeld, 1985).

The final N-glycosylation moiety depends on many factors. As indicated in Table 1, different species possess differing glycosylation abilities, but, even within species, cell lines derived from different tissues produce different glycosylation patterns. The physiological condition of the cell also plays an important role. The availability of glucose (Glc) and other sugars, lipids, and nucleotides can change the glycosylation patterns of a cell (Jenkins et al., 1996). Furthermore, the presence of serum or ammonia can also affect glycosylation (Jenkins et al., 1996). Cells glycosylate different proteins to different extents, and even the individual sites within a glycoprotein can be glycosylated differently for various reasons including steric hindrance (Warren, 1993).

Therefore, populations of several different glycosylation moieties can be generated on a particular glycosylation site. The biantennary N-glycan moiety terminating in sialic acid residues is the benchmark for complex N-glycosylation (Figure 1A). However, the final glycan can have further modifications including additional branching (Takeuchi and Kobata, 1991), extensive fucosylation (Tollefsen and Rosenblum, 1988), or terminal sulfation (Bergwerff et al., 1995).

1.4. N-GLYCOSYLATION IN INSECT CELLS

Initial studies of N-glycosylation in insects were done using widely studied species such as *Drosophila* (Parker et al., 1991) and the mosquito (Hsieh and Robbins, 1984). While the initial processing events are similar to those in mammals, insect cells do not achieve the complex glycosylation patterns observed in mammals (Butters and Hughes, 1981; Hsieh and Robbins, 1984; Parker et al., 1991). Because of the interest in the baculovirus expression system in biotechnology, considerable research has gone into identifying the N-glycans produced by insect cell culture. Studies of the native and recombinant glycoproteins of baculovirus-compatible cell lines show a prevalence of high-mannose

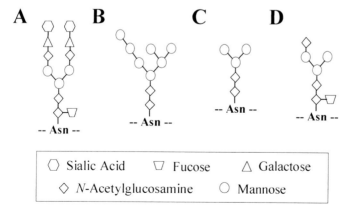

Figure 1. N-glycan types. (A) Complex. (B) High-mannose. (C) Paucimannosidic. (D) Hybrid.

structures, which include large numbers of mannose (Man) residues (Figure 1B), and paucimannosidic structures, which terminate in the three (Figure 1C) or even two "core" mannose residues (Jarvis and Summers, 1989; Kuroda *et al.*, 1990; Grabenhorst *et al.*, 1993; Yeh *et al.*, 1993; Kubelka *et al.*, 1994; Wagner *et al.*, 1996; Lopez *et al.*, 1997; Kulakosky *et al.*, 1998). Occasionally hybrid structures, including those terminating in *N*-acetylglucosamine (GlcNAc; Kubelka *et al.*, 1994; Wagner *et al.*, 1996) or even galactose (Gal; Ogonah *et al.*, 1996; Hsu *et al.*, 1997) on one branch (Figure 1D), have been identified. Isolated instances of complex structures (Figure 1A) such as sialylated *N*-glycans have been reported (Davidson and Castellino, 1991; Davis and Wood, 1995). As in mammalian cells, glycosylation in insect cells depends on the cell line, the target glycoprotein, and the cell culture conditions. However, insect cells generally produce glycoproteins with *N*-glycans that are less complex than those produced by mammalian cell lines.

1.5. THE SIALIC ACIDS AND GLYCOSYLATION

The sialic acids are a family of nine carbon 2-keto-3-deoxy sugars that have over 40 members, and their presence can significantly affect glycoprotein properties (Schauer *et al.*, 1995). They occupy the termini of glycoproteins and are somewhat unique sugars as they have pKa's of approximately 2.6 (De Bruyn and Michelson, 1979) and therefore are negatively charged at physiological pH. The most common sialic acid is *N*-acetyl-neuraminic acid (Neu5Ac) which contains an *N*-acetyl group at the 5-carbon position (Figure 2A). This group is often oxidized, particularly in non-human mammals (Much-more *et al.*, 1998), to an *N*-glycolyl group forming *N*-glycolylneuraminic acid (Neu5Gc; Figure 2B). Deaminated sialic acid, 2-keto-3-deoxy-D-*glycero*-D-*galacto*-nonoic acid (KDN; Figure 2C), was first identified in 1986 (Nadano *et al.*, 1986). Other common sialic acids have *N*-acetyl groups substituted at multiple positions (Schauer *et al.*, 1995).

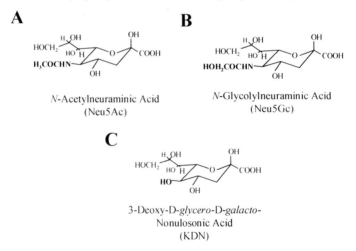

Figure 2. The sialic acids. (A) *N*-Acetylneuraminic acid. (B) *N*-Glycolylneuraminic acid. (C) 3-Deoxy-D-*glycero*-D-*galacto*-nonulosonic acid. Substituents at the 5-carbon position are shown in bold type.

The presence of sialic acids has been indicted in a number of biological recognition events of interest involving cancer, metastasis, microbe pathogenicity, and development (Schauer *et al.*, 1995). The effects of sialic acid on *in vivo* circulatory half-life are of particular interest to the pharmaceutical industry. Glycoproteins in the bloodstream that do not terminate in sialic acid residues are filtered by receptors in the liver and consequently have lower efficacies (Grossmann *et al.*, 1997). For example, recombinant human thyrotropin expressed in insect cells had three times the *in vitro* activity than that produced in the mammalian CHO cell line (Grossmann *et al.*, 1997). However, the CHO derived protein had much more *in vivo* activity than the recombinant insect thyrotropin. The lower *in vivo* activity was likely caused by the absence of terminal sialic acid residues resulting in removal from the circulatory system by asialoglycoprotein and mannose receptors in the liver (Grossmann *et al.*, 1997).

2.   Metabolic Engineering of the Glycosylation Pathways

A process for producing the complex glycoproteins similar to mammalian cells with the baculovirus expression system would have significant commercial value. Such a system would require a metabolic engineering approach in which the normal metabolic functions of an insect cell are "mammalianized." This can be accomplished by procedures as simple as adding sugars to the cell growth medium to stimulate beneficial pathways or by adding inhibitors to shunt counterproductive pathways. More complex efforts would involve genetic engineering. Recombinant enzymes catalyzing desirable pathways can be introduced, and deleterious native enzymes down-regulated by knockout mutations or anti-sense inhibition.

Before describing previous efforts in metabolic engineering with insect cells, the existing glycosylation machinery for generating the benchmark complex glycoprotein, a biantennary glycoprotein terminating in sialic acid (Figure 1A), is described. The process of mammalian *N*-glycosylation has been well characterized and is the result of glycosidases that selectively remove sugars (Figure 3A) and glycosyltransferases that selectively add sugars (Figure 3B) with specific linkages (Kornfeld and Kornfeld, 1985).

After the *en bloc* transfer of the $Glc_3Man_9GlcNAc_2$ residue (Figure 3A) from the dolichol-phosphate donor, the glucose residues are removed by the enzymes α-glucosidase I and II in the ER (Kornfeld and Kornfeld, 1985). Trimming of mannose residues by α-mannosidase I leaves a $Man_5GlcNAc_2$ residue to which GlcNAc is transferred to the α-1,3 mannose branch by GlcNAc transferase I. The activities of α-mannosidase II and core fucosyltransferases in insect cell lines heavily depend on the presence of the initial GlcNAc (Altmann and Marz, 1995; Altmann *et al.*, 1993). After α-mannosidase II removes the two mannose (Man) residues from the α-1,6 branch, GlcNAc transferase II adds a second GlcNAc residue to the $GlcNAcMan_3GlcNAc_2\pm Fuc$ acceptor. The action of galactosyltransferases and sialyltransferases successively complete the complex glycan structure. Many of these enzyme activities have been examined in a variety of baculovirus-compatible insect cell lines, and select findings are summarized in Table 2. Of the transferases required for complex glycosylation, low levels of GlcNAc transferase II and Gal transferase activities exist in insect cells as well

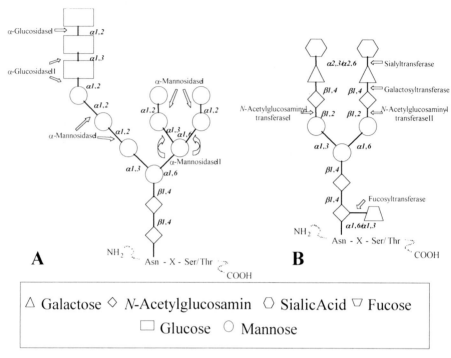

Figure 3. N-Glycan transferase and glycosidase activities. N-Glycan sugar linkages of (A) the initial glycan structure transferred from dolichol phosphate donor and (B) the mature complex biantennary glycan terminating in sialic acid. The sugar linkages removed by glycosidases (A) and added by glycosyltransferases (B) are also illustrated.

as a total absence of sialyltransferase activity. The results of the enzymatic activity assays explain the findings of N-glycan structural analysis in insect cells. While the initial N-glycan processing enzymes are present and similar to their mammalian counterparts (Altmann *et al.*, 1999), the enzymes of the later N-glycan processing pathway that give rise to complex N-glycans are present in conspicuously low levels if at all.

The activity of the mannosidases should also be noted. In mammalian cells, several different mannosidases have been identified, each with slightly different substrate specificities (reviewed by Moreman *et al.*, 1994). Often, an α-1,2-mannosidase is reported in the ER which cleaves the first Man residue with a Golgi α-1,2-mannosidase removing the three remaining α-1,2-Man linkages (Kornfeld and Kornfeld, 1985). In addition, endomannosidases that cleave Man residues within the glycan structure and other specialized mannosidases have been described (Moreman *et al.*, 1994). Redundant mannosidase pathways apparently also exist in insects as a cloned gene originally identified as an α-mannosidase II (Jarvis *et al.*, 1997) appears to have an alternate activity (Kawar *et al.*, 2000). Furthermore, unusual N-glycans missing the core α-1,3-Man branch have been frequently documented in Sf9 cells (Altmann *et al.*, 1999). Whether this Man

Table 2. Summary of native insect cell N-glycan processing enzymes. Bm and Mb indicate *Bombyx mori* and *Mamestra brassicae* cell lines, respectively.

Enzyme	Notes	Cell Line(s)	References
α-Glucosidase I and II	Standard inhibitors used to demonstrate processing	Sf9	Marchal *et al.*, 1999
α-Mannosidase I	Increased activity during baculovirus infection	Sf21	Davidson *et al.*, 1991
	Enzyme purification	Sf21	Ren *et al.*, 1995
	Cloning of gene	Sf9	Kawar *et al.*, 1997
GlcNAc Transferase I	Increased activity during baculovirus infection	Sf21	Velardo *et al.*, 1993
	High preference for Man₅ substrate compared to mammalian cell line	Sf9, Sf21, Bm, Mb	Altmann *et al.*, 1993
α-Mannosidase II	Requires addition of first GlcNAc for activity	Bm, Sf21, Mb	Altmann and Marz, 1995
	Purification	Sf21	Ren *et al.*, 1997
Fucosyltransferase	α-1,3 and α-1,6 activities not differentiated; Activity increases with existing GlcNAc residues	Sf9, Sf21, Mb, Bm	Altmann *et al.*, 1993
	α-1,3 and α-1,6 fucose transferred by different enzymes; α-1,6 activity occurs first		Staudacher and Marz, 1998
N-Acetylglucosaminidase	Membrane bound	Sf21, Mb, Bm	Altmann *et al.*, 1995
GlcNAc Transferase II	Much lower specific activity than GlcNAc transferase I	Sf9, Mb, Bm	Altmann *et al.*, 1993
Galactosyltransferase	Very low specific activity	High 5, Sf9, Mb	Van Die *et al.*, 1996
Sialyltransferase	ABSENCE of activity	Sf9	Hooker *et al.*, 1999

removal is the result of genuine enzymatic activity or represents glycan degradation with time remains unknown (Altmann *et al.*, 1999).

Another interesting aspect of insect cell glycosylation that strongly affects the ultimate glycan structure is the presence of a *N*-acetylglucosaminidase activity (Altmann *et al.*, 1995). The enzyme efficiently and specifically removes the GlcNAc residue transferred by GlcNAc transferase I after removal of two mannose residues by α-mannosidase II (Figure 4). The result is a large quantity of *N*-glycans with a Man₃GlcNAc₂ structure which may or may not include fucose (Fuc; Altmann *et al.*, 1999). GlcNAc transferase II has a high specificity for glycans with the initial GlcNAc (Altmann *et al.*, 1993), and removal of the first GlcNAc residue, therefore, precludes additional processing and complex *N*-glycosylation (Figure 4). This *N*-acetylglucosaminidase activity will probably need to be inhibited in order to obtain significant fractions of glycoproteins with complex *N*-glycans from insect cell culture.

To obtain these complex *N*-glycans in insect cells, certain metabolic pathways must be enhanced and others reduced. Each glycosyltransferase reaction, however, follows a similar reaction scheme by requiring three components: a glycosyltransferase enzyme,

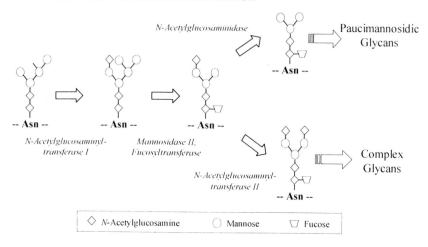

Figure 4. Effect of N-acetylglucosaminidase on final glycan structure. The action of *N*-cetylglucosaminidase can prevent the addition of a second GlcNAc residue (lower pathway) and consequently give rise to *N*-glycans of paucimannosidic form (upper pathway).

the glycan acceptor, and the sugar nucleotide donor (Figure 5). Efforts in the metabolic engineering of each of the three components are described.

3. Glycosyltransferases

Recombinant expression of glycosyltransferase enzymes in mammalian cell lines has been shown to alter the final glycoprotein composition. For example, recombinant expression of fucosyltransferase in BHK (Costa *et al.*, 1997) and porcine (Sepp *et al.*, 1997) cell lines, sialyltransferase in BHK (Grabenhorst *et al.*, 1995), and both sialyl-transferase and galactosyltransferase in CHO cells (Weikert *et al.*, 1999) improves glycosylation

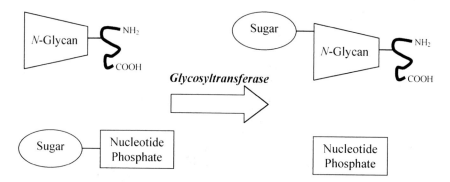

Figure 5. The glycosyltransferase reaction. Glycosyltransferases transfer sugars from their activated sugar nucleotide phosphate forms to specifically recognized glycans.

patterns. The absence of the transferase enzyme activity in the latter pathways of complex *N*-glycosylation in insect cells suggests interesting possibilities as insect cells may represent a clean slate for complex glycosylation. Glycans containing unwanted sugars and linkages found in mammalian cell culture, such as $\alpha1,3$-Gal in many mouse lines and $\alpha2,3$-sialic acid in CHO and BHK cells (Jenkins *et al.*, 1996), are not problematic in insect culture as the responsible enzymes are not very active or may not even exist.

Insect cells have been used to express numerous foreign proteins including active glycosyltransferases for activity and purification studies. Selected examples include the purification of multiple active sialyltransferases (Williams *et al.*, 1995) and the cloning of active mung bean fucosyltransferase (Leiter *et al.*, 1999). Through a metabolic engineering approach, recombinant enzymes have been co-expressed with target glyco-proteins to "mammalianize" these target proteins. These efforts are summarized in Table 3. The first such attempt was the co-expression of human GlcNAc transferase I with influenza virus hemagglutinin. Using a dual baculovirus infection, a fourfold increase in glycans terminating in GlcNAc resulted (Wagner *et al.*, 1996). Nonetheless, most of the glycans terminated in mannose suggesting even overexpression of GlcNAc transferase I could not overcome the native *N*-acetylglucosaminidase activity (Altmann *et al.*, 1999).

Considerable work has been done with the overexpression of a bovine Gal transferase (Hollister *et al.*, 1998; Ailor *et al.*, 2000; Hollister and Jarvis, 2001). A detailed analysis of the *N*-glycan structures from a co-expressed target glycoprotein, human transferrin, revealed that recombinant expression of the Gal transferase led to galactosylation about 10% of the *N*-glycans compared to no galactosylation in glycans from cultures without the galactosyltransferase (Ailor *et al.*, 2000). More importantly, Gal transferase expression reduced paucimannosidic glycan levels suggesting that the Gal residues may cap GlcNAc residues and protect them from *N*-acetylglucosaminidase activity (Ailor *et al.*, 2000).

The galactosyltransferase studies listed in Table 3 are also important as the enzyme has been expressed under an *ie1* promoter. This intermediate/early promoter is active

Table 3. Attempts to modify insect cell N-glycosylation through recombinant glycosyltransferase expression.

Enzyme	Glycoprotein Analyzed	Notes	Cell Line	Reference
Human β-1,2-GlcNAc Transferase I	Fowl plague virus hemagglutinin	4x increase in GlcNAc terminating glycans	Sf9	Wagner *et al.*, 1996
Bovine β-1,4-Galactosyltransferase	gp64	*ie1* promoter used; Addition of terminal Gal residues	Sf9	Jarvis and Finn, 1996
	gp64; Human t-PA	Stable cell lines; Addition of terminal Gal residues	Sf9	Hollister *et al.*, 1998
	Human transferrin	Multigene baculovirus; Addition of terminal Gal residues	High 5	Ailor *et al.*, 2000
Bovine β-1,4-Galactosyltransferase + Rat α-2,6-Sialyltransferase	gp64	Stable cell lines; Addition of terminal sialic acid residues	Sf9	Hollister and Jarvis, 2001

before the stronger promoters such as *polh* that are typically used for high-level target protein expression. Using this promoter strategy, processing enzymes are ideally in place when the target glycoprotein enters the secretory pathway. In addition, stably transformed insect cells are available with transferases expressed under the *ie1* promoter (Hollister *et al.*, 1998; Hollister and Jarvis, 2001). Genes under this promoter are constitutively expressed so these cell lines have transferases present at the time of infection (Hollister *et al.*, 1998). With the co-expression of galactosyltransferases and sialyltransferases, sialylated glycoproteins are even recovered from Sf9 culture grown in serum-containing medium as determined by lectin blotting and anion exchange chromatography (Hollister and Jarvis, 2001). However, serum presents many problems including variability, expense, and the potential for pathogen contamination (Altmann *et al.*, 1999).

4. Acceptor Glycan Structure

In the latter stages of complex *N*-glycosylation, the acceptor carbohydrate structure is the product of the previous glycosyltransferase reaction. However, the action of glycosidases which reverse progress toward "complex" glycosylation is a relevant problem in insect cell culture. As previously mentioned, insect cells have a particularly active *N*-acetylglucosaminidase or hexosaminidase.

This hexosaminidase activity is reported in *Spodoptera frugiperda* (Altmann *et al.*, 1995; Wagner *et al.*, 1996), and has been proposed to exist in *Mamestra brassicae* and *Bombyx mori* cell lines as well (Kubelka *et al.*, 1994). *Estigmene acrea* cells, however, have reduced hexosaminidase activity and consequently higher fractions of *N*-glycans terminating in GlcNAc residues (Ogonah *et al.*, 1996; Wagner *et al.*, 1996). Since insects represent such a vast store of biodiversity, cell lines that natively allow more complex glycosylation patterns may be identified as better starting points for further metabolic engineering.

In order to produce glycoproteins with more complex glycosylation patterns, inhibitors of glucosaminidases have been used. The inhibitor 2-acetamido-1,2,5-trideoxy-1,5-imino-D-glucitol in insect cell culture has been used successfully to obtain *N*-glycans with higher fractions of GlcNAc residues in Sf9 (Wagner *et al.*, 1996). Another inhibitor, 6-acetamido-6-deoxycastanospermine, has been shown to inhibit hexosaminidase activity *in vitro* (Altmann *et al.*, 1993). While these inhibitors are important tools in metabolic engineering, particularly for elucidating the metabolic pathways, their cost and safety considerations may be a limitation to their use on a large scale.

Consequently, alternative measures will likely be required. Overexpression of Glc NAc transferase I has been attempted as previously discussed with partially successful results in Sf9 cells (Wagner *et al.*, 1996). However, the use of gene-knockout and antisense RNA technologies maybe required to obtain the complete effect. Other potentially problematic enzymes include the possibility of "overactive" mannosidases leading to $GlcNAc_2Man_2$ structures and $\alpha 1,3$-fucosyltransferase. The $\alpha 1,3$-fucose linkage is immunogenic in mammals (Prenner *et al.*, 1992) so the transferase(s) involved in forming these structures in insect cells may need to be disabled (Altmann *et al.*, 1999).

Table 4. Summary of sugar nucleotides used in glycosyltransferase reactions. (Voet & Voet, 1995).

UDP	GDP	CMP
Galactose	Fucose	Sialic Acids (Neu5Ac,
Glucose	Mannose	Neu5Gc, KDN)
N-Acetylglucosamine		
N-Acetylgalactosamine		

5. Sugar Nucleotide Donors

The sugars involved in extending a glycan must be enzymatically activated to a sugar nucleotide form that is recognized by glycosyltransferases. The sugar nucleotides involved in glycosyltransferase reactions are listed in Table 4. Most sugars are activated by coupling with a nucleotide diphosphate. The sialic acids are somewhat unusual in that a monophosphate form, cytidine monophosphate (CMP), of the sugar nucleotide is used as a donor.

Changes in the intracellular sugar nucleotide levels and even the final glycosylation moieties have been shown to occur when the cell growth medium is supplemented by carbohydrates and their respective nucleotide bases. For example, the addition of N-acetylmannosamine has been repeatedly shown to increase sialylation in mammalian cells (Gu and Wang, 1998; Keppler et al., 1999), and mannosamine feeding has increased GlcNAc addition to glycans in insect cells (Donaldson et al., 1999). Addition of uridine and cytidine with different radiolabeled sugars caused increases in sugar incorporation in glycans of rat hepatocytes (Pels Rijcken et al., 1995).

The content of these sugar nucleotides has been measured in many common cell lines. Insect cell lines have comparable, if not higher levels, of UDP-N-acetylhexosamine (UDP-GlcNAc + UDP-GlcNAc), UDP-hexose (UDP-Glc + UDP-Gal), GDP-Man, and GDP-fucose (Hooker et al., 1999) than CHO cells. In another study in which the UDP sugar levels were determined separately, insect cells again had comparable and often higher levels of most sugar nucleotides when compared to CHO cells (Tomiya et al., 2001). The one glaring exception, however, is the CMP-sialic acid content as CMP-sialic acids were not detectable in insect cells grown on serum free medium (Hooker et al., 1999; Tomiya et al., 2001). The lack of CMP-sialic acids and few reports of native sialylated glycoproteins produced in serum-free cultures suggest insect cells may not have the capacity to produce sialic acids or that this capacity is at least tightly regulated. This lack of CMP-sialic acids in insect cells grown in serum-free cell culture may be one reason for the inability of insect cells to produce complex, sialylated glycoproteins.

The presence of sialic acid metabolism in insects has been debated in the literature. The presence of sialylated glycoconjugates has been used as evidence of endogenous sialic acid metabolism in at least some species. Sialylated structures have been detected in cicada (Malykh et al., 2000) and Drosophila (Roth et al., 1992). In addition, very low activity levels of the UDP-GlcNAc epimerase/ManNAc kinase enzyme, which is involved in sialic acid metabolism, have been detected (Effertz et al., 1999).

Nonetheless, the enzymatic activities and sialic acid levels are low in all cases. Opponents of the presence of sialic acid metabolism in insect cells have used evolutionary arguments. According to these arguments, deuterostomes, including echinoderms and mammals, developed sialic acid metabolism after the split from protostomes, which include insects (Corfield and Schauer, 1982; Angata and Varki, 2000). In support of this model, the development of siglecs, lectins which bind to sialylated structures, is believed to parallel the evolution of sialic acid metabolism in deuterostomes (Angata and Varki, 2000). The understanding of genomes, such as the recently released *Drosophila* genome, may hold answers to this debate. Nonetheless, the pathways for the synthesis, activation, and transfer of sialic acid in common insect cell lines are minimally active at best and possibly non-existent.

Achieving sialylated glycoproteins in insect cell culture therefore requires the metabolic engineering of a complete metabolic pathway for CMP-sialic acid production. The major pathways of sialic acid metabolism in bacteria and mammals are shown in Figure 6 starting with the synthesis of ManNAc, the rate-limiting step in sialic acid production in mammals (Keppler *et al.*, 1999). In mammalian tissues, ManNAc synthesis is primarily accomplished through the action of the UDP-GlcNAc epimerase

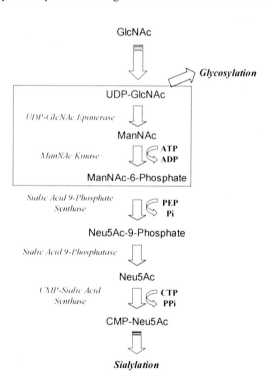

Figure 6. Mammalian Sialic Acid Metabolic Pathways. Mammalian sialic acid metabolic pathways with bifunctional epimerase enzyme activities enclosed in dashed box. UDP-GlcNAc and CMP-Neu5Ac are used for *N*-glycosylation by the appropriate transporters and transferases.

enzyme. (Comb and Roseman, 1958). In eucaryotes, this activity is performed by a bifunctional enzyme that catalyzes both the epimerase and the following activity, phosphorylation of ManNAc to give ManNAc-6-phosphate (Hinderlich *et al.*, 1997). The mammalian pathway uses phosphate intermediates, starting with ManNAc-6-phosphate, for the production of the most common sialic acid, Neu5Ac. Sialic acid phosphate synthase condenses phosphoenolpyruvate (PEP) with ManNAc-6-phosphate to produce Neu5Ac-9-phosphate (Watson *et al.*, 1966). Sialic acid phosphatase finally removes the phosphate group to give Neu5Ac (Jourdian *et al.*, 1964). In bacteria, sialic acid synthesis is achieved by a single enzyme, sialic acid synthase, which directly condenses PEP with ManNAc to form Neu5Ac (Vann *et al.*, 1997). Both bacteria and mammalian cells have an aldolase activity that condenses ManNAc with pyruvate (Brunetti *et al.*, 1962). However, the aldolase activity is primarily degradative as the enzyme catalyzes an equilibrium reaction favoring ManNAc production (Brunetti *et al.*, 1962; Vimr, 1992). Neu5Ac is then activated to CMP-Neu5Ac by CMP-sialic acid synthase in both bacteria and mammalian cells (Kean, 1991). CMP-Neu5Ac is the sugar nucleotide recognized by sialyltransferases for transfer to glycoconjugates.

While the activities of the sialic acid metabolic enzymes have been described since the early 1960's, the genes encoding the enzymes of these pathways have only recently been identified. Many of these enzymes have already been actively expressed in insect cell systems with promising results. The bifunctional UDP-GlcNAc epimerase/Man NAc kinase enzyme has been cloned from rat (Stäsche *et al.*, 1997) and subsequently expressed in Sf9 using a baculovirus vector (Effertz *et al.*, 1999). Notably, slight native activity for this enzyme was identified in Sf9 cells, although at levels 30 times less than those seen in rat hepatocyte cells (Effertz *et al.*, 1999).

The human sialic acid phosphate synthase enzyme has also been cloned (Lawrence *et al.*, 2000) based on homology to the known bacterial sialic acid synthesis gene, *neuB* (Annunziato *et al.*, 1995). When expressed in Sf9 cells through baculovirus infection, high levels of Neu5Ac production are observed only when ManNAc is supplemented into the cell culture medium. Again, insects apparently do not have the ability to produce significant levels of ManNAc since feeding is required. However, ManNAc kinase and Neu5Ac phosphatase activities, possibly non-specific, are natively present in sufficient levels so that only the sialic acid phosphate synthase has to be provided recombinantly to obtain Neu5Ac with ManNAc medium supplementation (Lawrence *et al.*, 2000). Interestingly, the production of KDN, a deaminated Neu5Ac, is observed upon infection with the baculovirus encoding sialic acid synthase even without ManNAc feeding. Evidently, this enzyme uses mannose-6-phosphate naturally present in insect cells cultured in serum-free medium to produce KDN-9-phosphate (Lawrence *et al.*, 2000). The human sialic acid phosphate synthase, unlike the activity of recently cloned mouse enzyme (Nakata *et al.*, 2000), has specificity for multiple substrates although Neu5Ac synthesis is greatly preferred over KDN (Lawrence *et al.*, 2000). A murine CMP-sialic acid synthase has also been recently cloned (Munster *et al.*, 1998), and several bacterial CMP-sialic acid synthases have also been identified (Zapata *et al.*, 1989; Ganguli *et al.*, 1994; Tullius *et al.*, 1996) although the activity of these enzymes expressed recombinantly in insect cells has not yet been demonstrated.

6. Conclusions

By using metabolic engineering, significant progress towards the goal of producing sialylated glycoproteins in insect cells grown on serum-free medium has been made. Achieving this goal requires the merging of several parallel efforts focusing on the three components required for sialylation: the CMP-sialic acid donor substrate, acceptor *N*-glycans terminating in galactose, and the sialyltransferase enzyme catalyzing the reaction. While independent efforts in generating each of these components have been successful, the major remaining challenge is to unite these efforts in an insect cell system that can produce sialylated glycoproteins.

The generation of each component requires the recombinant expression of at least one enzyme. Generating CMP-sialic acids will likely require a sialic acid phosphate synthase and a CMP-sialic acid synthase. Furthermore, a CMP-sialic acid transporter may be required to transport the CMP-sialic acid into the Golgi apparatus, and an enzyme with GlcNAc epimerase activity is required to avoid costly ManNAc feeding. Galacto-syltransferase expression increases the fraction of the second requirement: *N*-glycans terminating in galactose (Jarvis and Finn, 1996; Ailor *et al.*, 2000). However, galactose residues are currently limited primarily to the α-1,3 branch (Ailor *et al.*, 2000). Therefore, further steps must be taken to generate biantennary *N*-glycans terminating in two galactose residues. Finally, insect cells require an active sialyltransferase activity through the recombinant expression of another protein.

The production of sialylated glycoproteins in insect cells may require the expression of a number of genes. Multiple genes can be expressed by multiple viral infections or by the infection using a single virus with multiple genes (O'Reilly *et al.*, 1992). Insect cell lines stably transformed with foreign genes represent another possible method. Combinations of methods including multiple infections, baculoviruses containing multiple genes, and stably transformed cell lines may ultimately be used for the goal of sialylation. Cell growth in suspension, infectability, and expression levels of the target glycoprotein are likely criteria for assessing the best strategy.

The current results suggest the possibility of generating complex, sialylated glycoproteins in serum-free insect cell culture which could have a significant impact on biotechnology. If sialylated glycoproteins can be produced in insect cells by metabolic engineering, such products may also be possible in other cell lines, such as yeast or plants, using a similar metabolic engineering strategy. Cost considerations rather than pharmacological properties of the resulting glycoprotein would then dictate the choice of cell line used for recombinant glycoprotein production.

7. References

Ailor, E., Takahashi, N., Tsukamoto, Y., Masuda, K., Rahman, B. A., Jarvis,D. L., Lee, Y. C. and Betenbaugh, M. J. (2000) *N*-glycan patterns of human transferrin produced in *Trichoplusia ni* insect cells: effects of mammalian galactosyltransferase. *Glycobiology* 10, 837-47.
Altmann, F., Kornfeld, G., Dalik, T., Staudacher, E. and Glossl, J. (1993) Processing of asparagine-linked oligosaccharides in insect cells. *N*-acetylglucosaminyltransferase I and II activities in cultured lepidopteran cells. *Glycobiology*, 3, 619-25.

Altmann, F. and Marz, L. (1995) Processing of asparagine-linked oligosaccharides in insect cells: evidence for alpha-mannosidase II. *Glycoconj. J.* 12, 150-5.

Altmann, F., Schwihla, H., Staudacher, E., Glossl, J. and Marz, L. (1995) Insect cells contain an unusual, membrane-bound beta-*N*-acetylglucosaminidase probably involved in the processing of protein *N*-glycans. *J. Biol. Chem.* 270, 17344-9.

Altmann, F., Staudacher, E., Wilson, I. B. and Marz, L. (1999) Insect cells as hosts for the expression of recombinant glycoproteins. *Glycoconj. J.* 16, 109-23.

Angata, T. and Varki, A. (2000) Cloning, characterization, and phylogenetic analysis of siglec-9, a new member of the CD33-related group of siglecs. Evidence for co-evolution with sialic acid synthesis pathways. *J. Biol. Chem.* 275, 22127-35.

Annunziato, P. W., Wright, L. F., Vann, W. F. and Silver, R. P. (1995) Nucleotide sequence and genetic analysis of the *neuD* and *neuB* genes in region 2 of the polysialic acid gene cluster of *Escherichia coli* K1. *J. Bacteriol.* 177, 312-9.

Bergwerff, A. A., Van Oostrum, J., Kamerling, J. P. and Vliegenthart, J. F. (1995) The major *N*-linked carbohydrate chains from human urokinase. The occurrence of 4-O-sulfated, (alpha 2-6)-sialylated or (alpha 1-3)-fucosylated *N*-acetylgalactosamine(beta 1-4)-*N*-acetylglucosamine elements. *Eur. J. Biochem.* 228, 1009-19.

Brunetti, P., Jourdian, G., and Roseman, S. (1962) The sialic acids. III. Distribution and properties of animal *N*-acetylneuraminic acid aldolase. *J. Biol. Chem.* 237, 2447-53.

Butters, T. D. and Hughes, R. C. (1981) Isolation and characterization of mosquito cell membrane glycoproteins. *Biochim. Biophys. Acta.* 640, 655-71.

Comb, D. G. and Roseman, S. (1958) *Biochim. Biophys. Acta.* 29, 653-4.

Corfield, A., and Schauer, R. (1982) Occurrence of sialic acids. In R. Schauer (ed.), *Sialic Acids: Chemistry. Metabolism. and Function.* Springer-Verlag, Vienna, pp. 5-55.

Costa, J., Grabenhorst, E., Nimtz, M. and Conradt, H. S. (1997) Stable expression of the Golgi form and secretory variants of human fucosyltransferase III from BHK-21 cells. Purification and characterization of an engineered truncated form from the culture medium. *J. Biol. Chem.* 272, 11613-21.

Davidson, D. J. and Castellino, F. J. (1991) Structures of the asparagine-289-linked oligosaccharides assembled on recombinant human plasminogen expressed in a *Mamestra brassicae* cell line (IZD-MBO503). *Biochemistry.* 30, 6689-96.

Davidson, D. J., Bretthauer, R. K. and Castellino, F. J. (1991) alpha-Mannosidase-catalyzed trimming of high-mannose glycans in noninfected and baculovirus-infected *Spodoptera frugiperda* cells (IPLB- SF-21AE). A possible contributing regulatory mechanism for assembly of complex-type oligosaccharides in infected cells. *Biochemistry.* 30, 9811-5.

Davis, T. R. and Wood, H. A. (1995) Intrinsic glycosylation potentials of insect cell cultures and insect larvae. *In Vitro Cell. Dev. Biol. Anim.* 31, 659-63.

De Bruyn, P. P. and Michelson, S. (1979) Changes in the random distribution of sialic acid at the surface of the myeloid sinusoidal endothelium resulting from the presence of diaphragmed fenestrae, *J. Cell. Biol.* 82, 708-14.

Donaldson, M., Wood, H. A., Kulakosky, P. C. and Shuler, M. L. (1999) Use of mannosamine for inducing the addition of outer arm *N*-acetylglucosamine onto *N*-linked oligosaccharides of recombinant proteins in insect cells. *Biotechnol. Prog.* 15, 168-73.

Effertz, K., Hinderlich, S. and Reutter, W. (1999) Selective loss of either the epimerase or kinase activity of UDP-*N*-acetylglucosamine 2-epimerase/*N*-acetylmannosamine kinase due to site-directed mutagenesis based on sequence alignments, *J. Biol. Chem.* 274, 28771-8.

Ganguli, S., Zapata, G., Wallis, T., Reid, C., Boulnois, G., Vann, W. F. and Roberts, I. S. (1994) Molecular cloning and analysis of genes for sialic acid synthesis in *Neisseria meningitidis* group B and purification of the meningococcal CMP-NeuNAc synthetase enzyme. *J. Bacteriol.* 176, 4583-9.

Goochee, C. F., Gramer, M. J., Andersen, D. C., Bahr, J. B. and Rasmussen, J. R. (1991) The oligosaccharides of glycoproteins: bioprocess factors affecting oligosaccharide structure and their effect on glycoprotein properties. *Bio/technology* 9, 1347-55.

Grabenhorst, E., Hofer, B., Nimtz, M., Jager, V. and Conradt, H. S. (1993) Biosynthesis and secretion of human interleukin 2 glycoprotein variants from baculovirus-infected Sf21 cells. Characterization of polypeptides and posttranslational modifications. *Eur. J. Biochem.* 215, 189-97.

Grabenhorst, E., Hoffmann, A., Nimtz, M., Zettlmeissl, G. and Conradt, H. S. (1995) Construction of stable BHK-21 cells coexpressing human secretory glycoproteins and human Gal(beta 1-4)GlcNAc-R alpha 2,6-sialyltransferase alpha 2,6-linked NeuAc is preferentially attached to the Gal(beta 1-4)GlcNAc(beta 1-2)Man(alpha 1-3)-branch of diantennary oligosaccharides from secreted recombinant beta-trace protein. *Eur. J. Biochem.* 232, 718-25.

Grossmann, M., Wong, R., Teh, N. G., Tropea, J. E., East-Palmer, J., Weintraub, B. D. and Szkudlinski, M. W. (1997) Expression of biologically active human thyrotropin (hTSH) in a baculovirus system: effect of insect cell glycosylation on hTSH activity *in vitro* and *in vivo. Endocrinology* 138, 92-100.

Gu, X. and Wang, D. I. (1998) Improvement of interferon-gamma sialylation in Chinese hamster ovary cell culture by feeding of *N*-acetylmannosamine. *Biotechnol. Bioeng.* 58, 642-8.

Hinderlich, S., Stasche, R., Zeitler, R. and Reutter, W. (1997) A bifunctional enzyme catalyzes the first two steps in *N*-acetylneuraminic acid biosynthesis of rat liver. Purification and characterization of UDP-*N*-acetylglucosamine 2-epimerase/*N*- acetylmannosamine kinase. *J. Biol. Chem.* 272, 24313-8.

Hollister, J. R., Shaper, J. H. and Jarvis, D. L. (1998) Stable expression of mammalian beta 1,4-galactosyltransferase extends the *N*-glycosylation pathway in insect cells. *Glycobiology* 8, 473-80.

Hollister, J. R. and Jarvis, D. L. (2001) Engineering lepidopteran insect cells for sialoglycoprotein production by genetic transformation with mammalian beta1,4-galactosyltransferase and alpha2,6-sialyltransferase genes. *Glycobiology* 11, 1-9.

Hooker, A. D., Green, N. H., Baines, A. J., Bull, A. T., Jenkins, N., Strange, P. G. and James, D. C. (1999) Constraints on the transport and glycosylation of recombinant IFN-gamma in Chinese hamster ovary and insect cells. *Biotechnol. Bioeng.* 63, 559-72.

Hsieh, P. and Robbins, P. W. (1984) Regulation of asparagine-linked oligosaccharide processing. Oligosaccharide processing in *Aedes albopict*us mosquito cells. *J. Biol. Chem.* 259, 2375-82.

Hsu, T. A., Takahashi, N., Tsukamoto, Y., Kato, K., Shimada, I., Masuda, K., Whiteley, E. M., Fan, J. Q., Lee, Y. C. and Betenbaugh, M. J. (1997) Differential *N*-glycan patterns of secreted and intracellular IgG produced in *Trichoplusia ni* cells. *J. Biol. Chem.* 272, 9062-70.

Jarvis, D. L. and Summers, M. D. (1989) Glycosylation and secretion of human tissue plasminogen activator in recombinant baculovirus-infected insect cells. *Mol. Cell. Biol.* 9, 214-23.

Jarvis, D. L. and Finn, E. E. (1996) Modifying the insect cell *N*-glycosylation pathway with immediate early baculovirus expression vectors. *Nat. Biotechnol.* 14, 1288-92.

Jarvis, D. L., Bohlmeyer, D. A., Liao, Y. F., Lomax, K. K., Merkle, R. K., Weinkauf, C. and Moremen, K. W. (1997) Isolation and characterization of a class II alpha-mannosidase cDNA from lepidopteran insect cells. *Glycobiology* 7, 113-27.

Jenkins, N., Parekh, R. B. and James, D. C. (1996) Getting the glycosylation right: implications for the biotechnology industry. *Nat. Biotechnol.* 14, 975-81.

Jourdian, G. W., Swanson, A. L., Watson, D., and Roseman, S. (1964) Isolation of sialic acid 9-phosphatase from human erythrocytes. *J. Biol. Chem.* 239, 2714-5.

Kawar, Z., Herscovics, A. and Jarvis, D. L. (1997) Isolation and characterization of an alpha 1,2-mannosidase cDNA from the lepidopteran insect cell line Sf9. *Glycobiology* 7, 433-43.

Kawar, Z., Romero, P. A., Herscovics, A. and Jarvis, D. L. (2000) *N*-Glycan processing by a lepidopteran insect alpha1,2-mannosidase. *Glycobiology* 10, 347-55.

Kean, E. L. (1991) Sialic acid activation. *Glycobiology* 1, 441-7.

Keppler, O. T., Hinderlich, S., Langner, J., Schwartz-Albiez, R., Reutter, W. and Pawlita, M. (1999) UDP-GlcNAc 2-epimerase: a regulator of cell surface sialylation. *Science* 284, 1372-6.

Kornfeld, R. and Kornfeld, S. (1985) Assembly of asparagine-linked oligosaccharides. *Ann. Rev. Biochem.* 54, 631-64.

Kubelka, V., Altmann, F., Kornfeld, G. and Marz, L. (1994) Structures of the *N*-linked oligosaccharides of the membrane glycoproteins from three lepidopteran cell lines (Sf-21, IZD-Mb-0503, Bm-N). *Arch. Biochem. Biophys.* 308, 148-57.

Kulakosky, P. C., Hughes, P. R. and Wood, H. A. (1998) *N*-Linked glycosylation of a baculovirus-expressed recombinant glycoprotein in insect larvae and tissue culture cells. *Glycobiology* 8, 741-5.

Kuroda, K., Geyer, H., Geyer, R., Doerfler, W. and Klenk, H. D. (1990) The oligosaccharides of influenza virus hemagglutinin expressed in insect cells by a baculovirus vector. *Virology* 174, 418-9.

Lawrence, S. M., Huddleston, K. A., Pitts, L. R., Nguyen, N., Lee, Y. C., Vann, W. F., Coleman, T. A. and Betenbaugh, M. J. (2000) Cloning and expression of the human *N*-acetylneuraminic acid phosphate synthase gene with 2-keto-3-deoxy-D-*glycero*- D-*galacto*-nononic acid biosynthetic ability. *J. Biol. Chem.* 275, 17869-77.

Leiter, H., Mucha, J., Staudacher, E., Grimm, R., Glossl, J. and Altmann, F. (1999) Purification, cDNA cloning, and expression of GDP-L-Fuc:Asn-linked GlcNAc alpha1,3-fucosyltransferase from mung beans. *J. Biol. Chem.* 274, 21830-9.

Lopez, M., Coddeville, B., Langridge, J., Plancke, Y., Sautiere, P., Chaabihi, H., Chirat, F., Harduin-Lepers, A., Cerutti, M., Verbert, A. and Delannoy, P. (1997) Microheterogeneity of the oligosaccharides carried by the recombinant bovine lactoferrin expressed in *Mamestra brassicae* cells. *Glycobiology* 7, 635-51.

Malykh, Y. N., Krisch, B., Gerardy-Schahn, R., Lapina, E. B., Shaw, L. and Schauer, R. (1999) The presence of *N*-acetylneuraminic acid in Malpighian tubules of larvae of the cicada *Philaenus spumarius. Glycoconj. J.* 16, 731-9.

Marchal, I., Mir, A. M., Kmiecik, D., Verbert, A. and Cacan, R. (1999) Use of inhibitors to characterize intermediates in the processing of *N*-glycans synthesized by insect cells: a metabolic study with Sf9 cell line. *Glycobiology* 9, 645-54.

Moremen, K. W., Trimble, R. B. and Herscovics, A. (1994) Glycosidases of the asparagine-linked oligosaccharide processing pathway. *Glycobiology* 4, 113-25.

Muchmore, E. A., Diaz, S. and Varki, A. (1998) A structural difference between the cell surfaces of humans and the great apes. *Am. J. Phys. Anthropol.* 107, 187-98.

Munster, A. K., Eckhardt, M., Potvin, B., Muhlenhoff, M., Stanley, P. and Gerardy-Schahn, R. (1998) Mammalian cytidine 5'-monophosphate *N*-acetylneuraminic acid synthetase: a nuclear protein with evolutionarily conserved structural motifs. *Proc. Natl. Acad. Sci. USA* 95, 9140-5.

Nadano, D., Iwasaki, M., Endo, S., Kitajima, K., Inoue, S. and Inoue, Y. (1986) A naturally occurring deaminated neuraminic acid, 3-deoxy-D-*glycero*-D-*galacto*-nonulosonic acid (KDN). Its unique occurrence at the nonreducing ends of oligosialyl chains in polysialoglycoprotein of rainbow trout eggs. *J. Biol. Chem.* 261, 11550-7.

Nakata, D., Close, B. E., Colley, K. J., Matsuda, T. and Kitajima, K. (2000) Molecular cloning and expression of the mouse *N*-acetylneuraminic acid 9-phosphate synthase which does not have deaminoneuraminic acid (KDN) 9-phosphate synthase activity. *Biochem. Biophys. Res. Commun.* 273, 642-8.

O'Reilly, D. R., Miller, L. K. and Luckow, V. A. (1992) *Baculovirus Expression Vectors: A Laboratory Manual*, W. H. Freeman and Company, New York.

Oganah, O. W., Freedman, R.B., Jenkins, N., Patel, K., and Rooney, B.C. (1996) Isolation and characterization of an insect cell line able to perform complex *N*-linked glycosylation on recombinant proteins. *Bio/technology* 14, 197-202.

Parker, G. F., Williams, P. J., Butters, T. D. and Roberts, D. B. (1991) Detection of the lipid-linked precursor oligosaccharide of *N*-linked protein glycosylation in *Drosophila melanogaster. FEBS Lett.* 290, 58-60.

Pels Rijcken, W. R., Overdijk, B., Van den Eijnden, D. H. and Ferwerda, W. (1995) The effect of increasing nucleotide-sugar concentrations on the incorporation of sugars into glycoconjugates in rat hepatocytes. *Biochem J.* 305, 865-70.

Prenner, C., Mach, L., Glossl, J. and Marz, L. (1992) The antigenicity of the carbohydrate moiety of an insect glycoprotein, honey-bee (*Apis mellifera*) venom phospholipase A2. The role of alpha 1,3-fucosylation of the asparagine-bound *N*-acetylglucosamine. *Biochem. J.* 284, 377-80.

Prokop, A. (1991) Implications of cell biology in animal cell biotechnology, *Biotechnology* 17, 21-58.

Ren, J., Bretthauer, R. K. and Castellino, F. J. (1995) Purification and properties of a Golgi-derived (alpha 1,2)-mannosidase- I from baculovirus-infected lepidopteran insect cells (IPLB-SF21AE) with preferential activity toward mannose-6-*N*-acetylglucosamine2. *Biochemistry* 34, 2489-95.

Ren, J., Castellino, F. J. and Bretthauer, R. K. (1997) Purification and properties of alpha-mannosidase II from Golgi-like membranes of baculovirus-infected *Spodoptera frugiperda* (IPLB-SF-21AE) cells. *Biochem. J.* 324, 951-6.

Roth, J., Kempf, A., Reuter, G., Schauer, R. and Gehring, W. J. (1992) Occurrence of sialic acids in *Drosophila melanogaster. Science* 256, 673-5.

Schauer, R., Kelm, S., Reuter, G., Roggentin, P., and Shaw, L. (1995) Biochemistry and role of sialic acids. In A. Rosenberg (ed.), *Biology of the Sialic Acids*, Plenum Press, New York, pp. 7-67.

Sepp, A., Skacel, P., Lindstedt, R. and Lechler, R. I. (1997) Expression of alpha-1,3-galactose and other type 2 oligosaccharide structures in a porcine endothelial cell line transfected with human alpha-1,2-fucosyltransferase cDNA. *J. Biol. Chem.* 272, 23104-10.

Stasche, R., Hinderlich, S., Weise, C., Effertz, K., Lucka, L., Moormann, P. and Reutter, W. (1997) A bifunctional enzyme catalyzes the first two steps in *N*-acetylneuraminic acid biosynthesis of rat liver. Molecular cloning and functional expression of UDP-*N*-acetyl-glucosamine 2-epimerase/*N*-acetylmannosamine kinase. *J. Biol. Chem.* 272, 24319-24.

Staudacher, E. and Marz, L. (1998) Strict order of (Fuc to Asn-linked GlcNAc) fucosyltransferases forming core-difucosylated structures. *Glycoconj. J.* 15, 355-60.

Takeuchi, M. and Kobata, A. (1991) Structures and functional roles of the sugar chains of human erythropoietins. *Glycobiology* 1, 337-46.

Tollefsen, S. E. and Rosenblum, J. L. (1988) Role of terminal fucose residues in clearance of human glycosylated alpha-amylase. *Am. J. Physiol.* 255, 374-81.

Tomiya, N., Ailor, E., Lawrence, S., Betenbaugh, M, Lee, Y.C. (2001) Determination of nucleotides and sugar nucleotides involved in protein glycosylation by high-performance anion-exchange chromatography: sugar nucleotide contents in cultured insect cells and mammalian cells. *Anal. Biochem.* In Press.

Tullius, M. V., Munson, R. S., Jr., Wang, J. and Gibson, B. W. (1996) Purification, cloning, and expression of a cytidine 5'-monophosphate *N*-acetylneuraminic acid synthetase from *Haemophilus ducreyi*. *J. Biol. Chem.* 271, 15373-80.

van Die, I., van Tetering, A., Bakker, H., van den Eijnden, D. H. and Joziasse, D. H. (1996) Glycosylation in lepidopteran insect cells: identification of a beta 1->4-*N*-acetylgalactosaminyltransferase involved in the synthesis of complex-type oligosaccharide chains. *Glycobiology* 6, 157-64.

Vann, W. F., Tavarez, J. J., Crowley, J., Vimr, E. and Silver, R. P. (1997) Purification and characterization of the Escherichia coli K1 neuB gene product *N*-acetylneuraminic acid synthetase. *Glycobiology* 7, 697-701.

Velardo, M. A., Bretthauer, R. K., Boutaud, A., Reinhold, B., Reinhold, V. N. and Castellino, F. J. (1993) The presence of UDP-*N*-acetylglucosamine:alpha-3-D-mannoside beta 1,2-N- acetylglucosaminyltransferase I activity in *Spodoptera frugiperda* cells (IPLB-SF-21AE) and its enhancement as a result of baculovirus infection. *J. Biol. Chem.* 268, 17902-7.

Vimr, E. R. (1992) Selective synthesis and labeling of the polysialic acid capsule in *Escherichia coli* K1 strains with mutations in *nanA* and *neuB*, *J. Bacteriol.* 174, 6191-7.

Voet, D. and Voet, J. G. (1995) *Biochemistry*, 2nd edition, John Wiley and Sons, New York.

Wagner, R., Geyer, H., Geyer, R. and Klenk, H. D. (1996) *N*-acetyl-beta-glucosaminidase accounts for differences in glycosylation of influenza virus hemagglutinin expressed in insect cells from a baculovirus vector. *J. Virol.* 70, 4103-9.

Wagner, R., Liedtke, S., Kretzschmar, E., Geyer, H., Geyer, R. and Klenk, H. D. (1996) Elongation of the *N*-glycans of fowl plague virus hemagglutinin expressed in *Spodoptera frugiperda* (Sf9) cells by coexpression of human beta 1,2-*N*-acetylglucosaminyltransferase I. *Glycobiology* 6, 165-75.

Warren, C. E. (1993) Glycosylation. *Curr. Opin. Biotechnol.* 4, 596-602.

Watson, D. R., Jourdian, G. W. and Roseman, S. (1966) The sialic acids. 8. Sialic acid 9-phosphate synthetase. *J. Biol. Chem.* 241, 5627-36.

Weikert, S., Papac, D., Briggs, J., Cowfer, D., Tom, S., Gawlitzek, M., Lofgren, J., Mehta, S., Chisholm, V., Modi, N., Eppler, S., Carroll, K., Chamow, S., Peers, D., Berman, P. and Krummen, L. (1999) Engineering Chinese hamster ovary cells to maximize sialic acid content of recombinant glycoproteins. *Nat. Biotechnol.* 17, 1116-21.

Williams, M. A., Kitagawa, H., Datta, A. K., Paulson, J. C. and Jamieson, J. C. (1995) Large-scale expression of recombinant sialyltransferases and comparison of their kinetic properties with native enzymes. *Glycoconj. J.* 12, 755-61.

Yeh, J. C., Seals, J. R., Murphy, C. I., van Halbeek, H. and Cummings, R. D. (1993) Site-specific *N*-glycosylation and oligosaccharide structures of recombinant HIV-1 gp120 derived from a baculovirus expression system. *Biochemistry* 32, 11087-99.

Zapata, G., Vann, W. F., Aaronson, W., Lewis, M. S. and Moos, M. (1989) Sequence of the cloned *Escherichia coli* K1 CMP-*N*-acetylneuraminic acid synthetase gene. *J. Biol. Chem.* 264, 14769-74.

9. ASPARAGINE-LINKED GLYCOSYLATIONAL MODIFICATIONS IN YEAST

JYH-MING WU,[1] CHENG-KANG LEE,[1] TSU-AN HSU[2]*
[1]*Department of Chemical Engineering, National Taiwan University of Science and Technology, Taipei, 106, Taiwan*
[2] *Division of Biotechnology and Pharmaceutical Research, National Health Research Institutes, Taipei, 115, Taiwan*
* *Corresponding Author: E-mail: tsuanhsu@nhri.org.tw*

1. Introduction

Recombinant proteins produced in yeast can be modified post-translationally. Some modifications such as phosphorylation, glycosylation, etc., are important for the biological functions of the modified proteins. For glycoproteins, the glycans' structural or functional roles have gained more and more appreciation and have been extensively reviewed (Dennis et al., 1999; Durand and Seta, 2000; Bhatia and Mukhopadhyay, 1999; Lis and Sharon, 1993; Varki, 1993). Intracellularly, N-linked glycans may help nascent glycoproteins to fold and to target for transport to specific cellular organells (Helenius and Aebi, 2001; Helenius, 2001). When attached to mature glycoproteins, the carbohydrates may contribute to proteins' biological activities, conformation, stability, activity, immunogenecity, protease resistance, and pharmacokinetic properties in animals including human (Liu, 1992; Jenkins et al., 1996; Cumming, 1991; Takeuchi and Kobata, 1991; Bhatia and Mukhopadhyay, 1999; Umana et al., 1999).

Saccharomyces cerevisiae (S. cerevisiae) is the most frequently used yeast as an expression host for the production of recombinant proteins. *S. cerevisiae* is easy to manipulate by recombinant DNA technology in the laboratory and its genome sequence has been fully elucidated in 1996 (Goffeau et al., 1996; http://genome-www.stanford. edu/Saccharomyces/). It is now quite straightforward to generate genetic knock-out or knock-in mutants. In addition to *S. cerevisiae*, other yeast strains have all been used as production hosts for endogenous or heterologous protein over-expression in large scale. These include strains such as *Pichia pastoris, Pichia methanolica, Schizosaccharomyces pombe, Hansenula polymorpha, Kluyveromyces lactis*, and *Yarrowia lipolytica* (Gellissen and Hollenberg, 1997; Cereghino and Cregg, 2000; Juretzek et al., 2001; Muller et al., 1998; Fleer, 1992).

Many yeast strains including *S. cerevisiae* are considered as GRAS (generally recognized as safe) substances by the U.S. Food and Drug Administration regulation due to their long history of safety record in food and drug applications (http://www. cfsan.fda.gov/~rdb/opa-gras.html). Yeast is free of endotoxin and does not contain human viral or oncogenic DNA. These attributes ensure that yeast will remain as one of the main organisms of choice for protein expression in pharmaceutical industry. However, when the yeast-derived recombinant proteins are considered for parenteral

M. Al-Rubeai.(ed.). Cell Engineering. 215-232.

administration, it is important to know whether the post-translational modifications, especially the glycosylation of proteins, would effect or trigger unwanted responses or side effects.

The scope of this article is not to summarize detailed N- and O-linked glycosylational modifications in yeast that have already been reviewed extensively (Parodi, 1999; Gemmill and Trimble, 1999; Dean, 1999; Bretthauer and Castellino, 1999; Herscovics and Orlean, 1993). In this review, we focus on discussing N-linked glycosylational modifications of proteins expressed in commonly used yeast strains with some emphasis on pathway engineering toward glycoprotein biosynthesis in yeast cells.

2.   N-Linked Glycosylational Modifications in Yeasts

The endoplasmic reticulum (ER) and Golgi complex serve as the principal sites within a eukaryotic cell for the glycosylational modifications of secreted proteins to take place. In yeast or higher eukaryotes, the initial steps in the biosynthesis and trimming of N-linked oligosaccharides are well conserved (Kornfeld and Kornfeld, 1985). A core unit of tetradecasaccharide ($Glc_3Man_9GlcNAc_2$) is synthesized as a dolichyl diphosphate (oligosaccharide-PP-dolicol) precursor by enzymes located in the ER membrane. The core oligosaccharide structure is then transferred *en bloc* to an asparagine side chain by a membrane bound oligosaccharide transferase (Kelleher et al., 1992; Kelleher and Gilmore, 1994). The Asn acceptor of oligosaccharide is generally located in the context of Asn-Xaa-Ser/Thr, where Xaa can be any amino acid except Pro. However, some exceptions do exist; the N-linked glycosylation of protein C occurs at the sequence Asn–Xaa–Cys (Miletich and Broze, 1990).

After the tetradecasaccharide moiety is transferred to Ans, the maturation processes involve glycosylational modifications including removal of glucose and mannose residues and the addition of new sugar residues in both the ER and Golgi complex (Figure 1) (Kornfeld and Kornfeld, 1985). While the structures of the N-linked core oligosaccharide formed in the ER are remarkably similar among eukaryotic cells, the terminal modifications that occur in the Golgi can vary extensively among different organisms (Jenkins et al., 1996; Hsu et al., 1997). Although the same glycosylation machinery is available to all proteins that enter the secretory pathway in a given cell, most glycoproteins emerge with characteristic glycosylational patterns and hetero-geneous populations of glycans at each glycosylation site. These terminal modifications can also vary among proteins within a given organism as well as among different glycosylational sites within the same protein expressed (Andersen et al., 1994; Basco et al., 1993; Garcia et al., 2001; Joao and Dwek, 1993).

3.   *Saccharomyces cerevisiae*

In budding yeast *S. cerevisiae*, the N-linked core tetradecasaccharide can be trimmed to $Man_8GlcNAc_2$ through the glycosidase actions of glucosidase I & II and α-1,2-manno-sidase I in the ER (Figure 1). Subsequently, the N-linked oligosaccharide is extended

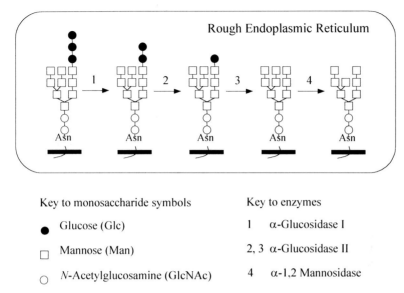

Figure 1.   The trimming of the N-linked core tetradecasaccharide to $Man_8GlcNAc_2$ through the glycosidase actions of glucosidase I & II and α-1,2-mannosidase I in the ER.

by various mannosyl transferases in the Golgi apparatus to form hyper-mannosylated glycoproteins with outer chains that may contain up to 50 to 200 mannose (Figure 2). These hyper-mannose structures consist of mainly a long backbone of α-1,6-linked residues with α-1,2-linked branches that terminate in α-1,3-linked residues (Figure 2) (Gemmill and Trimble, 1999; Jungmann and Munro, 1998; Jungmann et al., 1999; Kang et al., 1998; Herscovics and Orlean, 1993). It is important to note that the α-1,3-linked mannose residues are highly immunogenic and glycoproteins containing such mannose residues may not be ideal for therapeutic purposes (Jenkins et al., 1996). These hyper-mannose structures are thought to play important roles in cell wall integrity and in protection of cell wall surface glycoproteins from mechanical stress or undesirable enzymatic digestion.

The biological functions of glycans have been extensively explored by studying mutant strains with defects in the glycosylational pathway. These mutants can be generated by random mutagenesis coupled with a selection method to look for unique phenotype. Alternatively, gene function can be examined from mutants generated by genetic manipulation to delete or add the gene of interest. Here we have compiled a list of identified yeast genes encoding for proteins involve in glycosylational modifications (Table 1). In this table, mutant strains with defective genes are also listed.

4.   *Pichia pastoris*

Recently, *P. pastoris* has been used increasingly for recombinant protein production because of its extremely high expression levels and its ease of fermentation scale-up

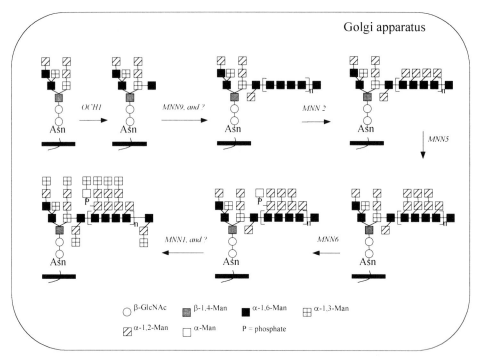

Figure 2. The extension of N-linked oligosaccharide by various mannosyltransferases in yeast (*S. cerevisiae*) Golgi apparatus to form hyper-mannosylated glycoproteins with outer chains that may contain up to 50 to 200 mannose. These hyper-mannose structures consist of mainly a long backbone of α-1,6-linked mannose residues with α-1,2-linked branches that terminate in α-1,3-linked mannose residues.

(Cereghino and Cregg, 2000; Cregg, 1999). In *P. pastoris*, the alcohol oxidase (*AOX1*) can be induced by methanol to extremely high levels; up to 30% of total cellular protein (Cregg, 1999). Consequently, the promoter of *AOX1* has been identified and successfully used for protein over-expression (Tschopp et al., 1987).

Glycosylation of *P. pastoris*-derived proteins has been reviewed recently (Bretthauer and Castellino, 1999; Gemmill and Trimble, 1999; Lundblad, 1999). Montesino R, et al. (1998), have reported the characterization of N-linked oligosaccharides on six foreign glycoproteins secreted from *P. pastoris*. Their results showed that glycoproteins secreted from *P. pastoris* are not subject to the hyper-mannosylation that often observed in *S. cerevisiae*. They found that the glycans on different proteins were observed to vary in size; with $Man_8GlcNAc_2$ and $Man_9GlcNAc_2$ structures being the most frequently observed species. Longer structures (up to $Man_{18}GlcNAc_2$) were common on aspartic protease and enterokinase. Phosphorylated oligosaccharides were observed only on one protein, aspartic protease. No terminal α-1,3-linked mannosylation was observed on any of the six *P. pastoris*-derived proteins.

An earlier study has elucidated the detailed structure of oligosaccharides of invertase of *S. cerevisiae* secreted from *P. pastoris* (Trimble et al., 1991). Heterologous invertase

Table 1. Yeast Glycosylation Genes and Mutants.

Gene	Mutant	Gene Product	Genebank or Protein Bank Accession No	References
Endoplasmic Reticulum				
ALG1	alg1	beta-1,4 mannosyltransferase	J05416	(Albright and Robbins, 1990)
ALG2	alg2	glucosyltransferase	X87947	(Jackson et al., 1993)
ALG3	alg3		P38179	(Aebi et al., 1996)
ALG5	alg5	UDP-glucose:dolichyl-phosphate glucosyltransferase	X77573	(Heesen et al., 1994)
ALG6	alg6	endoplasmic glucosyltransferase	-	(Reiss et al., 1996)
ALG7	alg7	UDP-N-acetylglucosamine-1-phosphate transferase	M15862	(Kukuruzinska and Robbins, 1987)
ALG8	alg8	glucosyltransferase	X75929	(Stagljar et al., 1994)
ALG9	alg9	mannosyltransferase	X96417	(Burda et al., 1996; Cipollo and Trimble, 2000)
ALG11	alg11	alpha -1,2- glucosyltransferase	U62941	(Cipollo et al., 2001; Umeda et al., 2000; Burda and Aebi, 1998)
CWH41	gls1	glucosidase I	U35669	(Esmon et al., 1984; Jiang et al., 1996)
DPM1		dolichyl-phosphate beta-D-mannosyltransferase	J04184	(Orlean et al., 1988)
MNS1		alpha 1,2 mannosidase	M63598	(Camirand et al., 1991)
OST1	ost1	oligosaccharyltransferase, alpha subunit	Z46719	(Silberstein et al., 1995b)
OST2	ost2	oligosaccharyltransferase, epsilon subunit	U32307	(Silberstein et al., 1995a)
OST3	ost3	oligosaccharyltransferase, gamma subunit	U25052	(Karaoglu et al., 1995; Kelleher and Gilmore, 1994)
OST4	ost4	oligosaccharyltransferase subunit	L42519	(Chi et al., 1996)
OST5	ost5	oligosaccharyltransferase, zeta subunit	X97545	(Reiss et al., 1997)
OST6	ost6	oligosaccharyltransferase subunit	Y08606	(Knauer and Lehle, 1999)
PMT1		dolichyl phosphate-D-mannose:protein O-D-mannosyltransferase	L19169	(Boskovic et al., 1996; Strahl-Bolsinger et al., 1993)
PMT3		dolichyl phosphate-D-mannose:protein O-D-mannosyltransferase	X83797	(Immervoll et al., 1995)
PMT4		dolichyl phosphate-D-mannose:protein O-D-mannosyltransferase	X83798	(Immervoll et al., 1995)
PMT5		dolichyl phosphate-D-mannose:protein O-D-mannosyltransferase	X92759	(Boskovic et al., 1996)
RHK1		mannosyltransferase	E15544	(Kimura et al., 1997; Kimura et al., 1999)
ROT2	rot2	glucosidase II, catalytic subunit	Z36098	(Feldmann et al., 1994; Simons et al., 1998)
SEC59		dolichol kinase	M25779	(Bernstein et al., 1989)
STT3	stt3	oligosaccharyltransferase subunit	D28952	(Yoshida et al., 1995)
SWP1		oligosaccharyltransferase,delta subunit	X67705	(te Heesen et al., 1993)
WBP1	wbp1	oligosaccharyltransferase, beta subunit	X61388	(te Heesen et al., 1992; te Heesen et al., 1991)
Golgi Apparatus				
ANP1	anp1	reguired for protein glycosylation in the golgi	S66114	(Melnick and Sherman,1993)
KRE2	kre2	alpha-1,2 mannosyltransferase	X62647	(Hausler and Robbins, 1992; Hill et al., 1992)
KTR1	ktr1	mannosyltransferase involved in N-linked and O-linked glycosylation	X62941	(Hill et al., 1992)

Table 1. Yeast Glycosylation Genes and Mutants. (cont.)

Gene	Mutant	Gene Product	Genebank or Protein Bank Accession No	References
Golgi Apparatus				
KTR2	ktr2	mannosyltransferase	L17083	(Lussier et al., 1993)
KTR3	ktr3	putative alpha-1,2 mannosyltransferase	Z36074	(Feldmann et al., 1994)
KTR4	ktr4	alpha-1,2 mannosyltransferase	Z36068	(Sipos et al., 1995)
KTR5	ktr5	putative mannosyltransferase	Z71305	(Lussier et al., 1997)
MNN1	mnn1	alpha-1,3 mannosyltransferase	L23753	(Wiggins and Munro, 1998; Yip et al., 1994)
MNN2-2		UDP-GlcNAc transporter (K. lactis)	AF106080	(Abeijon et al., 1996)
MNN4	mnn4	mannosylphosphate transferase	D83006	(Odani et al., 1996; Odani et al., 1997)
MNN6	mnn6	mannosylphosphate transferase	U43922	(Nakayama et al., 1998; Wang et al., 1997)
MNN9	mnn9	alpha-1,6 mannosyltransferase	L23752	(Hernandez et al., 1989; Yip et al., 1994)
MNN10	mnn10	mannosyltransferase complex component	L42540	(Dean and Poster, 1996)
MNT1	mnt1	alpha-1,2 mannosyltransferase	M81110	(Hausler and Robbins, 1992)
OCH1	och1	alpha-1,6 mannosyltransferase	D11095	(Nakayama et al., 1992)
VAN1	van1	phosphorylation	M33957	(Kanik-Ennulat et al., 1995; Kanik-Ennulat and Neff, 1990)
VRG4		GDP-mannose transporter	L33915	(Dean et al., 1997)
YEA4		UDP-GlcNAc transporter		(Roy et al., 2000)
YUR1	yur1	mannosyltransferase	X58099	(Foreman et al., 1991; Lussier et al., 1996)

from *P. pastoris* appeared on SDS-PAGE as an 85-90 kDa protein compared to invertase from *S. cerevisiae* of approximately 100 to 150 kDa on SDS-PAGE. After *endo-β-N-acetylglycosaminidase* H (*Endo* H) digestion to release glycans from *P. pastoris* invertase, carbohydrates were found to consist of $Man_{8-11}GlcNAc$ moieties; 75% were Man_{8-9} GlcNAc, 17% were $Man_{10}GlcNAc$, and 8% were $Man_{11}GlcNAc$. The lipid oligosaccharide precursor was found to consist of 5 % $Man_9GlcNAc$ structures and 95% of Man_8 GlcNAc isomer with a terminal α-1,6-linked mannose on the lower-arm α-1,3-core-linked residue. In this report, in accord with others' observation, immunogenic α-1,3-linked mannose was not found in *P. pastoris*-derived proteins (Montesino et al., 1998).

5.   Alternative Yeast Strains for Protein Over-Expression

With the development of gene expression in several yeast strains, it is now possible to evaluate protein expression in different yeast hosts. Yeast strains such as *Schizosaccharomyces pombe*, *Hansenula polymorpha*, *Kluyveromyces lactis*, and *Yarrowia lipolytica*, also bear attractive characteristics as hosts for recombinant protein production (Fleer, 1992; Muller et al., 1998; Muller et al., 1998; Romanos et al., 1992). In general, yeast is an economic host that can be grown in an inexpensive medium. The transformation systems for many yeast strains have been established. However, the biosynthesis of the N-linked oligosaccharides from these yeast strains is less well characterized than those derived from *S. cerevisiae* and *P. pastoris*.

Fission yeast *Sch. pombe* trims $Glc_3Man_9GlcNAc_2$ to $Man_9GlcNAc_2$ in the ER but does not trim $Man_9GlcNAc_2$ to $Man_8GlcNAc_2$ and no α-1,2-mannosidase activity could be detected in ER of *Sch. pombe* (Ziegler et al., 1994; Ballou et al., 1994). Consequently, *Sch. pombe* synthesizes large N-linked galacto-mannans based on the $Man_9GlcNAc_2$ core unit precursor for oligosaccharide elongation in contrast to the $Man_8GlcNAc_2$ precursor found in *S. cerevisiae* (Ziegler et al., 1994; Ziegler et al., 1999). Glycosylation mutants of *Sch. pombe* have been isolated and these mutants may be used to produce glycoproteins with reduced terminal mannose and galactose on both N- and O- linked glycans (Ballou and Ballou, 1995; Ballou et al., 1994; Huang and Snider, 1995). Analysis of N-linked oligosaccharides released from *Sch. pombe* glycoproteins revealed that there was no phosphate, sulfate, or acetate on the glycans but approximately six molecules of pyruvic acid were found on 98% of the oligosaccharides (Gemmill and Trimble, 1996).

The methylotrophic yeast *Hansenula polymorpha* has also been frequently used as a heterologous protein over-expression system (Gellissen et al., 1991; Gellissen et al., 1992; van Dijk et al., 2000; Diminsky et al., 1997; Janowicz et al., 1991; Bellu et al., 1999; Gellissen and Melber, 1996; Fellinger et al., 1991; Hollenberg and Gellissen, 1997; Gellissen et al., 1995; Gellissen and Hollenberg, 1997). This expression system is very similar to *P. pastoris* in many perspectives. A strong promoter to drive methanol oxidase (*MOX*) expression has been used to over-express recombinant proteins at extremely high levels (Fellinger et al., 1991). *H. polymorpha* has been specifically engineered and used as an economical expression system for production of phytase at the yield between 4.5 and 13.5 g/L (Mayer et al., 1999). Glycosylated carboxypeptidase Y (CPY) from *H. polymorpha* appeared as a polypeptide of approximately 62 kDa. After *endo* H treatment, the deglycosylated protein migrated at approximately 47 kDa on SDS-PAGE, indicating that glycans constitutes approximately 25% of apparent molecular weight of CPY (Bellu et al., 1999). It is important to note that SDS may not bind the glycans to confer mobility in SDS-polyacrylamide gel electrophoresis under denaturing conditions. Consequently, the apparent molecular weight (MW) on SDS-PAGE may not closely reflect the genuine MW of glycoproteins.

Kluyveromyces lactis also has a long history for protein expression (Buckholz and Gleeson, 1991; Fleer, 1992; Fleer et al., 1991). *K. lactis* has been used for many years to produce β-galactosidase (lactase) in food industry. Consequently, large-scale fermentation of this organism has been extensively developed in the production process. *Yarrowia lipolytica* has been developed for many industrial processes for the manufacturing of various materials including citric acid, erythritol, mannitol, and others (Muller et al., 1998; Juretzek et al., 2001; Buckholz and Gleeson, 1991; Romanos et al., 1992). It was found that tissue plasminogen activator produced from *Y. lipolytica* only contained 8-10 mannose residues on N-linked glycans (Buckholz and Gleeson, 1991). Not much detailed analysis for protein glycosylation of these two hosts have been reported in the literature yet.

3.   Transporter of Sugar Donors from Cytosol into ER and Golgi

The glycosylational modifications occur in the ER and Golgi. The core oligosaccharide moiety in a glycoprotein is derived from the dolichyl oligosaccharides. During the elongation process, most substrates of glycosyl transferases are nucleotide sugars. These nucleotide sugar precursors are synthesized in the cytosol that has to be transported into the lumen of ER or Golgi (Hirschberg et al., 1998). Provision of nucleotide sugar precursors for glycosyl transferases in the lumen of Golgi apparatus has been postulated as a critical step to effect glycosylational modifications. It was found that limited supply of nucleotide sugars mediates the glycosylational modifications in the Golgi lumen of mammalian cells (Guillen et al., 1998).

4.   Glycosylational Engineering

It is clear that yeast cells have their limitations as hosts for the production of mammalian-like glycoproteins as therapeutic agents. As has been discussed, the major limitation is in yeast's capacity for glycosylational modifications. The biochemical steps involving the biosynthesis of glycoproteins in yeast have been extensively explored and strains can be modified with respect to their glycosylational properties.

In *S. cerevisiae*, the N-linked oligosaccharide structure can be extended by an array of mannosyl transferases including OCH1, MNN1, MNN2, MNN9, etc. (Table 1) (Lehle et al., 1995; Nakayama et al., 1992; Nakanishi-Shindo et al., 1993; Gopal and Ballou, 1987). The addition of mannose residues catalyzed by mannosyl transferases results in hyper-mannosylation of most glycoproteins produced form *S. cerevisiae*. To alleviate the hyper-mannosylation problems, the undesired enzymatic activities could be disrupted or expressed conditionally by homologous recombination using an engineered linear DNA fragment to transform host cells. Concomitantly, additional enzymes involved in the elaboration of oligosaccharides can be expressed in yeast to render glycosylation in yeast mutants the same as that in mammalian cells.

The OCH1 gene encodes a type II membrane-bound α-1,6-mannosyltransferase that initiates α-1,6- polymannose outer chain elongation on the N-linked oligosaccharide of $Man_8GlcNAc_2$ in *S. cerevisiae* (Nakanishi-Shindo et al., 1993; Nakayama et al., 1992). The MNN1 gene is likely to encode the α-1,3-mannosyltransferase located in the Golgi complex (Yip et al., 1994; Wiggins and Munro, 1998). The MNN9 gene encodes a membrane-associate protein that plays role in the addition of the long α-1,6-mannose backbone of the complex mannans (Yip et al., 1994).

Several *S. cerevisiae* mutants have been successfully used to produce recombinant glycoproteins without the undesired hyper-mannosylation problems (Lehle et al., 1995; Nakanishi-Shindo et al., 1993). The Δoch1 mutant was found to have a defect in the initiation of the α-1,6-linked outer chain and displayed a temperature sensitivity in growth at 37°C (Lehle et al., 1995). The Δoch1 mnn1 double mutant was able to prevent hyper-mannosylation and produce a high mannose-type oligosaccharide structure, $Man_8Glc NAc_2$ (Nakanishi-Shindo et al., 1993).

The *S. cerevisiae* alg 1,2,3, and 4 mutants have been shown to accumulate lipid-linked $GlcNAc_2$, $Man_{1-2}GlcNAc_2$, $Man_5GlcNAc_2$, $Man_{1-8}GlcNAc_2$, respectively, and alg5 and alg6 mutants accumulated lipid-linked $Man_9GlcNAc_2$ (Huffaker and Robbins, 1983). The ALG9 gene is likely to encode a putative mannosyl transferase (Burda et al., 1996). The analysis of over-expressed invertase in Δalg9 mutant revealed the accumulation of lipid-linked $Man_{6-10}GlcNAc_2$ oligosaccharids in *S. cerevisiae* (Cipollo and Trimble, 2000). In alg3 mutant, $Man_5GlcNAc_2$- PP-Dol accumulates and *endo* H resistant carbohydrates are transferred to protein by the oligosaccharyl transferase complex (Aebi et al., 1996). The Δoch1 mnn1 alg3 triple mutants accumulated $Man_5GlcNAc_2$ and $Man_8 GlcNAc_2$ in total cell mannoprotein (Nakanishi-Shindo et al., 1993). These studies demonstrated the feasibility of engineering yeast cells to synthesize a glycoprotein which can mimic mammalian counterparts.

Although mutants in glycosylation can be acquired by various selection methods such as vanadate selection (Ballou et al., 1991), [^{3}H]mannose suicide selection (Huffaker and Robbins, 1983), lectin selection (Takegawa et al., 1996), and osmotical sensitivity (Huang and Snider, 1995), it is important to make genetic knock-out since mutants with point mutation have great potential to revert. Many of the yeast mutants with glycosylational defects are sensitive to shear stress, medium osomlality, and bear reduced growth rates. As a result, it may be a challenge to develop a large-scale production process using some of these sensitive mutants to produce glycoproteins with desired glycosylational patterns.

5.   Case Studies for Glycosylational Engineering in Yeast

Hepatitis B virus (HBV) vaccine contains mainly *S. cerevisiae*-derived recombinant HBV S antigen which is not glycosylated. It has been licensed and shown to be very effective toward prevention of HBV infection (West et al., 1994; Greenberg et al., 1996; Chang et al., 2000; Huang and Lin, 2000; Mahoney, 1999). However, a second-generation vaccine is needed to extend vaccine protection to those who do not respond well to the current vaccine. Consequently, preS2 + S antigens were considered to be the potential composition for the second-generation vaccine (Kniskern et al., 1994; Tron et al., 1989). PreS2 expressed from wild type *S. cerevisiae* was shown to contain N-linked polymannose residues and the immunogenicity of this antigen was reduced (Kniskern et al., 1994). When the preS2 polypeptide was expressed in *mnn9* mutant, the non-hyper-glycosylated preS2 + S polypeptide showed similar immunogenicity in mice compared to RECOMBIVAX HB® and the mammalian cell-derived GenHevac-B (Kniskern et al., 1994; Tron et al., 1989). The non-hyperglycosylated preS2 + S polypeptide was much more immunogenic in humans than its hyper-glycosylated counterpart (Kniskern et al., 1994). As a result, hyper-mannosylation-deficient recombinant yeast should be considered for the production of potential sub-unit vaccines to increase their efficacy and to diminish the failure rate due to compromised activities of hyper-mannosylated protein antigens.

Recently, the reaction mechanism for class I α-1,2-mannosidases has begun to be elucidated (Vallee et al., 2000a; Vallee et al., 2000b; Herscovics, 1999). The recombinant yeast α-1,2-mannosidase has proved to be a good model to study the catalytic mechanism of the family of α-1,2-mannosidases.

The crystal structure of a class I α-1,2- mannosidase has been solved (Vallee et al., 2000b). This α-1,2-mannosidase is a type II transmembrane protein with a short cytosolic amino-terminal fragment. Class I α-1,2- mannosidase trims $Man_9GlcNAc_2$ to Man_8 $GlcNAc_2$ in the ER of S. cerevisiae (Henrissat, 1991). This was the first structure of this class of mannosidase to provide molecular framework in elucidating the detailed enzymatic mechanisms. Insights into mannosidase substrate specificity have come from the X-ray crystallographic studies on complexes with oligosaccharide substrate. It was demonstrated that with a single site-directed mutation in the catalytic domain (Arg273 Leu) of the S. cerevisiae α -1,2-mannosidase altered its substrate specificity and rendered the engineered enzyme, expressed in recombinant P. pastoris, capable of trimming $Man_9GlcNAc$ to $Man_5GlcNAc$ in vitro (Romero et al., 2000).

Attempts have been made to generate a recombinant S. cerevisiae that can produce $Man_5GlcNAc_2$-oligosaccharides, the intermediate for hybrid-type and complex-type sugar chain (Chiba et al., 1998). In this mutant S. cerevisiae, a soluble form of α-1,2-mannosidase, derived from Aspergillus saitoi, fused with the "HDEL" ER retention signal was expressed and localized to the desired cellular organell in a Δoch1 mnn1 mnn4 triple mutant (Inoue et al., 1995). This S. cerevisiae mutant lacks glycosyltransferase activities of α-1,6-mannosyltransferase and α-1,3-mannosyltransferase encoded by OCH1 and MNN1 genes, respectively. Consequently, there is no outer chain elongation of mannosyl residues on the glycoproteins synthesized wherein.

Previously, the Δoch1 mnn1 mutant has been used to produce glycoproteins containing mammalian-like high mannose type oligosaccharides. This mutant is potentially useful for producing recombinant glycoproteins in S. cerevisiae without the antigenicity introduced by α-1,3-linked mannose (Nakanishi-Shindo et al., 1993). The Δoch1 mnn1 mutant can only produce the $Man_8GlcNAc_2$ without further trimming of mannose residues.

In the study reported by Chiba Y. et al. (1998), the oligosaccharide structures were analyzed for carboxypeptidase Y and cell surface glycoproteins produced from the engineered strain. It was shown that the recombinant yeast is capable of producing a series of high mannose-type sugar chains including $Man_5GlcNAc_2$, the intermediate for hybrid-type and complex-type sugar chain. The molar ratio of each glycoform along the sequential route was $Man_5GlcNAc_2$: $Man_6GlcNAc_2$: $Man_7GlcNAc_2$: $Man_8GlcNAc_2$ = 10 : 13 : 16 : 61 for cell wall mannoproteins and = 27 : 22 : 22 : 29 for carboxypeptidase Y (CPY) glycoprotein. A schematic description of their engineering strategy is shown in Figure 3.

5.   Concluding Remarks

Several yeast strains have been well developed for protein over-expression. There are many successful examples of producing therapeutic proteins from yeast including hepatitis B surface antigen, human proinsulin, human serum albumin, etc. Although glycosylational modifications in yeast are not perfect to make therapeutic glycoproteins,

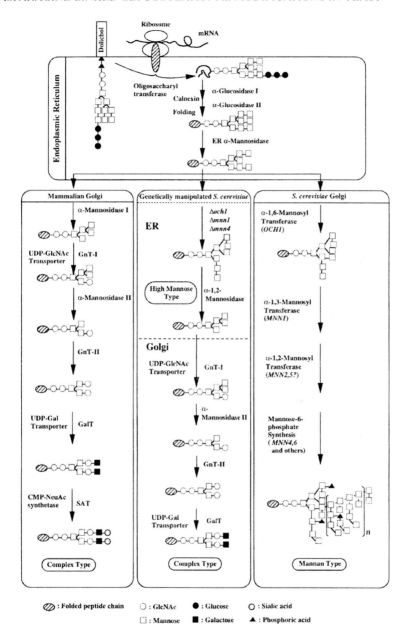

Figure 3. Strategy for glycosylational engineering in *S. cerevisiae* to produce mammalian-like complex type oligosaccharide. Reproduced by permission from American Society for Biochemistry and Molecular , Inc. (Chiba Y. et al., 1998).

efforts have been made to mediate the glycosylational machinery in yeast. These modulations can be achieved either by random selection of mutants generated by various methods or by rational genetic manipulation to add or delete specific enzymatic activities. Furthermore, enzymes involved in glycosylatinal pathway can be engineered to alter their substrate specificity to confer novel biological activities as has been discussed in previous sections. The aim would be to develop the technique to alter protein glycosylation in yeast as easy as protein backbones can be engineered through site directed mutagenesis. The success of glycosylational engineering in yeast ought to offer a very economic means to produce desired glycoprotein therapeutics to meet current and future medical needs. The feasibility to produce specific glycoforms in yeast will also be helpful in glycobiology reseaches to elucidate the structure and function relationships of glycan moiety attached to proteins.

6.   Acknowledgements

The authors thank Dr. Michael J. Betenbaugh at The Johns Hopkins University for encouragement and comments on the manuscript.

7.   References

Abeijon, C., Mandon, E. C., and Hirschberg, C. B. (1997) Transporters of nucleotide sugars, nucleotide sulfate and ATP in the Golgi apparatus. *Trends Biochem. Sci.* 22, 203-7.

Abeijon, C., Mandon, E. C., Robbins, P. W., and Hirschberg, C. B. (1996) A mutant yeast deficient in Golgi transport of uridine diphosphate N- acetylglucosamine. *J. Biol. Chem.* 271, 8851-4.

Aebi, M., Gassenhuber, J., Domdey, H., and te Heesen, S. (1996) Cloning and characterization of the ALG3 gene of Saccharomyces cerevisiae. *Glycobiology* 6 , 439-44.

Albright, C. F., and Robbins, R. W. (1990) The sequence and transcript heterogeneity of the yeast gene ALG1,an essential mannosyltransferase involved in N-glycosylation. *J. Biol. Chem.* 265, 7042-9.

Andersen, D. C., Goochee, C. F., Cooper, G., and Weitzhandler, M. (1994) Monosaccharide and oligosaccharide analysis of isoelectric focusing- separated and blotted granulocyte colony-stimulating factor glycoforms using high-pH anion-exchange chromatography with pulsed amperometric detection. *Glycobiology* 4, 459-67.

Ballou, C. E., Ballou, L., and Ball, G. (1994) Schizosaccharomyces pombe glycosylation mutant with altered cell surface properties. *Proc. Natl. Acad. Sci. USA* 91, 9327-31.

Ballou, L., and Ballou, C. (1995) Schizosaccharomyces pombe mutants that are defective in glycoprotein galactosylation. *Proc. Natl. Acad. Sci. USA* 92, 2790-4.

Ballou, L., Hitzeman, R. A., Lewis, M. S., and Ballou, C. E. (1991) Vanadate-resistant yeast mutants are defective in protein glycosylation. *Proc. Natl. Acad. Sci. USA* 88, 3209-12.

Basco, R. D., Munoz, M. D., Hernandez, L. M., Vazquez de Aldana, C., and Larriba, G. (1993) Reduced efficiency in the glycosylation of the first sequon of Saccharomyces cerevisiae exoglucanase leads to the synthesis and secretion of a new glycoform of the molecule. *Yeast* 9, 221-34.

Bellu, A. R., van der Klei, I. J., Rechinger, K. B., Yavuz, M., Veenhuis, M., and Kiel, J. A. (1999) Characterization of the Hansenula polymorpha CPY gene encoding carboxypeptidase Y. *Yeast* 15, 181-9.

Bernstein, M., Kepes, F., and Schekman, R. (1989) Sec59 encodes a membrane protein required for core glycosylation in Saccharomyces cerevisiae. *Mol. Cell. Biol.* 9, 1191-9.

Bhatia, P. K., and Mukhopadhyay, A. (1999) Protein glycosylation: implications for in vivo functions and therapeutic applications. *Adv. Biochem. Eng. Biotechnol.* 64, 155-201.

Boskovic, J., Soler-Mira, A., Garcia-Cantalejo, J. M., Ballesta, J. P., Jimenez, A., and Remacha, M. (1996) The sequence of a 16,691 bp segment of Saccharomyces cerevisiae chromosome IV identifies the DUN1, PMT1, PMT5, SRP14 and DPR1 genes, and five new open reading frames. *Yeast* 12, 1377-84.

Bretthauer, R. K., and Castellino, F. J. (1999) Glycosylation of Pichia pastoris-derived proteins. *Biotechnol. Appl. Biochem.* 30, 193-200.

Buckholz, R. G., and Gleeson, M. A. (1991) Yeast systems for the commercial production of heterologous proteins. *Biotechnology (N Y)* 9, 1067-72.

Burda, P., and Aebi, M. (1998) The ALG10 locus of Saccharomyces cerevisiae encodes the alpha-1,2 glucosyltransferase of the endoplasmic reticulum: the terminal glucose of the lipid-linked oligosaccharide is required for efficient N-linked glycosylation. *Glycobiology* 8, 455-62.

Burda, P., te Heesen, S., Brachat, A., Wach, A., Dusterhoft, A., and Aebi, M. (1996) Stepwise assembly of the lipid-linked oligosaccharide in the endoplasmic reticulum of Saccharomyces cerevisiae: identification of the ALG9 gene encoding a putative mannosyl transferase. *Proc. Natl. Acad. Sci. USA* 93, 7160-5.

Camirand, A., Heysen, A., Grondin, B., and Herscovics, A. (1991) Glycoprotein biosynthesis in Saccharomyces cerevisiae. Isolation and characterization of the gene encoding a specific processing alpha- mannosidase. *J. Biol. Chem.* 266. 15120-7.

Cereghino, J. L., and Cregg, J. M. (2000) Heterologous protein expression in the methylotrophic yeast Pichia pastoris. *FEMS Microbiol.Rev.* 24, 45-66.

Chang, M. H., Shau, W. Y., Chen, C. J., Wu, T. C., Kong, M. S., Liang, D. C., Hsu, H. M., Chen, H. L., Hsu, H. Y., and Chen, D. S. (2000) Hepatitis B vaccination and hepatocellular carcinoma rates in boys and girls. *JAMA* 284, 3040-2.

Chi, J. H., Roos, J., and Dean, N. (1996) The OST4 gene of Saccharomyces cerevisiae encodes an unusually small protein required for normal levels of oligosaccharyltransferase activity. *J. Biol. Chem.* 271, 3132-40.

Chiba, Y., Suzuki, M., Yoshida, S., Yoshida, A., Ikenaga, H., Takeuchi, M., Jigami, Y., and Ichishima, E. (1998) Production of human compatible high mannose-type (Man5GlcNAc2) sugar chains in Saccharomyces cerevisiae. *J. Biol. Chem.* 273, 26298-304.

Cipollo, J. F., and Trimble, R. B. (2000) The accumulation of Man(6)GlcNAc(2)-PP-dolichol in the Saccharomyces cerevisiae Deltaalg9 mutant reveals a regulatory role for the Alg3p alpha1,3-Man middle-arm addition in downstream oligosaccharide-lipid and glycoprotein glycan processing. *J. Biol. Chem.* 275, 4267-77.

Cipollo, J. F., Trimble, R. B., Chi, J. H., Yan, Q., and Dean, N. (2001) The yeast ALG11 gene specifies addition of the terminal {alpha}1,2-Man to the Man5GlcNAc2-PP-dolichol N-glycosylation intermediate formed on the cytosolic side of the endoplasmic reticulum. *J. Biol. Chem.*(in press)

Cregg, J. M. (1999) Expression in the Methylotrophic Yeast Pichia pastoris.In Gene Expression Systems: Using Nature for the Art of Expression. J. M. Fernandez, and J. P. Hoefflereds.(Academic Press) pp.157-191.

Cumming, D. A. (1991) Glycosylation of recombinant protein therapeutics: control and functional implications. *Glycobiology* 1, 115-30.

Dean, N. (1999) Asparagine-linked glycosylation in the yeast Golgi. *Biochim. Biophys. Acta.* 1426, 309-22.

Dean, N., and Poster, J. B. (1996) Molecular and phenotypic analysis of the S. cerevisiae MNN10 gene identifies a family of related glycosyltransferases. *Glycobiology* 6, 73-81.

Dean, N., Zhang, Y. B., and Poster, J. B. (1997) The VRG4 gene is required for GDP-mannose transport into the lumen of the Golgi in the yeast, Saccharomyces cerevisiae. *J. Biol. Chem.* 272, 31908-14.

Dennis, J. W., Granovsky, M., and Warren, C. E. (1999) Protein glycosylation in development and disease. *BioEssays* 21, 412-21.

Deutscher, S. L., Nuwayhid, N., Stanley, P., Briles, E. I. & Hirschberg, C. B. (1984) Translocation across Golgi vesicle membranes: a CHO glycosylation mutant deficient in CMP-sialic acid transport. *Cell* 39, 295-9.

Diminsky, D., Schirmbeck, R., Reimann, J., and Barenholz, Y. (1997) Comparison between hepatitis B surface antigen (HBsAg) particles derived from mammalian cells (CHO) and yeast cells (Hansenula polymorpha): composition, structure and immunogenicity. *Vaccine* 15, 637-47.

Durand, G., and Seta, N. (2000) Protein glycosylation and diseases: blood and urinary oligosaccharides as markers for diagnosis and therapeutic monitoring. *Clin. Chem.* 46, 795-805.

Esmon, B., Esmon, P. C., and Schekman, R. (1984) Early steps in processing of yeast glycoproteins. *J. Biol. Chem.* 259, 10322-7.

Feldmann, H., Aigle, M., Aljinovic, G., Andre, B., Baclet, M. C., Barthe, C., Baur, A., Becam, A. M., Biteau, N., Boles, E., and et al. (1994) Complete DNA sequence of yeast chromosome II. *EMBO J.* 13, 5795-809.

Fellinger, A. J., Verbakel, J. M., Veale, R. A., Sudbery, P. E., Bom, I. J., Overbeeke, N., and Verrips, C. T. (1991) Expression of the alpha-galactosidase from Cyamopsis tetragonoloba (guar) by Hansenula polymorpha. *Yeast* 7, 463-73.

Fleer, R. (1992) Engineering yeast for high level expression. *Curr. Opin. Biotechnol.* 3, 486-96.

Fleer, R., Chen, X. J., Amellal, N., Yeh, P., Fournier, A., Guinet, F., Gault, N., Faucher, D., Folliard, F., Fukuhara, H., and et al. (1991) High-level secretion of correctly processed recombinant human interleukin-1 beta in Kluyveromyces lactis. *Gene* 107, 285-95.

Foreman, P. K., Davis, R. W., and Sachs, A. B. (1991) The Saccharomyces cerevisiae RPB4 gene is tightly linked to the TIF2 gene. *Nucleic Acids Res.* 19, 2781.

Garcia, R., Cremata, J. A., Quintero, O., Montesino, R., Benkestock, K., and Stahlberg, J. (2001) Characterization of protein glycoforms with N-linked neutral and phosphorylated oligosaccharides: studies on the glycosylation of endoglucanase 1 (Cel7B) from Trichoderma reesei. *Biotechnol. Appl. Biochem.* 33, 141-52.

Gellissen, G., and Hollenberg, C. P. (1997) Application of yeasts in gene expression studies: a comparison of Saccharomyces cerevisiae, Hansenula polymorpha and Kluyveromyces lactis - a review. *Gene* 190, 87-97.

Gellissen, G., Hollenberg, C. P., and Janowicz, Z. A. (1995) Gene expression in methylotrophic yeasts. *Bioprocess. Technol.* 22, 195-239.

Gellissen, G., Janowicz, Z. A., Merckelbach, A., Piontek, M., Keup, P., Weydemann, U., Hollenberg, C. P., and Strasser, A. W. (1991) Heterologous gene expression in Hansenula polymorpha: efficient secretion of glucoamylase. *Biotechnology (N Y)* 9, 291-5.

Gellissen, G., and Melber, K. (1996) Methylotrophic yeast hansenula polymorpha as production organism for recombinant pharmaceuticals. *Arzneimittelforschung* 46, 943-8.

Gellissen, G., Melber, K., Janowicz, Z. A., Dahlems, U. M., Weydemann, U., Piontek, M., Strasser, A. W., and Hollenberg, C. P. (1992) Heterologous protein production in yeast. *Antonie Van Leeuwenhoek* 62, 79-93.

Gemmill, T. R., and Trimble, R. B. (1996) Schizosaccharomyces pombe produces novel pyruvate-containing N-linked oligosaccharides. *J. Biol. Chem.* 271, 25945-9.

Gemmill, T. R., and Trimble, R. B. (1999) Overview of N- and O-linked oligosaccharide structures found in various yeast species. *Biochim. Biophys. Acta.* 1426, 227-37.

Goffeau, A., Barrell, B. G., Bussey, H., Davis, R. W., Dujon, B., Feldmann, H., Galibert, F., Hoheisel, J. D., Jacq, C., Johnston, M., *et al.* (1996) Life with 6000 Genes. *Science* 274, 546-567.

Gopal, P. K., and Ballou, C. E. (1987) Regulation of the protein glycosylation pathway in yeast: structural control of N-linked oligosaccharide elongation. *Proc. Natl. Acad. Sci. USA* 84, 8824-8.

Greenberg, D. P., Vadheim, C. M., Wong, V. K., Marcy, S. M., Partridge, S., Greene, T., Chiu, C. Y., Margolis, H. S., and Ward, J. I. (1996) Comparative safety and immunogenicity of two recombinant hepatitis B vaccines given to infants at two, four and six months of age. *Pediatr. Infect. Dis. J.* 15, 590-6.

Guillen, E., Abeijon, C., and Hirschberg, C. B. (1998) Mammalian Golgi apparatus UDP-N-acetylglucosamine transporter: molecular cloning by phenotypic correction of a yeast mutant. *Proc. Natl. Acad. Sci. USA* 95, 7888-92.

Hausler, A., and Robbins, P. W. (1992) Glycosylation in Saccharomyces cerevisiae: cloning and characterization of an alpha-1.2-mannosyltransferase structural gene. *Glycobiology* 2, 77-84.

Heesen, S., Lehle, L., Weissmann, A., and Aebi, M. (1994) Isolation of the ALG5 locus encoding the UDP-glucose:dolichyl-phosphate glucosyltransferase from Saccharomyces cerevisiae. *Eur. J. Biochem.* 224, 71-9.

Helenius, A. (2001) Quality control in the secretory assembly line. *Philos. Trans. R. Soc. Lond., Ser. B. Biol. Sci.* 356, 147-50.

Helenius, A., and Aebi, M. (2001) Intracellular functions of N-linked glycans. *Science* 291, 2364-9.

Henrissat, B. (1991) A classification of glycosyl hydrolases based on amino acid sequence similarities. *Biochem. J.* 280, 309-16.

Hernandez, L. M., Ballou, L., Alvarado, E., Tsai, P. K., and Ballou, C. E. (1989) Structure of the phosphorylated N-linked oligosaccharides from the mnn9 and mnn10 mutants of Saccharomyces cerevisiae. *J. Biol. Chem.* 264, 13648-59.

Herscovics, A. (1999) Processing glycosidases of Saccharomyces cerevisiae. *Biochim. Biophys. Acta.* 1426, 275-85.

Herscovics, A., and Orlean, P. (1993) Glycoprotein biosynthesis in yeast. *FASEB J.* 7, 540-50.

Hill, K., Boone, C., Goebl, M., Puccia, R., Sdicu, A. M., and Bussey, H. (1992) Yeast KRE2 defines a new gene family encoding probable secretory proteins, and is required for the correct N-glycosylation of proteins. *Genetics* 130, 273-83.

Hirschberg, C. B., Robbins, P. W., and Abeijon, C. (1998) Transporters of nucleotide sugars, ATP, and nucleotide sulfate in the endoplasmic reticulum and Golgi apparatus. *Annu. Rev. Biochem.* 67, 49-69.

Hollenberg, C. P., and Gellissen, G. (1997) Production of recombinant proteins by methylotrophic yeasts. *Curr. Opin. Biotechnol.* 8, 554-60.

Hsu, T. A., Takahashi, N., Tsukamoto, Y., Kato, K., Shimada, I., Masuda, K., Whiteley, E. M., Fan, J. Q., Lee, Y. C., and Betenbaugh, M. J. (1997) Differential N-glycan patterns of secreted and intracellular IgG produced in Trichoplusia ni cells. *J. Biol. Chem.* 272, 9062-70.

Huang, K., and Lin, S. (2000) Nationwide vaccination: a success story in Taiwan. *Vaccine* 18, S35-8.

Huang, K. M., and Snider, M. D. (1995) Isolation of protein glycosylation mutants in the fission yeast Schizosaccharomyces pombe. *Mol. Biol. Cell.* 6, 485-96.

Huffaker, T. C., and Robbins, P. W. (1983) Yeast mutants deficient in protein glycosylation. *Proc. Natl. Acad. Sci. USA* 80, 7466-70.

Immervoll, T., Gentzsch, M., and Tanner, W. (1995) PMT3 and PMT4, two new members of the protein-O-mannosyltransferase gene family of Saccharomyces cerevisiae. *Yeast* 11, 1345-51.

Inoue, T., Yoshida, T., and Ichishima, E. (1995) Molecular cloning and nucleotide sequence of the 1,2-alpha-D- mannosidase gene, msdS, from Aspergillus saitoi and expression of the gene in yeast cells. *Biochim. Biophys. Acta.* 1253, 141-5.

Jackson, B. J., Kukuruzinska, M. A., and Robbins, P. (1993) Biosynthesis of asparagine-linked oligosaccharides in Saccharomyces cerevisiae: the alg2 mutation. *Glycobiology* 3, 357-64.

Janowicz, Z. A., Melber, K., Merckelbach, A., Jacobs, E., Harford, N., Comberbach, M., and Hollenberg, C. P. (1991) Simultaneous expression of the S and L surface antigens of hepatitis B, and formation of mixed particles in the methylotrophic yeast, Hansenula polymorpha. *Yeast* 7, 431-43.

Jenkins, N., Parekh, R. B., and James, D. C. (1996) Getting the glycosylation right: implications for the biotechnology industry. *Nat. Biotechnol.* 14, 975-81.

Jiang, B., Sheraton, J., Ram, A. F., Dijkgraaf, G. J., Klis, F. M., and Bussey, H. (1996) CWH41 encodes a novel endoplasmic reticulum membrane N-glycoprotein involved in beta 1,6-glucan assembly. *J. Bacteriol.* 178, 1162-71.

Joao, H. C., and Dwek, R. A. (1993) Effects of glycosylation on protein structure and dynamics in ribonuclease B and some of its individual glycoforms. *Eur. J. Biochem.* 218, 239-44.

Jungmann, J., and Munro, S. (1998) Multi-protein complexes in the cis Golgi of Saccharomyces cerevisiae with alpha-1,6-mannosyltransferase activity. *EMBO J.* 17, 423-34.

Jungmann, J., Rayner, J. C., and Munro, S. (1999) The Saccharomyces cerevisiae protein Mnn10p/Bed1p is a subunit of a Golgi mannosyltransferase complex. *J. Biol. Chem.* 274, 6579-85.

Juretzek, T., Le Dall, M., Mauersberger, S., Gaillardin, C., Barth, G., and Nicaud, J. (2001) Vectors for gene expression and amplification in the yeast Yarrowia lipolytica. *Yeast* 18, 97-113.

Kang, H. A., Sohn, J. H., Choi, E. S., Chung, B. H., Yu, M. H., and Rhee, S. K. (1998) Glycosylation of human alpha 1-antitrypsin in Saccharomyces cerevisiae and methylotrophic yeasts. *Yeast* 14, 371-81.

Kanik-Ennulat, C., Montalvo, E., and Neff, N. (1995) Sodium orthovanadate-resistant mutants of Saccharomyces cerevisiae show defects in Golgi-mediated protein glycosylation, sporulation and detergent resistance. *Genetics* 140, 933-43.

Kanik-Ennulat, C., and Neff, N. (1990) Vanadate-resistant mutants of Saccharomyces cerevisiae show alterations in protein phosphorylation and growth control. *Mol. Cell. Biol.* 10, 898-909.

Karaoglu, D., Kelleher, D. J., and Gilmore, R. (1995) Functional characterization of Ost3p. Loss of the 34-kD subunit of the Saccharomyces cerevisiae oligosaccharyltransferase results in biased underglycosylation of acceptor substrates. *J. Cell. Biol.* 130, 567-77.

Kelleher, D. J., and Gilmore, R. (1994) The Saccharomyces cerevisiae oligosaccharyltransferase is a protein complex composed of Wbp1p, Swp1p, and four additional polypeptides. *J. Biol. Chem.* 269, 12908-17.

Kelleher, D. J., Kreibich, G., and Gilmore, R. (1992) Oligosaccharyltransferase activity is associated with a protein complex composed of ribophorins I and II and a 48 kd protein. *Cell* 69, 55-65.

Kimura, T., Kitamoto, N., Kito, Y., Iimura, Y., Shirai, T., Komiyama, T., Furuichi, Y., Sakka, K., and Ohmiya, K. (1997) A novel yeast gene, RHK1, is involved in the synthesis of the cell wall receptor for the HM-1 killer toxin that inhibits beta-1,3-glucan synthesis. *Mol. Gen. Genet.* 254, 139-47.

Kimura, T., Komiyama, T., Furuichi, Y., Iimura, Y., Karita, S., Sakka, K., and Ohmiya, K. (1999) N-glycosylation is involved in the sensitivity of Saccharomyces cerevisiae to HM-1 killer toxin secreted from Hansenula mrakii IFO 0895. *Appl. Microbiol. Biotechnol.* 51, 176-84.

Knauer, R., and Lehle, L. (1999) The oligosaccharyltransferase complex from Saccharomyces cerevisiae. Isolation of the OST6 gene, its synthetic interaction with OST3, and analysis of the native complex. *J. Biol. Chem.* 274, 17249-56.

Kniskern, P. J., Hagopian, A., Burke, P., Schultz, L. D., Montgomery, D. L., Hurni, W. M., Ip, C. Y., Schulman, C.A., Maigetter, R. Z., Wampler, D. E., and et al. (1994) Characterization and evaluation of a recombinant hepatitis B vaccine expressed in yeast defective for N-linked hyperglycosylation. *Vaccine* 12, 1021-5.

Kornfeld, R., and Kornfeld, S. (1985) Assembly of asparagine-linked oligosaccharides. *Ann. Rev. Biochem.* 54, 631-64.

Kukuruzinska, M. A., and Robbins, P. W. (1987) Protein glycosylation in yeast: transcript heterogeneity of the ALG7 gene. *Proc. Natl. Acad. Sci. USA* 84, 2145-9.

Lehle, L., Eiden, A., Lehnert, K., Haselbeck, A., and Kopetzki, E. (1995) Glycoprotein biosynthesis in Saccharomyces cerevisiae: ngd29, an N- glycosylation mutant allelic to och1 having a defect in the initiation of outer chain formation. *FEBS Lett.* 370, 41-5.

Lis, H., and Sharon, N. (1993) Protein glycosylation. Structural and functional aspects. *Eur. J. Biochem.* 218, 1-27.

Liu, D. T. (1992) Glycoprotein pharmaceuticals: scientific and regulatory considerations, and the US Orphan Drug Act. *Trends Biotechnol.* 10, 114-20.

Lundblad, R. L. (1999) Glycosylation in Pichia pastoris. *Biotechnol. Appl. Biochem.* 30, 191-2.

Lussier, M., Camirand, A., Sdicu, A. M., and Bussey, H. (1993) KTR2: a new member of the KRE2 mannosyltransferase gene family. *Yeast* 9, 1057-63.

Lussier, M., Sdicu, A. M., Camirand, A., and Bussey, H. (1996) Functional characterization of the YUR1, KTR1, and KTR2 genes as members of the yeast KRE2/MNT1 mannosyltransferase gene family. *J. Biol. Chem.* 271, 11001-8.

Lussier, M., Sdicu, A. M., Winnett, E., Vo, D. H., Sheraton, J., Dusterhoft, A., Storms, R. K., and Bussey, H. (1997) Completion of the Saccharomyces cerevisiae genome sequence allows identification of KTR5, KTR6 and KTR7 and definition of the nine-membered KRE2/MNT1 mannosyltransferase gene family in this organism. *Yeast* 13, 267-74.

Mahoney, F. J. (1999) Update on diagnosis, management, and prevention of hepatitis B virus infection. *Clin. Microbiol. Rev.* 12, 351-66.

Mayer, A.F., Hellmuth, K., Schlieker, H., Lopez-Ulibarri, R., Oertel, S., Dahlems, U., Strasser, A.W.M., G., A.P., and Loon, M. v. (1999) An expression system matures: A highly efficient and cost-effective process for phytase production by recombinant strains of Hansenula polymorpha. *Biotechnol. Bioeng.* 63, 373-81.

Melnick, L., and Sherman, F. (1993) The gene clusters ARC and COR on chromosomes 5 and 10, respectively, of Saccharomyces cerevisiae share a common ancestry. *J. Mol. Biol.* 233, 372-88.

Miletich, J. P., and Broze. GJ. Jr. (1990) Beta protein C is not glycosylated at asparagine 329. The rate of translation may influence the frequency of usage at asparagine-X-cysteine sites. *J. Biol. Chem.* 265, 11397-404.

Montesino, R., Garcia, R., Quintero, O., and Cremata, J. A. (1998) Variation in N-linked oligosaccharide structures on heterologous proteins secreted by the methylotrophic yeast Pichia pastoris. *Protein. Expr. Purif.* 14, 197-207.

Muller, S., Sandal, T., Kamp-Hansen, P., and Dalboge, H. (1998) Comparison of expression systems in the yeasts Saccharomyces cerevisiae, Hansenula polymorpha, Klyveromyces lactis, Schizosaccharomyces pombe and Yarrowia lipolytica. Cloning of two novel promoters from Yarrowia lipolytica. *Yeast* 14, 1267-83.

Nakanishi-Shindo, Y., Nakayama, K. I., Tanaka, A., Toda, Y., and Jigami, Y. (1993) Structure of the N-linked oligosaccharides that show the complete loss of alpha-1,6-polymannose outer chain from och1, och1 mnn1, and och1 mnn1 alg3 mutants of Saccharomyces cerevisiae. *J. Biol. Chem.* 268, 26338-45.

Nakayama, K., Feng, Y., Tanaka, A., and Jigami, Y. (1998) The involvement of mnn4 and mnn6 mutations in mannosylphosphorylation of O-linked oligosaccharide in yeast Saccharomyces cerevisiae. *Biochim. Biophys. Acta.* 1425, 255-62.

Nakayama, K., Nagasu, T., Shimma, Y., Kuromitsu, J., and Jigami, Y. (1992) OCH1 encodes a novel membrane bound mannosyltransferase: outer chain elongation of asparagine-linked oligosaccharides. *EMBO J.* 11, 2511-9.

Odani, T., Shimma, Y., Tanaka, A., and Jigami, Y. (1996) Cloning and analysis of the MNN4 gene required for phosphorylation of N-linked oligosaccharides in Saccharomyces cerevisiae. *Glycobiology* 6, 805-10.

Odani, T., Shimma, Y., Wang, X. H., and Jigami, Y. (1997) Mannosylphosphate transfer to cell wall mannan is regulated by the transcriptional level of the MNN4 gene in Saccharomyces cerevisiae. *FEBS Lett.* 420, 186-90.

Oelmann, S., Stanley, P., and Gerardy-Schahn, R. (2001) Point mutations identified in Lec8 CHO glycosylation mutants that inactivate both the UDP-galactose and the CMP-sialic acid transporters. *J. Biol. Chem.* 276 (in press)

Orlean, P., Albright, C., and Robbins, P. W. (1988) Cloning and sequencing of the yeast gene for dolichol phosphate mannose synthase, an essential protein. *J. Biol. Chem.* 263, 17499-507.

Parodi, A. J. (1999) Reglucosylation of glycoproteins and quality control of glycoprotein folding in the endoplasmic reticulum of yeast cells. *Biochim. Biophys. Acta.* 1426, 287-95.

Praetorius-Ibba, M., Monnet, G., Meyhack, B., Kielland-Brandt, M., Nilsson-Tillgren, T., and Hinnen, A. (1997) Homologous recombination partly restores the secretion defect of underglycosylated acid phosphatase in yeast. *Curr. Genet.* 32, 190-6.

Reiss, G., te Heesen, S., Gilmore, R., Zufferey, R., and Aebi, M. (1997) A specific screen for oligosaccharyltransferase mutations identifies the 9 kDa OST5 protein required for optimal activity in vivo and in vitro. *EMBO J.* 16, 1164-72.

Reiss, G., te Heesen, S., Zimmerman, J., Robbins, P. W., and Aebi, M. (1996) Isolation of the ALG6 locus of Saccharomyces cerevisiae required for glucosylation in the N-linked glycosylation pathway. *Glycobiology* 6, 493-8.

Romanos, M. A., Scorer, C. A., and Clare, J. J. (1992) Foreign gene expression in yeast: a review. *Yeast* 8, 423-88.

Romero, P. A., Vallee, F., Howell, P. L., and Herscovics, A. (2000) Mutation of Arg(273) to Leu alters the specificity of the yeast N- glycan processing class I alpha1,2-mannosidase. *J. Biol. Chem.* 275, 11071-4.

Roy, S. K., Chiba, Y., Takeuchi, M., and Jigami, Y. (2000) Characterization of Yeast Yea4p, a uridine diphosphate-N-acetylglucosamine transporter localized in the endoplasmic reticulum and required for chitin synthesis. *J. Biol. Chem.* 275, 13580-7.

Silberstein, S., Collins, P. G., Kelleher, D. J., and Gilmore, R. (1995a) The essential OST2 gene encodes the 16-kD subunit of the yeast oligosaccharyltransferase, a highly conserved protein expressed in diverse eukaryotic organisms. *J. Cell. Biol.* 131, 371-83.

Silberstein, S., Collins, P. G., Kelleher, D. J., Rapiejko, P. J., and Gilmore, R. (1995b) The alpha subunit of the Saccharomyces cerevisiae oligosaccharyltransferase complex is essential for vegetative growth of yeast and is homologous to mammalian ribophorin I. *J. Cell. Biol.* 128, 525-36.

Simons, J. F., Ebersold, M., and Helenius, A. (1998) Cell wall 1,6-beta-glucan synthesis in Saccharomyces cerevisiae depends on ER glucosidases I and II, and the molecular chaperone BiP/Kar2p. *EMBO J.* 17, 396-405.

Sipos, G., Puoti, A., and Conzelmann, A. (1995) Biosynthesis of the side chain of yeast glycosylphosphatidylinositol anchors is operated by novel mannosyltransferases located in the endoplasmic reticulum and the Golgi apparatus. *J. Biol. Chem.* 270, 19709-15.

Stagljar, I., te Heesen, S., and Aebi, M. (1994) New phenotype of mutations deficient in glucosylation of the lipid- linked oligosaccharide: cloning of the ALG8 locus. *Proc. Natl. Acad. Sci. USA* 91, 5977-81.

Strahl-Bolsinger, S., Immervoll, T., Deutzmann, R., and Tanner, W. (1993) PMT1, the gene for a key enzyme of protein O-glycosylation in Saccharomyces cerevisiae. *Proc. Natl. Acad. Sci. USA* 90, 8164-8.

Takegawa, K., Tanaka, N., Tabuchi, M., and Iwahara, S. (1996) Isolation and characterization of a glycosylation mutant from Schizosaccharomyces pombe. *Biosci. Biotechnol. Biochem.* 60, 1156-9.

Takeuchi, M., and Kobata, A. (1991) Structures and functional roles of the sugar chains of human erythropoietins. *Glycobiology* 1, 337-46.

te Heesen, S., Janetzky, B., Lehle, L., and Aebi, M. (1992) The yeast WBP1 is essential for oligosaccharyl transferase activity in vivo and in vitro. *EMBO J.* 11, 2071-5.

te Heesen, S., Knauer, R., Lehle, L., and Aebi, M. (1993) Yeast Wbp1p and Swp1p form a protein complex essential for oligosaccharyl transferase activity. *EMBO J.* 12, 279-84.

te Heesen, S., Rauhut, R., Aebersold, R., Abelson, J., Aebi, M., and Clark, M. W. (1991) An essential 45 kDa yeast transmembrane protein reacts with anti-nuclear pore antibodies: purification of the protein, immunolocalization and cloning of the gene. *Eur. J. Cell Biol.* 56, 8-18.

Trimble, R. B., Atkinson, P. H., Tschopp, J. F., Townsend, R. R., and Maley, F. (1991) Structure of oligosaccharides on Saccharomyces SUC2 invertase secreted by the methylotrophic yeast Pichia pastoris. *J. Biol. Chem.* 266, 22807-17.

Tron, F., Degos, F., Brechot, C., Courouce, A. M., Goudeau, A., Marie, F. N., Adamowicz, P., Saliou, P., Laplanche, A., Benhamou, J. P., and et al. (1989) Randomized dose range study of a recombinant hepatitis B vaccine produced in mammalian cells and containing the S and PreS2 sequences. *J. Infect. Dis.* 160, 199-204.

Tschopp, J. F., Brust, P. F., Cregg, J. M., Stillman, C. A., and Gingeras, T. R. (1987) Expression of the lacZ gene from two methanol-regulated promoters in Pichia pastoris. *Nucleic Acids Res.* 15, 3859-76.

Uccelletti, D., Farina, F., Morlupi, A., and Palleschi, C. (1999) Mutants of Kluyveromyces lactis with altered protein glycosylation are affected in cell wall morphogenesis. *Res. Microbiol.* 150, 5-12.

Umana, P., Jean-Mairet, J., Moudry, R., Amstutz, H., and Bailey, J. E. (1999) Engineered glycoforms of an antineuroblastoma IgG1 with optimized antibody-dependent cellular cytotoxic activity. *Nat. Biotechnol.* 17, 176-80.

Umeda, K., Yoko-o, T., Nakayama, K., Suzuki, T., and Jigami, Y. (2000) Schizosaccharomyces pombe gmd3(+)/alg11(+) is a functional homologue of Saccharomyces cerevisiae ALG11 which is involved in N-linked oligosaccharide synthesis. *Yeast* 16, 1261-71.

Vallee, F., Karaveg, K., Herscovics, A., Moremen, K. W., and Howell, P. L. (2000a) Structural basis for catalysis and inhibition of N-glycan processing class I alpha 1,2-mannosidases. *J. Biol. Chem.* 275, 41287-98.

Vallee, F., Lipari, F., Yip, P., Sleno, B., Herscovics, A., and Howell, P. L. (2000b) Crystal structure of a class I alpha1,2-mannosidase involved in N- glycan processing and endoplasmic reticulum quality control. *EMBO J.* 19, 581-8.

van Dijk, R., Faber, K. N., Kiel, J. A., Veenhuis, M., and van der Klei, I. (2000) The methylotrophic yeast Hansenula polymorpha: a versatile cell factory. *Enzyme Microb. Technol.* 26, 793-800.

Varki, A. (1993) Biological roles of oligosaccharides: all of the theories are correct. *Glycobiology* 3, 97-130.

Wang, X. H., Nakayama, K., Shimma, Y., Tanaka, A., and Jigami, Y. (1997) MNN6, a member of the KRE2/MNT1 family, is the gene for mannosylphosphate transfer in Saccharomyces cerevisiae. *J. Biol. Chem.* 272, 18117-24.

West, D.J., Watson, B., Lichtman, J., Hesley, T.M., and Hedberg, K. (1994) Persistence of immunologic memory for twelve years in children given hepatitis B vaccine in infancy. *Pediatr. Infect. Dis. J.* 13, 745-7.

Wiggins, C.A., and Munro, S. (1998) Activity of the yeast MNN1 alpha-1,3-mannosyltransferase requires a motif conserved in many other families of glycosyltransferases. *Proc. Natl. Acad. Sci. USA* 95, 7945-50.

Yip, C. L., Welch, S. K., Klebl, F., Gilbert, T., Seidel, P., Grant, F. J., O'Hara, P. J., and MacKay, V. L. (1994) Cloning and analysis of the Saccharomyces cerevisiae MNN9 and MNN1 genes required for complex glycosylation of secreted proteins. *Proc. Natl. Acad. Sci. USA* 91, 2723-7.

Yoshida, S., Ohya, Y., Nakano, A., and Anraku, Y. (1995) STT3, a novel essential gene related to the PKC1/STT1 protein kinase pathway, is involved in protein glycosylation in yeast. *Gene* 164, 167-72.

Ziegler, F. D., Cavanagh, J., Lubowski, C., and Trimble, R. B. (1999) Novel Schizosaccharomyces pombe N-linked GalMan9GlcNAc isomers: role of the Golgi GMA12 galactosyltransferase in core glycan galactosylation. *Glycobiology* 9, 497-505.

Ziegler, F. D., Gemmill, T. R., and Trimble, R. B. (1994) Glycoprotein synthesis in yeast. Early events in N-linked oligosaccharide processing in Schizosaccharomyces pombe. *J. Biol. Chem.* 269, 12527-35.

Index

A

alcohol oxidase, 218
ammonia, 17; 60; 79-80; 199
Anion Exchange Chromatography, 33
antibody, 5; 11; 30; 76; 78; 81; 83; 85;
  93-4; 96; 99-105; 112; 116; 126;
  128; 135; 137; 139; 143-5; 177; 189;
  190

B

biotechnology, 134; 149; 192; 197-9;
  210
bisecting N-acetylglucosamine, 97; 99-
  100; 142

C

carbohydrate, 4-5; 13-4; 21; 24; 26; 28-
  30; 40; 45; 47; 61; 73-4; 76; 112;
  113; 123; 132-3; 142; 144-5; 149-
  54; 164; 174; 178; 186; 191-2; 206
cell culture, 77; 81-2; 100-1; 111; 113;
  114; 116; 131; 134; 186; 188-9; 197;
  199-200; 203; 205-8; 209-10
cell engineering, 94; 99-100; 126
cell surface engineering, 177-8; 190
chemical coupling partners, 177
chemoselective ligation, 173; 176-8;
  181; 184; 189-193
CHO, 10; 28; 71; 74-86; 93; 99-100;
  106-7; 112-3; 115; 121; 126; 128;
  131; 133-42; 144-5; 151-4; 201;
  204-5; 207
comparability, 101; 103
complex N-glycans, 134; 202-3

D

dissolved oxygen, 77; 81; 116
dolichol, 65; 74; 109; 131; 135; 198;
  201-2; 219

E

Electrospray, 13; 43; 133
Endo H, 220

F

flow cytometry, 137-8; 140; 179-80;
  183-4
Fluorescent Label, 34

G

galactosylation, 58; 81; 88; 97; 100-2;
  104; 111; 113; 126; 128; 142-4; 205
GDP-mannose transporter, 220
Gel Permeation Chromatography, 32
GlcNAc, 17; 19-20; 39; 62; 65; 67-8;
  72; 74-7; 84-5; 87; 98; 102; 111-3;
  117-8; 121; 123; 126; 128; 131; 145;
  150; 153-4; 156; 158-9; 161; 163-5;
  167; 174; 200-1; 203-10
glucose, 35-6; 69; 79; 101; 109; 134;
  139; 199; 201; 216
glucosidase I, 66; 201; 216-7; 219
glucosidase II, 66; 219
glycan heterogeneity, 145
glycobiology, 192; 226

Glycoform, 2; 81; 98
glycoprotein(s), 1-4; 7-8; 10; 12-5; 19-
  20; 22-4; 26-7; 30; 32-4; 36; 43; 47-
  8; 61-3; 66; 69; 72-6; 79; 83-4; 86-
  8; 94; 109; 111-4; 121; 128; 132-5;
  137-8; 141-2; 144-5; 149-54; 156;
  160; 167-8; 190; 192-3; 200-1; 203;
  206-8; 210; 215-8; 221-4
glycosylation, 1; 3-5; 10; 12-6; 18; 20;
  23-6; 28-34; 49; 61-3; 65-6; 69-71;
  73-83; 85-6; 88; 94-95; 97; 100-103;
  105; 131-6; 141; 143; 145; 149-50;

152-4; 156-8; 161-2; 166-7; 178-9; 192; 198-201; 203-8; 215-6; 219; 221-3; 226

glycosyltransferase(s), 62; 66; 68; 70; 72-3; 75; 77-8; 82; 84; 86-7; 102; 109; 111; 114-6; 126; 132-3, 136; 141; 149-50; 152; 154; 156-8; 160-8; 198; 201-7

Golgi, 61; 63; 66-8; 71; 74; 77-82; 85; 109; 114; 126; 132; 134; 136; 139; 149-50; 152; 157; 160-1; 163-7; 174; 202; 210; 216-20; 222

GPI anchor, 62

H

Hansenula polymorpha, 215; 220-1

HPAEC-PAD, 15; 114; 117-8; 121; 162

HPLC, 1-2; 4; 16; 18; 21-2; 25-6; 32-8; 40; 43; 47-8; 98-9; 133; 139; 142-3

hybrid glycans, 85

hydrazide, 19; 43; 178-80; 183-6; 189

I

insect cells, 4; 61; 71-3; 87; 131; 134; 199-203; 205-10

interferon-gamma, 87

K

ketone, 176-84; 188-91; 193

Kluyveromyces lactis, 215; 220-1

L

lectins, 76; 104; 132

lectin selection, 223

M

macroheterogeneity, 62; 79

MALDI-TOF, 36; 114; 116-8; 121; 126

mammalian, 11; 14; 62; 71; 73; 74; 76; 83; 86; 87; 109; 111-3; 115; 131; 134; 141; 149-51; 156-7; 167; 191; 198-205; 207-10; 222-5

mammalian cells, 11; 71; 73; 74; 83; 86; 109; 111; 115; 131; 141; 149-51; 191; 199-202; 207; 209; 222

mammalian expression systems, 150

mammalian host cells, 62; 83; 87; 149

ManLev, 178-86; 188-9

mannosidase, 39; 66; 74; 109; 160; 163; 201-2; 217; 219; 221; 224

mannosyl transferase, 217; 222-3

mannosyltransferase, 219; 220; 222; 224

mannosylphosphate transferase, 220

Mass Spectrometry, 12; 38; 40; 47

methanol oxidase, 221

methylotrophic, 221

microheterogeneity, 4; 8; 10; 62; 66; 111; 132-3; 149; 153; 168

monoclonal, 5; 8; 11; 25; 29; 76; 78; 81; 93; 105; 109; 112; 123; 126; 139

Monosaccharide Compositional Analysis, 14

N

N-acetylglucosaminyltransferase III, 76; 88; 111

N-glycans, 20; 38; 62; 71; 73; 74; 78; 84; 87; 111; 118; 121; 126; 132-3; 157; 158; 161199-200; 202-6; 210

N-linked oligosaccharides, 25; 27; 30; 31; 62; 66; 75; 84; 95; 123; 126; 131; 143; 158; 216; 218; 221

O

O-glycans, 20; 36; 62; 69; 72-4; 77

oligosaccharide processing, 71; 82; 152

oligosaccharide structure, 30; 63; 72; 75; 77-9; 84; 98; 102; 109; 117; 133-4; 143; 150; 152; 168; 216; 222; 224

Oligosaccharides, 22; 29-30; 38; 63; 98-9

overexpression, 109; 114; 115; 116; 118; 123; 126; 128; 205

P

pH, 1; 3-12; 27-8; 30; 32-4; 43; 48; 77-81; 83; 97; 116; 117; 134; 145; 180; 182-3; 200
pharmacokinetics, 49; 61; 121; 141-2
Pichia methanolica, 215
Pichia pastoris, 215; 217
plant cells, 71; 87
post-translational processing, 113
product quality, 2; 86; 113-5; 137

R

recombinant protein(s), 1; 25; 70-2; 74; 75; 77; 83; 99; 109; 111; 113; 116; 121; 123; 126; 128; 133-7; 141; 171; 192; 197; 198; 215; 217; 220-1

S

Saccharomyces cerevisiae, 215-6
Schizosaccharomyces pombe, 215
SiaLev, 176-86; 188-9; 190-1
sialic acid, 3; 7; 19; 23; 26; 29-30; 39; 41-2; 48; 62; 68; 72; 74-5; 79-80; 84-6; 97-8; 100-1; 111-3; 117-8; 121; 123; 128; 131; 133; 135; 137; 139; 141-3; 145; 173-8; 183; 186; 189-91; 199-202; 205; 207-10
sialic acid biosynthetic pathway, 173-5
sialylation, 3; 7; 10; 26; 70; 73; 75; 77-80; 82-4; 86; 97; 102; 111; 113; 123; 126; 128; 135-7; 140-5; 159-62; 164-5; 207; 210
sialyltransferase, 68; 73; 76; 82; 84-5; 100; 110; 114; 118; 134-5; 140; 145; 158-9; 164; 167; 174; 202; 204; 210

U

UDP-GlcNAc transporter, 220
UDP-glucose
    dolichyl-phosphate

glucosyltransferase, 219
UDP-N-acetylglucosamine-1-
    phosphate transferase, 219

V

vanadate selection, 223

Y

Yarrowia lipolytica, 215; 220
yeast, 61; 71; 73-4; 131; 192; 197; 210; 215-8; 220-4; 226

Cell Engineerings

1. M. Al-Rubeai (ed.): *Cell Engineering.* 1999   ISBN 0-7923-5790-6
2. M. Al-Rubeai (ed.): *Cell Engineering.* Vol. 2: Transient Expression. 2000
                                                                    ISBN 0-7923-6596-8
3. M. Al-Rubeai (ed.): *Cell Engineering.* Vol. 3: Glycosylation. 2002
                                                                    ISBN 1-4020-0733-7

Kluwer Academic Publishers – Dordrecht / Boston / London